Applied Probability and Statistics (Continued)

CHERNOFF and MOSES · Elementary Decision Theory
CHIANG · Introduction to Stochastic Processes in Biostatistics
CLELLAND, deCANI, BROWN, BURSK, and MURRAY · Basic Statistics with Business Applications
COCHRAN · Sampling Techniques, *Second Edition*
COCHRAN and COX · Experimental Designs, *Second Edition*
COX · Planning of Experiments
COX and MILLER · The Theory of Stochastic Processes
DAVID · Order Statistics
DEMING · Sample Design in Business Research
DODGE and ROMIG · Sampling Inspection Tables, *Second Edition*
DRAPER and SMITH · Applied Regression Analysis
ELANDT-JOHNSON · Probability Models and Statistical Methods in Genetics
GOLDBERGER · Econometric Theory
GUTTMAN, WILKS and HUNTER · Introductory Engineering Statistics, *Second Edition*
HAHN and SHAPIRO · Statistical Models in Engineering
HALD · Statistical Tables and Formulas
HALD · Statistical Theory with Engineering Applications
HOEL · Elementary Statistics, *Third Edition*
HUANG · Regression and Econometric Methods
JOHNSON and LEONE · Statistics and Experimental Design: In Engineering and the Physical Sciences, Volumes I and II
LANCASTER · The Chi Squared Distribution
MILTON · Rank Order Probabilities: Two-Sample Normal Shift Alternatives
PRABHU · Queues and Inventories: A Study of Their Basic Stochastic Processes
RAO and MITRA · Generalized Inverse of Matrices and Its Applications
SARD and WEINTRAUB · A Book of Splines
SARHAN and GREENBERG · Contributions to Order Statistics
SEAL · Stochastic Theory of a Risk Business
SEARLE · Linear Models
THOMAS · An Introduction to Applied Probability and Random Processes
WHITTLE · Optimization under Constraints
WILLIAMS · Regression Analysis
WOLD and JUREEN · Demand Analysis
WONNACOTT and WONNACOTT · Introduction to Econometric Methods
YOUDEN · Statistical Methods for Chemists
ZELLNER · An Introduction to Bayesian Inference in Econometrics

Tracts on Probability and Statistics

BILLINGSLEY · Ergodic Theory and Information
BILLINGSLEY · Convergence of Probability Measures
CRAMÉR and LEADBETTER · Stationary and Related Stochastic Processes
JARDINE and SIBSON · Mathematical Taxonomy
RIORDAN · Combinatorial Identities
TAKÁCS · Combinatorial Methods in the Theory of Stochastic Processes

Generalized Inverse of Matrices and its Applications

Generalized Inverse of Matrices and its Applications

C. RADHAKRISHNA RAO, Sc.D., F.N.A., F.R.S.
Director, Research and Training School
Indian Statistical Institute

SUJIT KUMAR MITRA, Ph.D.
Professor of Statistics
Indian Statistical Institute

JOHN WILEY & SONS, INC.
New York London Sydney Toronto

Library of Congress Catalog Card Number: 74-158528

ISBN 0-471-70821-6

Printed in the United States of America.

10 9 8 7 6 5 4 3 2 1

dedicated to
Professor P. C. Mahalanobis, F.R.S.
Founder, Indian Statistical Institute,
for his significant contributions
to the Statistical world.

Preface

This book is an attempt to bring together all the available results on "invertibility of singular matrices" under a unified theory and to discuss their applications.

It is well known that if **A** is a square non-singular matrix, then there exists a matrix **G**, such that $\mathbf{AG} = \mathbf{GA} = \mathbf{I}$, which is called the inverse of **A** and denoted by $\mathbf{A}^{-1}$. If **A** is a singular or a rectangular matrix, no such matrix **G** exists. However, Moore extended the notion of inverse to singular matrices in 1920 and discussed the concept at some length in 1935. Moore's definition of an inverse of **A** is equivalent to the existence of a matrix **G** such that

$$\mathbf{AG} = \mathbf{P_A}, \quad \mathbf{GA} = \mathbf{P_G}$$

where $\mathbf{P_X}$ stands for the projection operator onto $\mathscr{M}(\mathbf{X})$, the space generated by the columns of **X**. Unaware of Moore's work, Penrose defined in 1955 an inverse **G** of **A** as satisfying the conditions

$$\mathbf{AGA} = \mathbf{A}, \quad (\mathbf{AG})^* = \mathbf{AG}$$

$$\mathbf{GAG} = \mathbf{G}, \quad (\mathbf{GA})^* = \mathbf{GA}$$

which are equivalent to Moore's conditions (when the inner product between two vectors **x**, **y** is defined as $\mathbf{y}^*\mathbf{x}$, where * indicates conjugate transpose).

In three fundamental papers Tseng (1949a, 1949b and 1956) considered the problem of defining inverses of singular operators, which are more general than matrices. Attempts at defining and using an inverse of a singular matrix have been made from time to time (see Bjerhammer, 1951, 1957, 1958) but the results were less general or offered no systematic study.

In 1955 one of the authors, Rao, constructed an inverse of a singular matrix that occurs in normal equations in the least-squares theory, which he called a pseudoinverse, and showed that it serves the same purpose as the regular inverse of a nonsingular matrix in solving normal equations and also in computing standard errors of least-squares estimators. Rao's pseudoinverse did not satisfy all the conditions of Moore and Penrose, and the only property required of the inverse **G** was "that $\mathbf{x} = \mathbf{Gy}$ provides a solution of the equation $\mathbf{Ax} = \mathbf{y}$ for any **y**, such that $\mathbf{Ax} = \mathbf{y}$ is consistent." This is provided by a matrix **G** satisfying the only condition **AGA** = **A** in Penrose's definition. In 1962 Rao called a matrix **G** satisfying this single condition, **AGA** = **A**, a g-inverse (generalized inverse) of **A** and studied its properties in greater detail. In many practical applications, it is sufficient to work with a g-inverse satisfying this more general (weaker) definition, as demonstrated in two other publications by Rao in 1965 and 1966.

A g-inverse so defined is not unique and thus presents an interesting study in matrix algebra. In a publication in 1967, Rao showed how a variety of g-inverses could be constructed to suit different purposes and presented a classification (with nomenclature) of g-inverses.

The work was later pursued by Mitra (1968a and 1968b) who introduced some new classes of g-inverses. Further applications of g-inverses were considered in the joint publications by the authors (Mitra and Rao, 1968a, 1968b and 1969). The present book essentially describes the work of the authors and also within its framework brings all the important contributions by other authors on this subject up to date.

Some principal contributors to this subject since 1955 are Greville (1957), Bjerhammer (1957 and 1958), Ben-Israel and Charnes (1963), Chipman (1964), Chipman and Rao (1964) and Scroggs and Odell (1966). Bose (1959) mentions the use of g-inverse in his lecture notes on Analysis of Variance. Bott and Duffin (1953) introduced the concept of a constrained inverse of a square matrix, which is different from a g-inverse and is useful in some applications in network theory. Chernoff (1953) considered an inverse of a singular n.n.d. matrix, which is also not a g-inverse but is useful in discussing some problems in statistical estimation theory.

Chapter 1 of this book, *Generalized Inverse of Matrices and its Applications*, contains statements of certain results in matrix algebra which are used in the discussions of the later chapters. It also explains the notations used in the book. Properties of g-inverse based on the single condition **AGA** = **A**, and on the conditions **AGA** = **A**, **GAG** = **G** called reflexive g-inverse, are studied in Chapter 2. Solutions of some important matrix equations are also considered. It is shown that in many problems one needs a g-inverse satisfying the only condition **AGA** = **A**, and the other restraining conditions can only serve special purposes. Chapter 3 examines conditions on a g-inverse to obtain (i) minimum norm solution of a consistent equation **Ax** = **y**, (ii) least-squares solution of an inconsistent equation **Ax** = **y**, and (iii) minimum norm least-squares solution of an inconsistent equation **Ax** = **y**. It is shown that the g-inverse which produces a solution of the type (iii) is precisely the Moore–Penrose inverse. Other special types of g-inverses with reciprocal eigenvalue property, etc., are considered in Chapter 4. Chapter 5 contains a general discussion of projection operators and idempotent matrices and their explicit representations in terms of g-inverses of matrices. Simultaneous reduction of two hermitian forms when none of them need be positive definite is considered in Chapter 6. Applications of g-inverse in problems of estimation from linear models and robustness of statistical procedures under deviations from specified models are examined in Chapters 7 and 8. Very general results on the distribution of quadratic forms of random variables having a singular normal distribution are obtained in Chapter 9. Applications of g-inverse in network theory, mathematical programming and some problems in mathematical statistics (discriminant function when the dispersion matrix is singular and maximum likelihood estimation when the information matrix is singular) are considered in Chapter 10, while computational methods for obtaining a g-inverse are discussed in Chapter 11.

The new classes of constrained inverses introduced in Chapter 4 deserve special mention as they include all the types of g-inverses as special cases.

The material covered in the book relates to the research work done during the last 15 years and a number of unpublished results recently obtained by the authors. The applications of g-inverse are rapidly increasing; we have considered in some detail only a few of them. We hope that this full-length monograph on the subject will be of use to students and research workers in various fields. In the book we confine our attention to matrices only. Extension of the results on matrices to more general operators in abstract spaces offers a good scope for research work.

There is enough material in the book for one term course on g-inverse of matrices. In addition the book would be useful as supplementary material in a variety of courses such as Matrix Algebra, Network Theory, Mathematical Statistics, Optimization Problems and Numerical Analysis.

It gives us great pleasure to thank Mr. Arun Das and Mr. Mehar Lal, who have undertaken the heavy burden of typing the manuscript for the press at various stages of preparation, and to P. Bhimasankaram for reading the manuscript and making helpful comments.

Calcutta, India
April 1971

C. R. RAO
S. K. MITRA

Contents

Generalized Inverse of Matrices and its Applications

CHAPTER 1

Notations and Preliminaries

In this chapter we introduce the notations and some of the preliminary results on matrices needed elsewhere in the text. The proofs of these results will be found in standard textbooks on matrix algebra and are therefore omitted here. See, for instance, books by Gantmacher (1959), Householder (1964), Pease (1965), and Perlis (1952). A fairly complete discussion of properties of matrices, with special reference to applications in mathematical statistics, is contained in Chapter 1 of Rao (1965).

Matrices are denoted by boldface capital letters such as $\mathbf{A}$, $\mathbf{B}$, $\mathbf{\Sigma}$. Boldface lower-case letters $\mathbf{x}$, $\mathbf{y}$, ... denote column vectors, used synonymously for matrices with only one column. A null matrix or a null vector is denoted by $\mathbf{0}$. Unless otherwise stated, we shall consider only those matrices and vectors with elements defined over the field of complex numbers. In Chapters 7, 8, and 9, however, we consider only real matrices and vectors.

1.1 ROW AND COLUMN SPACES OF A MATRIX, SUBSPACES AND ORTHOGONAL COMPLEMENT, PROJECTION OPERATOR

Vector Spaces

For a matrix $\mathbf{A}$ of order $m \times n$ the linear space spanned by the columns of $\mathbf{A}$ is called the column space of $\mathbf{A}$ and denoted by the symbol $\mathcal{M}(\mathbf{A})$. Row space of $\mathbf{A}$, defined analogously, can therefore be denoted by $\mathcal{M}(\mathbf{A}')$; $\mathscr{E}^n$ and $\mathscr{R}^n$ denote the vector spaces of all n-tuples with complex and real coordinates, respectively.

Notice that $\mathcal{M}(\mathbf{A})$ consists of precisely those vectors in $\mathscr{E}^m$ which can be expressed as $\mathbf{Ax}$ for some $\mathbf{x}$ in $\mathscr{E}^n$. It is convenient to think of a matrix $\mathbf{A}$ as a linear transformation $\mathscr{E}^n \xrightarrow{\mathbf{A}} \mathscr{E}^m$, in which case $\mathcal{M}(\mathbf{A})$ is the range of the transformation $\mathbf{A}$. The null space of $\mathbf{A}$ is, on the other hand, the set of all vectors in $\mathscr{E}^n$ that are mapped into the null vector in $\mathscr{E}^m$ under this transformation.

Basis. Any set of linearly independent vectors spanning a given vector space (which may be a subspace) is called a basis of the vector space.

Dimension. The dimension of a vector space $\mathscr{V}$, denoted by $d[\mathscr{V}]$, is the number of vectors in a basis of $\mathscr{V}$.

Linear functional. A linear functional on a complex vector space $\mathscr{V}$ (i.e., a vector space on the field of complex numbers) is a complex-valued, additive homogeneous function ξ defined on $\mathscr{V}$; that is,

$$\xi(\mathbf{x} + \mathbf{y}) = \xi(\mathbf{x}) + \xi(\mathbf{y}) \quad \forall\, \mathbf{x}, \mathbf{y} \in \mathscr{V},$$

$$\xi(\alpha\mathbf{x}) = \alpha\xi(\mathbf{x}) \quad \forall\, \mathbf{x} \in \mathscr{V} \quad \text{and arbitrary complex number } \alpha.$$

Bilinear Functional. A bilinear functional on a complex vector space $\mathscr{V}$ is a complex-valued function ϕ, defined on the cartesian product of $\mathscr{V}$ with itself such that, if

$$\xi_{\mathbf{y}}(\mathbf{x}) = \eta_{\mathbf{x}}(\mathbf{y}) = \phi(\mathbf{x}, \mathbf{y}),$$

then, for each $\mathbf{y} \in \mathscr{V}$, $\xi_{\mathbf{y}}$ is a linear functional on $\mathscr{V}$ and, for each $\mathbf{x} \in \mathscr{V}$, $\eta_{\mathbf{x}}$ is a conjugate linear functional.

Inner Product and Orthogonality. An inner product in a complex vector space $\mathscr{V}$ denoted by $(\mathbf{x}, \mathbf{y})$ is a symmetric, strictly positive, bilinear functional on $\mathscr{V}$, that is, a bilinear functional satisfying (i) $(\mathbf{x}, \mathbf{y}) = \overline{(\mathbf{y}, \mathbf{x})}$, and (ii) $(\mathbf{x}, \mathbf{x}) > 0\ \forall$ non-null vectors $\mathbf{x} \in \mathscr{V}$. An inner product space is a complex vector space with an agreed inner product definition. In an inner product space, vectors $\mathbf{x}$ and $\mathbf{y}$ are said to be mutually orthogonal, if $(\mathbf{x}, \mathbf{y}) = 0$.

Reciprocal Bases (Dual Bases). Consider a vector space of finite dimension k and let $\boldsymbol{\alpha}_1, \ldots, \boldsymbol{\alpha}_k$ and $\boldsymbol{\beta}_1, \ldots, \boldsymbol{\beta}_k$ be two alternative bases. One is called the reciprocal basis of the other if $(\boldsymbol{\alpha}_i, \boldsymbol{\beta}_j) = \delta_{ij}$, where δ_{ij} is the Kronecker symbol.

Note that if $\mathbf{A}$ is a nonsingular matrix of order k, (a) the columns of $\mathbf{A}$ and the columns of $(\mathbf{A}^{-1})^*$ provide reciprocal bases of $\mathscr{E}^k$ if $(\mathbf{x}, \mathbf{y}) = \mathbf{y}^*\mathbf{x}$, (b) the columns of $\mathbf{A}$ and the columns of $\mathbf{\Lambda}^{-1}(\mathbf{A}^{-1})^*$ provide reciprocal bases of $\mathscr{E}^k$ if $(\mathbf{x}, \mathbf{y}) = \mathbf{y}^*\mathbf{\Lambda}\mathbf{x}$.

Intersection, Sum, and Direct Sum of Subspaces. If $\mathscr{S}$ and $\mathscr{T}$ are subspaces of a vector space $\mathscr{V}$, the set of vectors common to both $\mathscr{S}$ and $\mathscr{T}$ forms a subspace of $\mathscr{V}$. This subspace, called the intersection of $\mathscr{S}$ and $\mathscr{T}$, is denoted by the symbol $\mathscr{S} \cap \mathscr{T}$. Here, the set of all vectors in $\mathscr{V}$ which can be expressed as $\boldsymbol{\alpha} + \boldsymbol{\beta}$ with $\boldsymbol{\alpha} \in \mathscr{S}$ and $\boldsymbol{\beta} \in \mathscr{T}$ also forms a subspace of $\mathscr{V}$, called the sum of $\mathscr{S}$ and $\mathscr{T}$ and denoted by the symbol $\mathscr{S} + \mathscr{T}$. We have the dimensional relation

$$d(\mathscr{S} + \mathscr{T}) + d(\mathscr{S} \cap \mathscr{T}) = d(\mathscr{S}) + d(\mathscr{T}). \qquad (1.1.1)$$

Since the null vector is a necessary constituent of every subspace, $\mathscr{S} \cap \mathscr{T}$ will always contain the null vector. If $\mathscr{S} \cap \mathscr{T}$ is a single vector set consisting only of the null vector, $\mathscr{S}$ and $\mathscr{T}$ are said to be virtually disjoint and the sum of $\mathscr{S}$ and $\mathscr{T}$ is called the direct sum and denoted by the symbol $\mathscr{S} \oplus \mathscr{T}$.

Orthogonal Subspace. Let $\mathscr{V}$ be a vectorspace with a proper inner product defined for all pairs of vectors in $\mathscr{V}$. If $\mathscr{S}$ is a subspace of $\mathscr{V}$, the set of all vectors in $\mathscr{V}$ that are orthogonal to every vector in $\mathscr{S}$ forms a subspace $\mathscr{S}^\perp$ called the orthogonal complement of $\mathscr{S}$ (in $\mathscr{V}$). We have the dimensional equality

$$d(\mathscr{S}) + d(\mathscr{S}^\perp) = d(\mathscr{V}). \qquad (1.1.2)$$

The orthogonal complement of $\mathscr{M}(\mathbf{A})$ in $\mathscr{E}^m$ is denoted by $\mathscr{O}(\mathbf{A})$. $\mathbf{A}^\perp$ denotes a matrix such that

$$\mathscr{M}(\mathbf{A}^\perp) = \mathscr{O}(\mathbf{A}). \qquad (1.1.3)$$

Unless it is otherwise clear from the context, the columns of $\mathbf{A}^\perp$ are assumed to be linearly independent.

Projection Operator

A general treatment of projection operators is given in Chapter 5. However, a special class of projection operators needed in the discussion of generalized inverse of a matrix considered in Chapter 3 is described here. Let $\mathbf{A}$ be $m \times n$ matrix. We shall call a matrix $\mathbf{P_A}$ a projection operator onto $\mathscr{M}(\mathbf{A})$ with respect to a n.n.d. matrix $\mathbf{M}$ iff

$$\mathbf{P_A x} \in \mathscr{M}(\mathbf{A}), \qquad \forall\, \mathbf{x} \in \mathscr{E}^m$$

$$(\mathbf{x} - \mathbf{P_A x})^* \mathbf{M} (\mathbf{x} - \mathbf{P_A x}) \le (\mathbf{x} - \mathbf{Ay})^* \mathbf{M} (\mathbf{x} - \mathbf{Ay}), \qquad \forall\, \mathbf{x} \in \mathscr{E}^m,\ \mathbf{y} \in \mathscr{E}^n. \qquad (1.1.4)$$

It is easy to see that the conditions in (1.1.4) are equivalent to

$$\mathbf{P_A^* M P_A} = \mathbf{M P_A}, \qquad \mathbf{M P_A A} = \mathbf{MA}, \quad \text{and} \quad R(\mathbf{P_A}) = R(\mathbf{A}), \qquad (1.1.5)$$

where $R(\mathbf{X})$ indicates the rank of the matrix $\mathbf{X}$. If $\mathbf{M} = \mathbf{I}$, the conditions (1.1.5) reduce to

$$\begin{aligned} \mathbf{P_A^2} &= \mathbf{P_A}, & \mathbf{P_A^*} &= \mathbf{P_A}, \\ \mathbf{P_A A} &= \mathbf{A}, & R(\mathbf{P_A}) &= R(\mathbf{A}). \end{aligned} \qquad (1.1.6)$$

If $\mathbf{M}$ is p.d., the conditions (1.1.5) reduce to

$$\begin{aligned} \mathbf{P}_{\mathbf{A}}^2 &= \mathbf{P}_{\mathbf{A}}, & (\mathbf{M}\mathbf{P}_{\mathbf{A}})^* &= \mathbf{M}\mathbf{P}_{\mathbf{A}}, \\ \mathbf{P}_{\mathbf{A}}\mathbf{A} &= \mathbf{A}, & R(\mathbf{P}_{\mathbf{A}}) &= R(\mathbf{A}). \end{aligned} \tag{1.1.7}$$

Adjoint Matrix (Transformation)

Let $\mathbf{A}$ be a $m \times n$ matrix or a transformation mapping $\mathscr{E}^n$ into $\mathscr{E}^m$. If $\mathbf{x} \in \mathscr{E}^n$, then the transformation is written $\mathbf{y} = \mathbf{A}\mathbf{x}$, $\mathbf{y} \in \mathscr{E}^m$. Let $(\cdot, \cdot)_m$ and $(\cdot, \cdot)_n$ denote inner products in $\mathscr{E}^m$ and $\mathscr{E}^n$, respectively. The adjoint matrix of $\mathbf{A}$, denoted by $\mathbf{A}^{\#}$, is defined by the relation

$$(\mathbf{A}\mathbf{x}, \mathbf{z})_m = (\mathbf{x}, \mathbf{A}^{\#}\mathbf{z})_n \text{ for all } \mathbf{x} \in \mathscr{E}^n, \ \mathbf{z} \in \mathscr{E}^m. \tag{1.1.8}$$

By definition, if $\mathbf{A}$ is a $m \times n$ matrix, then $\mathbf{A}^{\#}$ is a $n \times m$ matrix.

Note that, if $(\mathbf{y}_1, \mathbf{y}_2)_m = \mathbf{y}_2^*\mathbf{M}\mathbf{y}_1$, $(\mathbf{x}_1, \mathbf{x}_2)_n = \mathbf{x}_2^*\mathbf{N}\mathbf{x}_1$, where $\mathbf{M}$ and $\mathbf{N}$ are p.d. matrices, then $\mathbf{A}^*\mathbf{M} = \mathbf{N}\mathbf{A}^{\#}$ or $\mathbf{A}^{\#} = \mathbf{N}^{-1}\mathbf{A}^*\mathbf{M}$. (If $\mathbf{M}$ and $\mathbf{N}$ are identity matrices of order m and n, respectively, then $\mathbf{A}^{\#} = \mathbf{A}^*$.)

From (1.1.8)

$$(\mathbf{A}^{\#}\mathbf{z}, \mathbf{x})_n = (\mathbf{z}, \mathbf{A}\mathbf{x})_m \quad \text{so that } (\mathbf{A}^{\#})^{\#} = \mathbf{A}. \tag{1.1.9}$$

1.2 CANONICAL FORMS OF MATRICES

In this section we consider canonical reduction of matrices into simpler forms by post- and premultiplication. For proofs of most of the results mentioned reference may be made to Section 1b.2 of Rao (1965).

Hermite Canonical Form

A square matrix $\mathbf{H}$ is said to be in Hermite canonical form if its principal diagonal consists of only zeros and unities and all subdiagonal elements are zero such that when a diagonal element is zero the entire row is zero, and when a diagonal element is unity the rest of the elements in the column are zero. Alternatively $\mathbf{H}$ is in the Hermite canonical form if there exists a permutation matrix $\mathbf{P}$ such that

$$\mathbf{P}\mathbf{H}\mathbf{P}' = \begin{pmatrix} \mathbf{I}_r & \mathbf{B} \\ \mathbf{0} & \mathbf{0} \end{pmatrix},$$

where $r = R(\mathbf{H})$ and $\mathbf{B}$ could be arbitrary. A matrix $\mathbf{H}$ in Hermite canonical form is necessarily idempotent (i.e., $\mathbf{H}^2 = \mathbf{H}$).

Let $\mathbf{A}$ be a square matrix of order m. Then there exists a nonsingular matrix $\mathbf{C}$ of order m such that

$$\mathbf{C}\mathbf{A} = \mathbf{H}, \tag{1.2.1}$$

where $\mathbf{H}$ is in Hermite canonical form.

Diagonal Reduction

Let **A** be $m \times n$ matrix of rank r. Then there exist nonsingular square matrices **B** and **C** such that

$$\mathbf{BAC} = \begin{pmatrix} \mathbf{I}_r & \mathbf{0} \\ \mathbf{0} & \mathbf{0} \end{pmatrix}, \tag{1.2.2}$$

which give the representations

$$\mathbf{A} = \mathbf{B}^{-1} \begin{pmatrix} \mathbf{I}_r & \mathbf{0} \\ \mathbf{0} & \mathbf{0} \end{pmatrix} \mathbf{C}^{-1}$$

$$\mathbf{A} = \mathbf{DE} = \boldsymbol{\partial}_1 \boldsymbol{\varepsilon}'_1 + \cdots + \boldsymbol{\partial}_r \boldsymbol{\varepsilon}'_r, \tag{1.2.3}$$

where **D** is $m \times r$ matrix of rank r consisting of the first r column vectors $\boldsymbol{\partial}_1, \ldots, \boldsymbol{\partial}_r$ of $\mathbf{B}^{-1}$ and **E** is $r \times n$ matrix of rank r consisting of the first r row vectors $\boldsymbol{\varepsilon}'_1, \ldots, \boldsymbol{\varepsilon}'_r$ of $\mathbf{C}^{-1}$. The representation (1.2.3) is called the *rank factorization* of **A**.

Householder's Transformation

Triangular Reduction. Let **A** be a $m \times n$ matrix and $m \geq n$. Then there exists a unitary matrix **B** of order m such that

$$\mathbf{BA} = \begin{pmatrix} \mathbf{T} \\ \mathbf{0} \end{pmatrix}, \tag{1.2.4}$$

where **T** is an upper triangular matrix of order n and **0** is a null matrix of order $(m - n) \times n$.

Bidiagonalization. Let **A** be $m \times n$ matrix and $m \geq n$. Then there exist unitary matrices **B** and **C** such that **BAC** is in bidiagonal form, that is, all the elements of **BAC** are zero except possibly those in the main diagonal and the one above (or below) it.

Spectral Decomposition

Hermitian Matrix. Let **A** be $k \times k$ hermitian matrix (i.e., $\mathbf{A} = \mathbf{A}^*$). Then there exists a unitary matrix **U** such that

$$\mathbf{A} = \mathbf{U}^* \boldsymbol{\Delta} \mathbf{U}, \tag{1.2.5}$$

where $\boldsymbol{\Delta}$ is a real diagonal matrix. If $\delta_1, \ldots, \delta_k$ are diagonal elements of $\boldsymbol{\Delta}$, (1.2.5) can also be written as

$$\mathbf{A} = \delta_1 \mathbf{P}_1 + \cdots + \delta_k \mathbf{P}_k, \tag{1.2.6}$$

where $\mathbf{P}_i^2 = \mathbf{P}_i$, $\mathbf{P}_i \mathbf{P}_j = \mathbf{0}$ for $i \neq j$ and $\mathbf{P}_1 + \cdots + \mathbf{P}_k = \mathbf{I}$.

Normal Matrix. Let $\mathbf{A}$ be $n \times n$ normal matrix (i.e., $\mathbf{AA}^* = \mathbf{A}^*\mathbf{A}$). Then there exists a unitary matrix $\mathbf{U}$ such that

$$\mathbf{A} = \mathbf{U}^*\mathbf{\Delta U}, \tag{1.2.7}$$

where $\mathbf{\Delta}$ is a diagonal matrix. If $\delta_1, \ldots, \delta_k$ are distinct diagonal elements of $\mathbf{\Delta}$, then (1.2.7) can also be written as

$$\mathbf{A} = \delta_1\mathbf{P}_1 + \cdots + \delta_k\mathbf{P}_k, \tag{1.2.8}$$

where $\mathbf{P}_i^2 = \mathbf{P}_i$, $\mathbf{P}_i\mathbf{P}_j = \mathbf{0}$ for $i \neq j$ and $\mathbf{P}_1 + \cdots + \mathbf{P}_k = \mathbf{I}$. Thus a normal matrix is unitary congruent to a diagonal matrix.

Commuting Hermitian Matrices. Let $\mathbf{A}_1$ and $\mathbf{A}_2$ be two hermitian matrices such that $\mathbf{A}_1\mathbf{A}_2 = \mathbf{A}_2\mathbf{A}_1$. Then there exists a unitary matrix $\mathbf{U}$ such that

$$\mathbf{A}_1 = \mathbf{U}^*\mathbf{\Delta}_1\mathbf{U} \qquad \mathbf{A}_2 = \mathbf{U}^*\mathbf{\Delta}_2\mathbf{U}, \tag{1.2.9}$$

where $\mathbf{\Delta}_1$ and $\mathbf{\Delta}_2$ are diagonal matrices.

Singular Value Decomposition. Let $\mathbf{A}$ be a $m \times n$ matrix of rank r. Then there exist two unitary matrices $\mathbf{U}$ of order m and $\mathbf{V}$ of order n such that

$$\mathbf{U}^*\mathbf{AV} = \begin{pmatrix} \mathbf{\Delta} & \mathbf{0} \\ \mathbf{0} & \mathbf{0} \end{pmatrix}, \tag{1.2.10}$$

where $\mathbf{\Delta}$ is a diagonal matrix of rank r with real elements, all of which are positive.

From (1.2.10) we have

$$\mathbf{A} = \mathbf{E\Delta F}^*, \tag{1.2.11}$$

where $\mathbf{E}^*\mathbf{E} = \mathbf{I}_r = \mathbf{F}^*\mathbf{F}$. If $\delta_1, \ldots, \delta_r$ (not all of which need be distinct) are the diagonal elements of $\mathbf{\Delta}$, $\mathbf{e}_1, \ldots, \mathbf{e}_r$ are the columns of $\mathbf{E}$, and $\mathbf{f}_1, \ldots, \mathbf{f}_r$ are the columns of $\mathbf{F}$, then (1.2.11) can be written as

$$\mathbf{A} = \delta_1\mathbf{e}_1\mathbf{f}_1^* + \cdots + \delta_r\mathbf{e}_r\mathbf{f}_r^*. \tag{1.2.12}$$

We note that $\delta_1^2, \ldots, \delta_r^2$ are the common positive eigenvalues of $\mathbf{AA}^*$ and $\mathbf{A}^*\mathbf{A}$, $\mathbf{e}_i$ is the eigenvector of $\mathbf{AA}^*$ corresponding to δ_i^2, and $\mathbf{f}_i$ is the eigenvector of $\mathbf{A}^*\mathbf{A}$ corresponding to δ_i^2. The vectors $\mathbf{e}_1, \ldots, \mathbf{e}_r$ are orthonormal and so are $\mathbf{f}_1, \ldots, \mathbf{f}_r$.

Simultaneous Singular Value Decomposition. If $\mathbf{A}_1$ and $\mathbf{A}_2$ are two $m \times n$ matrices, there exist two unitary matrices $\mathbf{U}$ and $\mathbf{V}$ such that

$$\mathbf{U}^*\mathbf{A}_1\mathbf{V} = \begin{pmatrix} \mathbf{\Delta}_1 & \mathbf{0} \\ \mathbf{0} & \mathbf{0} \end{pmatrix}, \qquad \mathbf{U}^*\mathbf{A}_2\mathbf{V} = \begin{pmatrix} \mathbf{\Delta}_2 & \mathbf{0} \\ \mathbf{0} & \mathbf{0} \end{pmatrix}, \tag{1.2.13}$$

where $\mathbf{\Delta}_1$ and $\mathbf{\Delta}_2$ are both diagonal matrices with real elements; $\mathbf{\Delta}_1$ has no negative elements if and only if $\mathbf{A}_1\mathbf{A}_2^*$ and $\mathbf{A}_1^*\mathbf{A}_2$ are both hermitian matrices.

***Singular Value Decomposition with Respect to* M *and* N.** Let **A** be a $m \times n$ matrix of rank r and **M** and **N** be p.d. matrices of orders m and n, respectively. Then **A** can be expressed in the form

$$\mathbf{MAN} = \mu_1 \boldsymbol{\xi}_1 \boldsymbol{\eta}_1^* + \cdots + \mu_r \boldsymbol{\xi}_r \boldsymbol{\eta}_r^*, \tag{1.2.14}$$

where $\boldsymbol{\xi}_i^* \mathbf{M}^{-1} \boldsymbol{\xi}_j = 0$ for $i \neq j$ and $= 1$ for $i = j$, and $\boldsymbol{\eta}_i^* \mathbf{N}^{-1} \boldsymbol{\eta}_j = 0$ for $i \neq j$ and $= 1$ for $i = j$.

In (1.2.14) $\mu_1^2, \ldots, \mu_r^2$ are the nonzero eigenvalues of $\mathbf{A}^*\mathbf{MA}$ with respect to $\mathbf{N}^{-1}$ or of $\mathbf{ANA}^*$ with respect to $\mathbf{M}^{-1}$, $\boldsymbol{\xi}_i$ is the eigenvector of $\mathbf{ANA}^*$ with respect to $\mathbf{M}^{-1}$ corresponding to the eigenvalue μ_i^2, and $\boldsymbol{\eta}_i$ is the eigenvector of $\mathbf{A}^*\mathbf{MA}$ with respect to $\mathbf{N}^{-1}$ corresponding to the eigenvalue μ_i^2.

Polar Reduction. Let **A** be a square matrix. Then there exists a n.n.d. matrix **G** such that

$$\mathbf{A} = \mathbf{HG}, \qquad \mathbf{H} \text{ unitary}. \tag{1.2.15}$$

In fact, **G** is the hermitian square root of $\mathbf{A}^*\mathbf{A}$. **H** is unique if $|\mathbf{A}| \neq 0$.

Similarly, $\mathbf{A} = \mathbf{FH}$, where **H** is unitary and **F** is the hermitian square root of $\mathbf{AA}^*$.

Polar Representations. A complex orthogonal matrix **M** can always be represented in the form

$$\mathbf{M} = \mathbf{R}e^{i\mathbf{K}}, \tag{1.2.16}$$

where **R** is real orthogonal and **K** is real antisymmetric. A unitary matrix **U** can always be represented as

$$\mathbf{U} = \mathbf{R}e^{i\mathbf{S}}, \tag{1.2.17}$$

where **R** is real orthogonal and **S** is real symmetric.

1.3 CHARACTERISTIC FUNCTION, MINIMUM POLYNOMIALS

With every square matrix **A** of order n is associated the matric polynomial

$$\lambda\mathbf{I} - \mathbf{A}, \tag{1.3.1}$$

which is called the characteristic matrix of **A**. Its determinant is a polynomial in λ:

$$f(\lambda) = |\lambda\mathbf{I} - \mathbf{A}| = \lambda^n + c_{n-1}\lambda^{n-1} + \cdots + c_0 \tag{1.3.2}$$

called the characteristic function of **A**. The equation $f(\lambda) = 0$ is called the characteristic equation of **A**. Its roots are known as the eigenvalues (characteristic roots, latent roots) of **A**. In (1.3.2)

$$c_{n-r} = (-1)^r tr_{(r)}\mathbf{A}, \tag{1.3.3}$$

where $tr_{(r)}(\mathbf{A})$ is the sum of the determinants of all the $\binom{n}{r}$ principal minors of order r of **A**.The trace of **A**, $tr_{(1)}(\mathbf{A})$, is the sum of the diagonal elements of **A** and $tr_{(n)}(\mathbf{A}) = |\mathbf{A}|$, the determinant of **A**.

Cayley–Hamilton Theorem

Every square matrix satisfies its own characteristic equation; that is, if **A** is a square matrix of order n and (1.3.2) is its characteristics function, then

$$f(\mathbf{A}) = \mathbf{A}^n + c_{n-1}\mathbf{A}^{n-1} + \cdots + c_0\mathbf{I} = \mathbf{0}. \tag{1.3.4}$$

Minimum Polynomial

For a matrix **A** of order n let m be the smallest integer for which the powers **I**, **A**, . . . , $\mathbf{A}^m$ are linearly dependent. Then we have an equation

$$\mathbf{A}^m + d_{m-1}\mathbf{A}^{m-1} + \cdots + d_0\mathbf{I} = \mathbf{0} \tag{1.3.5}$$

in which the coefficient d_m of $\mathbf{A}^m$ is not zero, hence may be taken as 1. The polynomial $\lambda^m + d_{m-1}\lambda^{m-1} + \cdots + d_0$ is known as the minimum polynomial of **A**.

1.4 EQUIVALENCE AND SIMILARITY

Smith's Canonical Matrices

Let $\mathscr{F}$ be a field and $\mathscr{F}(\lambda)$ the polynomial domain consisting of all polynomials in λ with coefficients chosen from $\mathscr{F}$. We consider the usual elementary operations on matrix with scalars from $\mathscr{F}(\lambda)$. An elementary matrix is obtained when an elementary operation is carried out on the rows (or columns) of a unit matrix **I**. Let **A** and **B** be matrices over $\mathscr{F}(\lambda)$; **B** is said to be *equivalent* to **A** over $\mathscr{F}(\lambda)$ iff $\mathbf{B} = \mathbf{PAQ}$, where **P** and **Q** are products of elementary matrices.

Each matrix **A** of order $m \times n$ and rank r with elements in $\mathscr{F}(\lambda)$ is equivalent over $\mathscr{F}(\lambda)$ to a matrix $\mathbf{B} = \{b_{ij}\}$ of the same order such that $b_{ii} = f_i(\lambda)$ if $i = 1, 2, \ldots, r$, $b_{ij} = 0$; otherwise, where $f_i(\lambda)$ are certain monic polynomials and $f_i(\lambda)$ divides $f_{i+1}(\lambda)$, $i = 1, 2, \ldots, (r-1)$. The matrix **B** is uniquely determined by the given matrix and is said to be in *Smith's canonical form*.

Further

$$f_t(\lambda) = \frac{g_t(\lambda)}{g_{t-1}(\lambda)}, \qquad t = 1, 2, \ldots, r \tag{1.4.1}$$

where $g_0(\lambda) = 1$ and $g_t(\lambda)$ is the greatest common divisor of all the $\binom{m}{t}\binom{n}{t}$ subdeterminants of **A**. The *invariant factors* of a matrix **A** over $\mathscr{F}(\lambda)$ are defined as the polynomials $f_i(\lambda)$, $i = 1, 2, \ldots, r$, appearing in the diagonal of the Smith canonical matrix equivalent to **A**.

N.S. Condition for Equivalence. Two $m \times n$ matrices over $\mathscr{F}(\lambda)$ are equivalent over $\mathscr{F}(\lambda)$ iff they have the same invariant factors.

If **A** and **B** are square matrices of order n over $\mathscr{F}$, **B** is said to be *similar* to **A** over $\mathscr{F}$ if there is a nonsingular matrix **P** with elements in $\mathscr{F}$, such that

$$\mathbf{B} = \mathbf{P}^{-1}\mathbf{A}\mathbf{P}. \tag{1.4.2}$$

N.S. Condition for Similarity. Matrices **A** and **B** of order n over $\mathscr{F}$ are similar iff $\lambda\mathbf{I} - \mathbf{A}$ and $\lambda\mathbf{I} - \mathbf{B}$ have the same invariant factors over $\mathscr{F}(\lambda)$, hence iff $\lambda\mathbf{I} - \mathbf{A}$ and $\lambda\mathbf{I} - \mathbf{B}$ are equivalent over $\mathscr{F}(\lambda)$.

The similarity invariants of a $n \times n$ matrix **A** are defined to be the invariant factors of $\lambda\mathbf{I} - \mathbf{A}$ over $\mathscr{F}(\lambda)$. The similarity invariant of **A**, which has the highest degree, namely, $f_n(\lambda)$, provides the minimum polynomial of **A**.

Let us write the characteristic polynomial of **A** as $f(\lambda) = [p_1(\lambda)]^{e_1} \times [p_2(\lambda)]^{e_2} \cdots [p_s(\lambda)]^{e_s}$, where $p_1(\lambda), \ldots, p_s(\lambda)$ are distinct monic polynomials which are irreducible over the scalar field $\mathscr{F}$ where e_j's are positive integers. Note that if $\mathscr{F}$ is the field of complex numbers each $p_j(\lambda)$ is necessarily of degree 1, that is, of the form $\lambda - \lambda_j$ and e_j is the algebraic multiplicity of λ_j. Clearly, $f_i(\lambda) = [p_1(\lambda)]^{e_{i1}}[p_2(\lambda)]^{e_{i2}} \ldots [p_s(\lambda)]^{e_{is}}$, where some of the e_{ij}'s could be zero, but if e_{ij} is a positive integer $e_{(i+1)j}$ is also positive and $e_{(i+1)j} \geq e_{ij}$. These polynomials $[p_j(\lambda)]^{e_{ij}}$ with $e_{ij} > 0$ which appear as factors in the similarity invariants are called the elementary divisors of **A** over $\mathscr{F}$.

If $p(\lambda)$ is an irreducible polynomial and $g(\lambda) = [p(\lambda)]^e = \lambda^k - a_1\lambda^{k-1} - \cdots - a_k$, the matrix

$$\mathbf{C}(g) = \begin{pmatrix} 0 & 0 & \ldots & 0 & a_k \\ 1 & 0 & \cdots & 0 & a_{k-1} \\ 0 & 1 & \cdots & 0 & a_{k-2} \\ \cdot & \cdot & \cdots & \cdot & \cdots \\ 0 & 0 & & 1 & a_1 \end{pmatrix} \tag{1.4.3}$$

is called the companion matrix associated with the polynomial $g(\lambda)$. For the companion matrix the characteristic and minimum polynomials are both equal to $[p(\lambda)]^e$.

Matrix Similar to **A.** Every matrix **A** of order n is similar to the direct sum of the companion matrices of its elementary divisors; that is, **A** is similar to diag $(\mathbf{C}_1, \ldots, \mathbf{C}_u)$, where $\mathbf{C}_1, \mathbf{C}_2, \ldots, \mathbf{C}_u$ are the companion matrices of its elementary divisors.

If $p(\lambda)$ is of degree t, so that $k = te$, the partitioned matrix of order $k \times k$,

$$\mathbf{C}_e(p) = \begin{pmatrix} \mathbf{C}(p) & \mathbf{0} & \cdots & \mathbf{0} & \mathbf{0} \\ \mathbf{N} & \mathbf{C}(p) & \cdots & \mathbf{0} & \mathbf{0} \\ \mathbf{0} & \mathbf{N} & \cdots & \mathbf{0} & \mathbf{0} \\ \cdot & \cdot & \cdots & \cdot & \cdot \\ \mathbf{0} & \mathbf{0} & \cdots & \mathbf{N} & \mathbf{C}(p) \end{pmatrix} \tag{1.4.4}$$

is called the hypercompanion matrix of $[p(\lambda)]^e$, where $\mathbf{C}(p)$, of order t, is as defined above and $\mathbf{N} = \{n_{ij}\}$ is a matrix of order t, with $n_{1t} = 1$ as its only nonzero element. The effect of the matrices **N** is to produce an unbroken line of 1 just below the main diagonal of $\mathbf{C}_e(p)$.

Jordan Canonical Form. Every matrix of order n is similar to the direct sum of the hypercompanion matrices of its elementary divisors.

When t is 1 and $p(\lambda) = \lambda - a$, $\mathbf{C}_e(p)$ is the well-known lower Jordan matrix,

$$\mathbf{J}_e(a) = \begin{pmatrix} a & 0 & \cdots & 0 & 0 \\ 1 & a & \cdots & 0 & 0 \\ 0 & 1 & \cdots & 0 & 0 \\ \cdot & \cdot & \cdots & \cdot & \cdot \\ 0 & 0 & \cdots & 1 & a \end{pmatrix}. \tag{1.4.5}$$

A matrix **A** of order n is said to be *semisimple* if it is similar to a diagonal matrix. Clearly, **A** is semisimple iff its eigenvectors span $\mathscr{E}^n$. The following theorem is also of interest.

N.S. Condition for **A** ***to be Semisimple.*** A matrix **A** of order n is semisimple iff the elementary divisors of **A** are of degree 1 (or, equivalently, the minimum polynomial of **A** is the product *of* distinct linear factors).

1.5 SPECIAL PRODUCTS OF MATRICES

Kronecker Product

Let $\mathbf{A} = (a_{ij})$ and $\mathbf{B} = (b_{ij})$ be $m \times n$ and $p \times q$ matrices, respectively. Then the Kronecker product

$$\mathbf{A} \otimes \mathbf{B} = (a_{ij}\mathbf{B}) \tag{1.5.1}$$

is a $mp \times nq$ matrix expressible as a partitioned matrix with $a_{ij}\mathbf{B}$ as the (i, j)th partition, $i = 1, \ldots, m$; $j = 1, \ldots, n$.

The formal rules for operating with Kronecker products are as follows:

(i) $\mathbf{0} \otimes \mathbf{A} = \mathbf{A} \otimes \mathbf{0} = \mathbf{0}$
(ii) $(\mathbf{A}_1 + \mathbf{A}_2) \otimes \mathbf{B} = (\mathbf{A}_1 \otimes \mathbf{B}) + (\mathbf{A}_2 \otimes \mathbf{B})$
(iii) $\mathbf{A} \otimes (\mathbf{B}_1 + \mathbf{B}_2) = \mathbf{A} \otimes \mathbf{B}_1 + \mathbf{A} \otimes \mathbf{B}_2$
(iv) $\alpha\mathbf{A} \otimes \beta\mathbf{B} = \alpha\beta(\mathbf{A} \otimes \mathbf{B})$
(v) $\mathbf{A}_1\mathbf{A}_2 \otimes \mathbf{B}_1\mathbf{B}_2 = (\mathbf{A}_1 \otimes \mathbf{B}_1)(\mathbf{A}_2 \otimes \mathbf{B}_2)$
(vi) $(\mathbf{A} \otimes \mathbf{B})^{-1} = \mathbf{A}^{-1} \otimes \mathbf{B}^{-1}$ (if the inverses exist).

It is seen from the definition of a generalized inverse (Chapter 2) that the more general result

(vii) $(\mathbf{A} \otimes \mathbf{B})^{-} = \mathbf{A}^{-} \otimes \mathbf{B}^{-}$($\mathbf{A}^{-}$ and $\mathbf{B}^{-}$ are g-inverses) is true whatever **A** and **B** may be.

Let $\mathbf{x}$ be a $nr \times 1$ vector obtained from a $n \times r$ matrix $\mathbf{X}$ by writing the columns of $\mathbf{X}$ one below the other [the $(i + 1)$th below the ith]. Similarly, let $\mathbf{c}$ be a $ms \times 1$ vector derived from a matrix $\mathbf{C}$. Consider the equation in $\mathbf{X}$

$$\mathbf{AXB} = \mathbf{C}, \tag{1.5.2}$$

where $\mathbf{A}$ is a $m \times n$ matrix and $\mathbf{B}$ is a $r \times e$ matrix. Given $\mathbf{A}, \mathbf{B}, \mathbf{C}$, the problem is to solve for $\mathbf{X}$. Using Kronecker product of $\mathbf{A}$ and $\mathbf{B}$ we can write (1.5.2) in the usual form of linear equations

$$(\mathbf{A} \otimes \mathbf{B}')\mathbf{x} = \mathbf{c}, \tag{1.5.3}$$

solve for $\mathbf{x}$, and write the solution in the matrix form $\mathbf{X}$.

Hadamard Product

If $\mathbf{A} = (a_{ij})$ and $\mathbf{B} = (b_{ij})$ be each $m \times n$ matrices, their Hadamard product is the $m \times n$ matrix of elementwise products

$$\mathbf{A}*\mathbf{B} = (a_{ij}b_{ij}). \tag{1.5.4}$$

Some formal rules for operating with Hadamard products are as follows:

(i) $\mathbf{A}*\mathbf{0} = \mathbf{0}$

(ii) $\mathbf{A}*\mathbf{ee}' = \mathbf{A} = \mathbf{ee}'*\mathbf{A}$ where $\mathbf{e}' = (1, 1, \ldots, 1)$.
(iii) $\mathbf{A}*\mathbf{B} = \mathbf{B}*\mathbf{A}$
(iv) $(\mathbf{A} + \mathbf{B})*\mathbf{C} = (\mathbf{A}*\mathbf{C}) + (\mathbf{B}*\mathbf{C})$
(v) $\text{tr}\,\mathbf{AB} = \mathbf{e}'(\mathbf{A}*\mathbf{B}')\mathbf{e}$ where $\mathbf{A}$ and $\mathbf{B}$ are square matrices.

Schur's Lemma

When $\mathbf{A}$ and $\mathbf{B}$ are p.s.d. matrices, so is their Hadamard product $\mathbf{A}*\mathbf{B}$. When $\mathbf{A}$ and $\mathbf{B}$ are p.d., so also is $\mathbf{A}*\mathbf{B}$.

A possible generalization of Hadamard product is

$$\mathbf{A}*\mathbf{B} = (\mathbf{A}_{ij}\mathbf{B}_{ij}), \tag{1.5.5}$$

where $\mathbf{A}_{ij}$ and $\mathbf{B}_{ij}$ are matrix partitions of $\mathbf{A}$ and $\mathbf{B}$, respectively; the (i, j)th partition of $\mathbf{A}*\mathbf{B}$ is $\mathbf{A}_{ij}\mathbf{B}_{ij}$.

A New Type of Product (Khatri-Rao)

This new product was introduced in a paper by Khatri and Rao (1968), which, in a general form, is as follows. Let

$$\mathbf{A} = (\mathbf{A}_1 \vdots \cdots \vdots \mathbf{A}_k) \text{ and } \mathbf{B} = (\mathbf{B}_1 \vdots \cdots \vdots \mathbf{B}_k) \tag{1.5.6}$$

TABLE 1

Function	Symbol	Description
Transpose	$\mathbf{A}'$	Matrix with (i, j)th element $= a_{ji}$
Conjugate transpose	$\mathbf{A}^*$	Matrix with (i, j)th element $= \bar{a}_{ji}$
Rank	$R(\mathbf{A})$	The number of independent rows or columns of $\mathbf{A}$
Trace	$tr(\mathbf{A})$	Σa_{ii}
Column space	$\mathscr{M}(\mathbf{A})$	Vector space generated by the columns of $\mathbf{A}$
Orthogonal complement	$\mathbf{A}^{\perp}$	$\mathscr{O}(\mathbf{A}) = \mathscr{M}(\mathbf{A}^{\perp})$ is orthogonal complement of $\mathscr{M}(\mathbf{A})$
Adjoint	$\mathbf{A}^{\#}$	Matrix such that $(\mathbf{x}, \mathbf{Ay}) = (\mathbf{A}^{\#}\mathbf{x}, \mathbf{y})\ \forall\ \mathbf{x}, \mathbf{y}$ where $(\cdot\,, \cdot)$ is inner product
Parallel sum (1)	$\mathbf{A} \mp \mathbf{B}$	$(\mathbf{A}^+ + \mathbf{B}^+)^+$ | where $\mathbf{X}^+$ indicates
Parallel sum (2)	$\mathbf{A} \mp \mathbf{B}$	$\mathbf{A}(\mathbf{A} + \mathbf{B})^+\mathbf{B}$ | Moore-Penrose inverse
Singular value (principal value)		Square root of a non-null eigenvalue of $\mathbf{AA}^*$
Right singular vector		Eigenvector of $\mathbf{A}^*\mathbf{A}$
Left singular vector		Eigenvector of $\mathbf{AA}^*$
Spectrum (square matrices only)	$\sigma(\mathbf{A})$	The set of distinct eigenvalues of $\mathbf{A}$

be two partitioned matrices with the same number of partitions. The new product is

$$\mathbf{A} \odot \mathbf{B} = (\mathbf{A}_1 \otimes \mathbf{B}_1 \vdots \cdots \vdots \mathbf{A}_k \otimes \mathbf{B}_k), \tag{1.5.7}$$

where $\mathbf{A}_i \otimes \mathbf{B}_i$ is the Kronecker product of $\mathbf{A}_i$ and $\mathbf{B}_i$. The case in which each partition in **A** and **B** is a column has been studied in some detail by Khatri and Rao (1968) and Rao (1970).

The following may be easily verified:

$$(\mathbf{A} \odot \mathbf{B}) \odot \mathbf{C} = \mathbf{A} \odot (\mathbf{B} \odot \mathbf{C}) \tag{1.5.8}$$

$$(\mathbf{T}_1 \otimes \mathbf{T}_2)(\mathbf{A} \odot \mathbf{B}) = \mathbf{T}_1\mathbf{A} \odot \mathbf{T}_2\mathbf{B}. \tag{1.5.9}$$

1.6 NOTATIONS

Let $\mathbf{A} = (a_{ij})$ be a $m \times n$ matrix. Some functions of **A** and the symbols used are described in Table 1.

Definitions of special matrices are given in Table 2, where **A** is taken to be a square matrix of order n (except for partial isometry which could be defined for rectangular matrices also).

TABLE 2

Type of Matrix **A**	Definition
Symmetric	$\mathbf{A} = \mathbf{A}'$
Hermitian	$\mathbf{A} = \mathbf{A}^*$
Idempotent	$\mathbf{A}^2 = \mathbf{A}$
Nilpotent	$\mathbf{A}^k = \mathbf{0}$ for some integer k
Positive definite (p.d.)	$\mathbf{x}^*\mathbf{A}\mathbf{x} > 0$ for all non-null $\mathbf{x}$
Positive semi-definite (p.d.)	$\mathbf{x}^*\mathbf{A}\mathbf{x} \geq 0$ for all $\mathbf{x}$ $\mathbf{x}^*\mathbf{A}\mathbf{x} = 0$ for some non-null $\mathbf{x}$
Non-negative definite (n.n.d.)	$\mathbf{x}^*\mathbf{A}\mathbf{x} \geq 0$ for all $\mathbf{x}$
Normal	$\mathbf{A}\mathbf{A}^* = \mathbf{A}^*\mathbf{A}$
EPr	$\mathbf{A}\mathbf{x} = \mathbf{0} \Leftrightarrow \mathbf{A}^*\mathbf{x} = \mathbf{0}$
EP	$\mathbf{A}\mathbf{E}\mathbf{x} = \mathbf{0} \Leftrightarrow \mathbf{A}^*\mathbf{F}\mathbf{x} = 0$, for some **E**, **F** with columns providing reciprocal bases of $\mathscr{E}^n$
Semisimple	$\mathbf{A} = \mathbf{P}\mathbf{\Delta}\mathbf{P}^{-1}$ for some **P** and **Δ** diagonal
Orthogonal	$\mathbf{A}\mathbf{A}' = \mathbf{A}'\mathbf{A} = \mathbf{I}$
Unitary	$\mathbf{A}\mathbf{A}^* = \mathbf{A}^*\mathbf{A} = \mathbf{I}$
Unitary (general inner product)	$\mathbf{A}\mathbf{A}^{\#} = \mathbf{A}^{\#}\mathbf{A} = \mathbf{I}$
Partial isometry	$\mathbf{A}^{\#} = \mathbf{A}^{+}$

TABLE 3.1

GENERAL INVERSES

Reference to Section	Notation	Condition	Purpose ($\mathbf{Ax} = \mathbf{y}$ consistent)
2.1	$\mathbf{A}_L^{-1}$	$\mathbf{GA} = \mathbf{I}$	Solving $\mathbf{Ax} = \mathbf{y}$ when $R(\mathbf{A}) = n$
2.1	$\mathbf{A}_R^{-1}$	$\mathbf{AG} = \mathbf{I}$	Solving $\mathbf{Ax} = \mathbf{y}$ when $R(\mathbf{A}) = m$.
2.2	$\mathbf{A}^-$	$\mathbf{AGA} = \mathbf{A}$	Solving $\mathbf{Ax} = \mathbf{y}$
2.5	$\mathbf{A}_r^-$	$\mathbf{AGA} = \mathbf{A}, \mathbf{GAG} = \mathbf{G}$	Solving $\mathbf{Ax} = \mathbf{y}$
2.6[a]	$\mathbf{A}_b^-$	$\mathbf{E}_1\mathbf{G} = (\mathbf{AE}_1')_L^{-1}$ $\mathbf{E}_2\mathbf{GA} = \mathbf{0}$	Basic solution of $\mathbf{Ax} = \mathbf{y}$ in programming problems
2.7	$\mathbf{A}_c^-$	$\mathbf{AGA} = \mathbf{A}$, $R(\mathbf{G}) = \min(m, n)$	Inverse with maximum rank for testing consistency of equation

[a] In (2.6) $\mathbf{E}' = (\mathbf{E}_1' \vdots \mathbf{E}_2')$ is a permutation matrix.

TABLE 3.2

BASIC TYPES OF g-INVERSE

(conditions valid for any inner products in $\mathscr{E}^m, \mathscr{E}^n$)

Reference to Section	Notation	Equivalent Conditions	Type of Solution of $\mathbf{Ax} = \mathbf{y}$
3.1	$\mathbf{A}_m^-$	$\mathbf{GA} = \mathbf{P_A} \Leftrightarrow \mathbf{AGA} = \mathbf{A}, (\mathbf{GA})^{\#} = \mathbf{GA}$	minimum norm
3.2	$\mathbf{A}_l^-$	$\mathbf{AG} = \mathbf{P_A} \Leftrightarrow \mathbf{AGA} = \mathbf{A}, (\mathbf{AG})^{\#} = \mathbf{AG}$	least squares
3.3	$\mathbf{A}^+$	$\left.\begin{array}{l}\mathbf{AG} = \mathbf{P_A}\\ \mathbf{GA} = \mathbf{P_G}\end{array}\right\} \Leftrightarrow \left\{\begin{array}{l}\mathbf{AGA} = \mathbf{A}, (\mathbf{GA})^{\#} = \mathbf{GA}\\ \mathbf{GAG} = \mathbf{G}, (\mathbf{AG})^{\#} = \mathbf{AG}\end{array}\right.$	minimum norm least squares

TABLE 3.3

BASIC TYPES OF g-INVERSE

(conditions when $(\mathbf{x}, \mathbf{y})_m = \mathbf{y}^*\mathbf{Mx}$, $(\mathbf{x}, \mathbf{y})_n = \mathbf{y}^*\mathbf{Nx}$)

Reference to Section	Notation	Condition	Type of Solution of $\mathbf{Ax} = \mathbf{y}$
3.1	$\mathbf{A}_{m(\mathbf{N})}^-$	$\mathbf{AGA} = \mathbf{A}, (\mathbf{GA})^*\mathbf{N} = \mathbf{NGA}$	minimum **N**-norm
3.2	$\mathbf{A}_{l(\mathbf{M})}^-$	$\mathbf{AGA} = \mathbf{A}, (\mathbf{AG})^*\mathbf{M} = \mathbf{MAG}$	**M**-least squares
3.3	$\mathbf{A}_{\mathbf{MN}}^+$	$\mathbf{AGA} = \mathbf{A}, (\mathbf{AG})^*\mathbf{M} = \mathbf{MAG}$ $\mathbf{GAG} = \mathbf{G}, (\mathbf{GA})^*\mathbf{N} = \mathbf{NGA}$	minimum **N**-norm **M**-least squares

Classification of g-inverses are given in Tables 3.1–3.5. In the conditions **G** denotes a g-inverse of **A** of order $m \times n$ and rank r. $\mathbf{P_X}$ denotes the projection operator onto $\mathcal{M}(\mathbf{X})$.

In Table 3.3, **M** and **N** are p.d. matrices. Similar conditions can be written when **M** and **N** are n.n.d. (see Section 3.1–3.3). We use the notations $\mathbf{A}_m^-$, $\mathbf{A}_l^-$, $\mathbf{A}^+$ when **M**, **N** are not specified and generally when $\mathbf{M} = \mathbf{I}$, $\mathbf{N} = \mathbf{I}$.

TABLE 3.4
SPECIAL TYPES OF g-INVERSE

Reference to Section	Notation	Condition	Purpose
4.2	$\mathbf{A}_\chi^-$	$\mathcal{M}(\mathbf{G}) \subset \mathcal{M}(\mathbf{A})$	**A** is square $\ni R(\mathbf{A}) = R(\mathbf{A}^2)$ $(\mathbf{A}_\chi^-)^k$ is a g-inverse of $\mathbf{A}^k$ for all integral k
4.3	$\mathbf{A}_\rho^-$	$\mathcal{M}(\mathbf{G}^*) \subset \mathcal{M}(\mathbf{A}^*)$	**A** is square $\ni R(\mathbf{A}) = R(\mathbf{A}^2)$; $(\mathbf{A}_\rho^-)^k$ is a g-inverse of $\mathbf{A}^k$ for all integral k
4.3	$\mathbf{A}_{\rho\chi}^-$	$\mathcal{M}(\mathbf{G}) \subset \mathcal{M}(\mathbf{A})$ $\mathcal{M}(\mathbf{G}^*) \subset \mathcal{M}(\mathbf{A}^*)$	**A** is square $\ni R(\mathbf{A}) = R(\mathbf{A}^2)$; $(\mathbf{A}_{\rho\chi}^-)^k$ is a g-inverse of $\mathbf{A}^k$ for all integral k
4.7	$\mathbf{A}_e^-$	$\mathbf{G} = \mathbf{X}\mathbf{D}^-\mathbf{X}^{-1}$ $\mathbf{A} = \mathbf{X}\mathbf{D}\mathbf{X}^{-1}$ **D** is in Jordan form, $\mathbf{D}^-$ as in (4.6.8)	Inverse with same eigenvectors and eigenvalues reciprocal to those of **A**

TABLE 3.5
CONSTRAINED INVERSES
(**F** and **E** are given and **V** and **U** are arbitrary matrices)

Reference to Theorem	Notation	Condition for Existence	Algebraic Expression
4.11.1	$\mathbf{A}_{cC}$	$R(\mathbf{AE}) = R(\mathbf{E})$	$\mathbf{E}(\mathbf{AE})^-$
4.11.3	$\mathbf{A}_{rR}$	$R(\mathbf{FA}) = R(\mathbf{F})$	$(\mathbf{FA})^-\mathbf{F}$
4.11.5	$\mathbf{A}_{cR}$	$R(\mathbf{FAE}) = R(\mathbf{F})$	$\mathbf{E}(\mathbf{FAE})^-\mathbf{F} + \mathbf{E}[\mathbf{I} - (\mathbf{FAE})^-\mathbf{FAE}]\mathbf{V}$
4.11.6	$\mathbf{A}_{rC}$	$R(\mathbf{FAE}) = R(\mathbf{E})$	$\mathbf{E}(\mathbf{FAE})^-\mathbf{F} + \mathbf{V}[\mathbf{I} - (\mathbf{FAE})(\mathbf{FAE})^-]\mathbf{F}$
4.11.7	$\mathbf{A}_{crCR}$	$R(\mathbf{FAE}) = R(\mathbf{E}) = R(\mathbf{F})$	$\mathbf{E}(\mathbf{FAE})^-\mathbf{F}$

The type of constraints on the inverse **G** of **A**, considered in Table 3.5, are

c: **G** maps vectors of $\mathscr{E}^m$ into $\mathscr{M}(\mathbf{E})$

r: **G*** maps vectors of $\mathscr{E}^n$ into $\mathscr{M}(\mathbf{F})$

C: **GA** is an identity in $\mathscr{M}(\mathbf{E})$

R: (**AG**)* is an identity in $\mathscr{M}(\mathbf{F})$

The subscript of **A** in the notation indicates the conditions satisfied by **G**.

TABLE 3.6

GLOSSARY OF SYMBOLS AND TERMINOLOGY

Symbols and Reference to Sections		Terminology used in the Book	Alternative Terminology Used in Literature
$\mathbf{A}_L^{-1}$	(2.1)	Left inverse	—
$\mathbf{A}_R^{-1}$	(2.1)	Right inverse	—
$\mathbf{A}^-$	(2.2)	Generalized inverse g-inverse	Pseudoinverse (Sheffield, 1958) Inverse (Bjerhammer, 1958)
$\mathbf{A}_r^-$	(2.5)	Reflexive g-inverse	Semi-inverse (Frame 1964) Reciprocal inverse (Bjerhammer, 1958) A particular case is called pseudoinverse (Rao, 1955)
$\mathbf{A}_m^-$	(3.1)	Minimum norm g-inverse	—
$\mathbf{A}_l^-$	(3.2)	Least-squares g-inverse	—
$\mathbf{A}_{mr}^-$	(3.1)	Minimum norm reflexive g-inverse	Weak generalized inverse (Goldman and Zelen, 1964)
$\mathbf{A}_{lr}^-$	(3.2)	Least-squares reflexive g-inverse	Normalized generalized inverse (Rohde, 1965)
$\mathbf{A}^+$	(3.3)	Minimum norm least-squares g-inverse Moore-Penrose inverse	General reciprocal (Moore, 1920, 1935) Generalized inverse (Penrose, 1955)
$\mathbf{A}_\rho^-$	(4.3)	ρ-inverse	—
A_χ^-	(4.2)	χ-inverse	—
$A_{\rho\chi}^-$	(4.3)	$\rho\chi$-inverse	Group-inverse (Erdelyi, 1967)
$\mathbf{A}_{\rho\chi^*}^-$	(4.10)	$\rho\chi^*$-inverse	—
$\mathbf{A}_{\rho^*\chi}^-$	(4.10)	$\rho^*\chi$-inverse	—
$(\mathbf{A}^m)_L$	(4.10)	Left power inverse of $\mathbf{A}^m$ (Cline, 1968)	—
$(\mathbf{A}^m)_R$	(4.10)	Right power inverse of A^m (Cline, 1968)	—

TABLE 3.6—*continued*
GLOSSARY OF SYMBOLS AND TERMINOLOGY

Symbols and Reference to Sections		Terminology used in the Book	Alternative Terminology Used in Literature
$\mathbf{A}_b^-$	(2.6)	g-inverse providing a basic solution (Rosen, 1964)	—
$\mathbf{A}_c^-$	(2.7)	Maximum rank g-inverse	—
A_{com}^-	(4.5)	Commuting g-inverse (Engelfield, 1966)	—
$\mathbf{A}_e^-$	(4.7)	g-inverse with eigen value property	—
$\mathbf{A}_X^{\oplus}$	(4.7)	$\mathbf{A}_e^-$ obtained from Jordan decomposition $\mathbf{A} = \mathbf{XDX}^{-1}$	A special choice of **X** gives a unique pseudo-inverse (Scroggs and Odell, 1966)
$\mathbf{A}_{com}$	(4.10)	Pseudoinverse with commuting property* (Drazin, 1958)	—
$\mathbf{A}_{cC}$	(4.11)	*cC*-constrained inverse *	—
$\mathbf{A}_{rR}$	(4.11)	*rR*-constrained inverse*	—
$\mathbf{A}_{cR}$	(4.11)	*cR*-constrained inverse*	—
$\mathbf{A}_{rC}$	(4.11)	*rC*-constrained inverse*	A special case is called constrained inverse (Bott and Duffin, 1963)
$\mathbf{A}_{crCR}$	(4.11)	*crCR*-constrained inverse *	—

* Not necessarily a g-inverse.

Note: The terminology and symbols for $\mathbf{A}^-$, $\mathbf{A}_r^-$ were developed in Rao (1962), for $\mathbf{A}_m^-$, $\mathbf{A}_l^-$, $\mathbf{A}_{mr}^-$, $\mathbf{A}_{lr}^-$, $\mathbf{A}_c^-$, $\mathbf{A}_e^-$ in Rao (1967), and for $\mathbf{A}_\rho^-$, $\mathbf{A}_\chi^-$ and $\mathbf{A}_{\rho\chi}^-$ in Mitra (1968).

Some of these symbols are standard and some are newly introduced by the authors. It is hoped that in the course of time future contributors to the subject will agree on a uniform list of notations and terminology to prevent the confusion that prevails. Table 3.6 is a step in that direction.

COMPLEMENTS

1. Let **A** and **B** be two $m \times n$ real matrices. Then $\mathbf{AA}' = \mathbf{BB}'$ if and only if $\mathbf{A} = \mathbf{BU}$ where **U** is orthogonal. (This is a very important result and has applications in multivariate statistical analysis.)

2. Any matrix **A** can be written, $\mathbf{A} = \mathbf{B} + i\mathbf{C}$, where **B** and **C** are hermitian. Indeed,

$$\mathbf{B} = \frac{1}{2}(\mathbf{A} + \mathbf{A}^*) \quad \text{and} \quad \mathbf{C} = \frac{1}{2i}(\mathbf{A} - \mathbf{A}^*)$$

3. Let **A** be a $m \times m$ matrix. Then there exist unitary transformations **U** and **V** such that $\mathbf{UAU} = \mathbf{A}^*$ and $\mathbf{V}^*\mathbf{A}^*\mathbf{AV} = \mathbf{AA}^*$.

4. $\mathbf{AA}^*$ and $\mathbf{A}^*\mathbf{A}$ have the same nonzero eigenvalues.

5. Let **B** and **C** be nonsingular matrices such that $\mathbf{BA} = \mathbf{H}_1$ and $\mathbf{CA} = \mathbf{H}_2$ are both in hermite canonical form. Then $\mathbf{H}_1 = \mathbf{H}_2$; that is, all hermite forms of **A** are the same. If **A** is nonsingular, its hermite form is **I**.

6. **A** and **B** have the same hermite form iff $\mathscr{M}(\mathbf{A}') = \mathscr{M}(\mathbf{B}')$.

7. **A** and $\mathbf{A}^*\mathbf{A}$ have the same hermite form. If $|\mathbf{B}| \neq 0$, then **BA** and **A** have the same hermite form.

8. Let **H** be the hermite form of **A**. If $i_1, \ldots, i_k$ diagonal elements of **H** are 1 and the remaining are 0, then the i_1th, …, i_kth columns of **A** are linearly independent.

Consider the tth column of **A**. Then it is a linear combination of i_1th, …, i_kth columns of **A** with coefficients that are the nonzero elements of the tth column of **H**.

9. In the following **A** is $m \times m$ real matrix.

(a) If every principal minor of **A** is not zero, then $\mathbf{A} = \mathbf{RT}$, where **R** is lower- and **T** is upper triangular.

(b) If **A** is p.d. or p.s.d., then $\mathbf{A} = \mathbf{T}'\mathbf{T}$, where **T** is unique (except for sign) upper triangular.

(c) If **A** is p.d., then $\mathbf{A} = \mathbf{T}'\mathbf{DT}$, where **D** is diagonal and **T** is upper triangular with $t_{ii} = 1$.

(d) If **A** is symmetric and every leading principal minor is nonzero, then $\mathbf{A} = \mathbf{T}'\mathbf{DT}$, where **T** is upper triangular and **D** is diagonal with elements ± 1.

(e) There exists a matrix **P** such that $\mathbf{P}^{-1}\mathbf{AP}$ is upper triangular, but **P** need not be real.

(f) There exists an orthogonal matrix **P** such that $\mathbf{PA} = \mathbf{T}$, where **T** is upper triangular with $t_{ii} \geq 0$.

10. Let **C** be any symmetric matrix. Then there exist two unique n.n.d. disjoint matrices **A** and **B** (i.e., $\mathbf{AB} = \mathbf{O}$) such that $\mathbf{C} = \mathbf{A} - \mathbf{B}$.

11. If **A** is a singular complex matrix of order n, the following statements are equivalent

(a) $R(\mathbf{A}) = R(\mathbf{A}^2)$

(b) The elementary divisors corresponding to the zero eigen value are linear.

(c) The range and the null space of **A** are virtually disjoint.

(d) **A** is an EP matrix. (Erdelyi, 1967b; Katz and Pearl, 1966.)

12. *Frobenius inequality:*

$$R(\mathbf{ABC}) + R(\mathbf{B}) \geq R(\mathbf{AB}) + R(\mathbf{BC}).$$

13. If $\mathbf{AB} = \mathbf{BA} = \mathbf{0}$ and $R(\mathbf{A}^2) = R(\mathbf{A})$, then $R(\mathbf{A} + \mathbf{B}) = R(\mathbf{A}) + R(\mathbf{B})$.

14. If $\mathbf{AB} = \mathbf{BA} = \mathbf{0}$, then for some integer k, $R(\mathbf{A}^k + \mathbf{B}^k) = R(\mathbf{A}^k) + R(\mathbf{B}^k)$.

CHAPTER 2

Generalized Inverse of a Matrix

2.1 MATRICES OF FULL RANK

Regular Inverse of a Square Matrix

Consider a square matrix $\mathbf{A}$ of order $m \times m$ with rank m. Then there exists a unique matrix $\mathbf{A}^{-1}$, called the inverse of $\mathbf{A}$, with the property

$$\mathbf{A}\mathbf{A}^{-1} = \mathbf{A}^{-1}\mathbf{A} = \mathbf{I}_m, \tag{2.1.1}$$

where $\mathbf{I}_m$ is the identity matrix of order m.

Right Inverse of a Rectangular Matrix

Let $\mathbf{A}$ be a rectangular $m \times n$ matrix and suppose that $R(\mathbf{A}) = m$, in which case $\mathbf{A}\mathbf{A}^*$ is of order $m \times m$ and rank m. The inverse $(\mathbf{A}\mathbf{A}^*)^{-1}$ exists and

$$\mathbf{I}_m = (\mathbf{A}\mathbf{A}^*)(\mathbf{A}\mathbf{A}^*)^{-1} = \mathbf{A}[\mathbf{A}^*(\mathbf{A}\mathbf{A}^*)^{-1}] = \mathbf{A}\mathbf{A}_R^{-1}, \tag{2.1.2}$$

that is, there exists a matrix $\mathbf{A}_R^{-1}$ defined in (2.1.2) which postmultiplies $\mathbf{A}$ to $\mathbf{I}$. Such a matrix is called a right inverse of $\mathbf{A}$.

Left Inverse of a Rectangular Matrix

Similarly, if $R(\mathbf{A}) = n$, there exists a matrix $\mathbf{A}_L^{-1}$ called the left inverse of $\mathbf{A}$ such that

$$\mathbf{A}_L^{-1}\mathbf{A} = \mathbf{I}_n. \tag{2.1.3}$$

One choice of $\mathbf{A}_L^{-1}$ is $(\mathbf{A}^*\mathbf{A})^{-1}\mathbf{A}^*$. It is clear that $\mathbf{A}_L^{-1}$ or $\mathbf{A}_R^{-1}$ exists only in special cases when the rank of the $m \times n$ matrix is n or m. If $n \neq m$, both $\mathbf{A}_L^{-1}$ and $\mathbf{A}_R^{-1}$ cannot exist. In either case the inverse is not unique as shown in Theorem 2.1.1.

THEOREM 2.1.1 (general representation of right and left inverses). *Let* $\mathbf{A}$ *be a matrix of order* $m \times n$. *A general solution of* $\mathbf{AX} = \mathbf{I}$ *when* $R(\mathbf{A}) = m$ *is*

$$\mathbf{X} = \mathbf{VA}^*(\mathbf{AVA}^*)^{-1}, \tag{2.1.4}$$

where $\mathbf{V}$ *is an arbitrary matrix such that* $R(\mathbf{AVA}^*) = R(\mathbf{A})$. *A general solution of* $\mathbf{XA} = \mathbf{I}$ *when* $R(\mathbf{A}) = n$ *is*

$$\mathbf{X} = (\mathbf{A}^*\mathbf{VA})^{-1}\mathbf{A}^*\mathbf{V}, \tag{2.1.5}$$

where $\mathbf{V}$ *is an arbitrary matrix such that* $R(\mathbf{A}^*\mathbf{VA}) = R(\mathbf{A})$.

Proof: It is clear that an $\mathbf{X}$ of the form (2.1.4) is a right inverse of $\mathbf{A}$. Conversely, given a right inverse $\mathbf{X}$, it can be exhibited in the form (2.1.4); for instance, one choice of $\mathbf{V}$ is $\mathbf{XX}^*$. Similarly, the form (2.1.5) for a left inverse of $\mathbf{A}$ can be established.

2.2 DEFINITION OF A GENERALIZED INVERSE

Now we consider a general matrix $\mathbf{A}$ of order $m \times n$ and rank k which may be less than $\min(m, n)$ and raise the question whether an inverse exists in some suitable sense. This naturally depends on the purpose for which such an inverse is needed.

If $\mathbf{A}$ is a nonsingular matrix of order m, the solution of the linear equation $\mathbf{Ax} = \mathbf{y}$, where $\mathbf{y}$ is a $m \times 1$ column vector, is given by $\mathbf{x} = \mathbf{A}^{-1}\mathbf{y}$ where $\mathbf{A}^{-1}$ is the inverse of $\mathbf{A}$ (i.e., $\mathbf{AA}^{-1} = \mathbf{A}^{-1}\mathbf{A} = \mathbf{I}$, the identity matrix). We ask the question whether a similar representation of the solution, that is, of the form $\mathbf{x} = \mathbf{Gy}$, is possible when $\mathbf{A}$ is a singular square or a rectangular matrix. If there exists a matrix $\mathbf{G}$ such that $\mathbf{x} = \mathbf{Gy}$ is a solution of $\mathbf{Ax} = \mathbf{y}$ for any $\mathbf{y}$ such that $\mathbf{Ax} = \mathbf{y}$ is a consistent equation, then $\mathbf{G}$ does the same job (or behaves) as the inverse of $\mathbf{A}$, hence may be called a *generalized inverse* (g-inverse) of $\mathbf{A}$. So we give a formal definition of a g-inverse.

Definition 1: Let $\mathbf{A}$ be an $m \times n$ matrix of arbitrary rank. A generalized inverse of $\mathbf{A}$ is an $n \times m$ matrix $\mathbf{G}$ such that $\mathbf{x} = \mathbf{Gy}$ is a solution of $\mathbf{Ax} = \mathbf{y}$ for any $\mathbf{y}$ which makes the equation consistent.

We establish the existence of $\mathbf{G}$, which we denote by $\mathbf{A}^-$, and investigate its properties.

LEMMA 2.2.1 $\mathbf{A}^-$ *exists* $\Leftrightarrow \mathbf{AA}^-\mathbf{A} = \mathbf{A}$.

Proof: The equation $\mathbf{Ax} = \mathbf{y}$ is obviously consistent if $\mathbf{y} = \mathbf{Az}$ where $\mathbf{z}$ is an arbitrary $n \times 1$ vector. Hence the existence of $\mathbf{A}^-$ implies $\mathbf{A}(\mathbf{A}^-\mathbf{Az}) = \mathbf{Az}, \forall\, \mathbf{z}$ which gives $\mathbf{AA}^-\mathbf{A} = \mathbf{A}$.

Conversely, assume that the equation $\mathbf{Ax} = \mathbf{y}$ is consistent. This implies the existence of a vector $\mathbf{w}$ such that $\mathbf{Aw} = \mathbf{y}$. Since $\mathbf{AA^-A} = \mathbf{A}$, $\mathbf{AA^-Aw} = \mathbf{Aw} \Rightarrow \mathbf{AA^-y} = \mathbf{y}$; that is, $\mathbf{A^-y}$ is a solution of the equation $\mathbf{Ax} = \mathbf{y}$. Lemma 2.2.1 suggests the following equivalent definition of a g-inverse.

Definition 2: A g-inverse of $\mathbf{A}$ of order $m \times n$ is a matrix $\mathbf{A}^-$ of order $n \times m$ such that

$$\mathbf{AA^-A} = \mathbf{A} \tag{2.2.1}$$

LEMMA 2.2.2 *(a)* $\mathbf{A}^-$ *exists* $\Leftrightarrow \mathbf{H} = \mathbf{A^-A}$ *is idempotent and* $R(\mathbf{H}) = tr\,\mathbf{H} = R(\mathbf{A})$, *(b)* $\mathbf{A}^-$ *exists* $\Leftrightarrow \mathbf{F} = \mathbf{AA^-}$ *is idempotent and* $R(\mathbf{F}) = tr\,\mathbf{F} = R(\mathbf{A})$.

Proof of (a): $\mathbf{A}^-$ exists $\Rightarrow \mathbf{AA^-A} = \mathbf{A} \Rightarrow \mathbf{A^-AA^-A} = \mathbf{A^-A} \Rightarrow \mathbf{H}^2 = \mathbf{H}$. Also $R(\mathbf{A}) \geq R(\mathbf{H}) \geq R(\mathbf{AH}) = R(\mathbf{A})$. Hence $R(\mathbf{A}) = R(\mathbf{H}) = tr\,\mathbf{H}$ since $\mathbf{H}$ is idempotent.

Conversely, suppose $\mathbf{H} = \mathbf{A^-A}$ is idempotent and $R(\mathbf{H}) = R(\mathbf{A})$. This implies $\mathscr{M}(\mathbf{H}^*) = \mathscr{M}(\mathbf{A}^*) \Rightarrow \mathscr{O}(\mathbf{H}^*) = \mathscr{O}(\mathbf{A}^*)$. Hence $\mathbf{H}(\mathbf{I} - \mathbf{H}) = \mathbf{0} \Rightarrow \mathbf{A}(\mathbf{I} - \mathbf{H}) = \mathbf{0} = \mathbf{A}(\mathbf{I} - \mathbf{A^-A})$ or $\mathbf{A} = \mathbf{AA^-A}$.

Proof of (b): This proof is similar. Lemma 2.2.2 provides the following equivalent definition of a g-inverse.

Definition 3: A g-inverse of $\mathbf{A}$ of order $m \times n$ is a matrix $\mathbf{A}^-$ of order $n \times m$ such that $\mathbf{A^-A}$ is idempotent and $R(\mathbf{A^-A}) = R(\mathbf{A})$ or, alternatively, $\mathbf{AA^-}$ is idempotent and $R(\mathbf{AA^-}) = R(\mathbf{A})$.

LEMMA 2.2.3 $\mathbf{A}^-$ *exists and* $R(\mathbf{A}^-) \geq R(\mathbf{A})$.

Proof: Let $R(\mathbf{A}) = r$. Consider a rank factorization of $\mathbf{A}$, $\mathbf{A} = \mathbf{CD}$, where $\mathbf{C}$ is $m \times r$ and $\mathbf{D}$ is $r \times n$ matrix, each of rank r. Then $\mathbf{A}^- = \mathbf{D}_R^{-1}\mathbf{C}_L^{-1}$ satisfies the definition 2, thus establishing the existence of $\mathbf{A}^-$. Trivially, for any $\mathbf{A}^-$ satisfying the definition 2, $R(\mathbf{A}^-) \geq R(\mathbf{A})$.

It would be useful to recognize the situations in which g-inverse behaves like a regular inverse. Some cases are considered in Lemmas 2.2.4–2.2.6.

LEMMA 2.2.4 (i) *A n.s. condition that*

$$\mathbf{BA^-A} = \mathbf{B} \tag{2.2.2}$$

is that $\mathscr{M}(\mathbf{B}') \subset \mathscr{M}(\mathbf{A}')$, *that is, there exists a matrix* $\mathbf{D}$ *such that* $\mathbf{B} = \mathbf{DA}$. *Similarly, if* $\mathbf{B} = \mathbf{AA^-B}$, *then it is n.s. that* $\mathbf{B} = \mathbf{AD}$ *for some* $\mathbf{D}$.

(ii) *Let* $\mathbf{A}$ *be* $n \times n$ *matrix and* $\mathbf{p}$ *be n-vector. Then* $\mathbf{p'A^-p}$ *is invariant for any choice of* $\mathbf{A}^-$ *if* $\mathbf{p} \in \mathscr{M}(\mathbf{A})$ *and* $\mathbf{p} \in \mathscr{M}(\mathbf{A}')$.

(iii) *Let* $\mathbf{A}$ *be* $m \times n$, $\mathbf{B}$ *be* $p \times n$ *and* $\mathbf{C}$ *be* $m \times q$ *matrices. Then* $\mathbf{BA^-C}$ *is invariant for any choice of* $\mathbf{A}^-$ *if* $\mathscr{M}(\mathbf{B}') \subset \mathscr{M}(\mathbf{A}')$ *and* $\mathscr{M}(\mathbf{C}) \subset \mathscr{M}(\mathbf{A})$.

Proof of (i): Sufficiency is obvious. To prove necessity choose $\mathbf{D} = \mathbf{BA}^-$ in the first case and $\mathbf{D} = \mathbf{A}^-\mathbf{B}$ in the later case. (ii) and (iii) are similarly proved.

LEMMA 2.2.5

(a) *If* $R(\mathbf{BC}) = R(\mathbf{B})$, *then* $\mathbf{C}(\mathbf{BC})^-$ *is a g-inverse of* $\mathbf{B}$.
(b) *If* $R(\mathbf{BC}) = R(\mathbf{C})$, *then* $(\mathbf{BC})^-\mathbf{B}$ *is a g-inverse of* $\mathbf{C}$.
(c) *If* $R(\mathbf{QAP}) = R(\mathbf{A})$, *then* $\mathbf{P}(\mathbf{QAP})^-\mathbf{Q}$ *is a g-inverse of* $\mathbf{A}$.
(d) $\mathbf{PWQ}$ *is a g-inverse of* $\mathbf{A}$ *iff* $R(\mathbf{QAP}) = R(\mathbf{A})$ *and* $\mathbf{W} = (\mathbf{QAP})^-$.

All the results of Lemma 2.2.5 follow from Lemma 2.2.2.

LEMMA 2.2.6 *The following results hold for any choice of g-inverse involved:*

(a) *One choice of* $(\mathbf{A}^*)^-$ *is* $(\mathbf{A}^-)^*$. (2.2.3)
(b) $\mathbf{A}(\mathbf{A}^*\mathbf{A})^-(\mathbf{A}^*\mathbf{A}) = \mathbf{A}$ *and* $(\mathbf{A}^*\mathbf{A})(\mathbf{A}^*\mathbf{A})^-\mathbf{A}^* = \mathbf{A}^*$. (2.2.4)
(c) $\mathbf{A}(\mathbf{A}^*\mathbf{VA})^-(\mathbf{A}^*\mathbf{VA}) = \mathbf{A}$ *and* $(\mathbf{A}^*\mathbf{VA})(\mathbf{A}^*\mathbf{VA})^-\mathbf{A}^* = \mathbf{A}^*$ *for any matrix* $\mathbf{V}$ *such that* $R(\mathbf{A}^*\mathbf{VA}) = R(\mathbf{A})$. (Note that $R(\mathbf{A}^*\mathbf{VA}) = R(\mathbf{A})$ is automatically satisfied if $\mathbf{V}$ is a p.d. matrix). (2.2.5)
(d) $\mathbf{A}(\mathbf{A}^*\mathbf{VA})^-\mathbf{A}^*$ *is invariant for any choice of* $(\mathbf{A}^*\mathbf{VA})^-$ *and is of rank equal to that of* $\mathbf{A}$, *provided* $R(\mathbf{A}^*\mathbf{VA}) = R(\mathbf{A})$. *Further, if* $\mathbf{A}^*\mathbf{VA}$ *is hermitian, so is* $\mathbf{A}(\mathbf{A}^*\mathbf{VA})^-\mathbf{A}^*$.
(e) $\mathbf{A}^*\mathbf{AGA} = \mathbf{A}^*\mathbf{A} \Leftrightarrow \mathbf{AGA} = \mathbf{A}$. (2.2.6)
(f) *If* $R(\mathbf{CAB}) = R(\mathbf{C}) = R(\mathbf{B})$, *then* $\mathbf{B}(\mathbf{CAB})^-\mathbf{CAB} = \mathbf{B}$ *and* $\mathbf{CAB}(\mathbf{CAB})^-\mathbf{C} = \mathbf{C}$.
(g) *If* $R(\mathbf{CAB}) = R(\mathbf{C}) = R(\mathbf{B})$, $\mathbf{B}(\mathbf{CAB})^-\mathbf{C}$ *is invariant for any choice of* $(\mathbf{CAB})^-$.

Proof: The result (a) follows from the definition of a g-inverse; (b) and (c) follow from Lemma 2.2.5. We give below an interesting alternative proof for (b) and (c). To prove (b), consider

$$[\mathbf{A}(\mathbf{A}^*\mathbf{A})^-\mathbf{A}^*\mathbf{A} - \mathbf{A}]^*[\mathbf{A}(\mathbf{A}^*\mathbf{A})^-\mathbf{A}^*\mathbf{A} - \mathbf{A}]$$
$$= [(\mathbf{A}^*\mathbf{A})(\mathbf{A}^*\mathbf{A})^{-*} - \mathbf{I}]\mathbf{A}^*[\mathbf{A}(\mathbf{A}^*\mathbf{A})^-\mathbf{A}^*\mathbf{A} - \mathbf{A}]$$
$$= [(\mathbf{A}^*\mathbf{A})(\mathbf{A}^*\mathbf{A})^{-*} - \mathbf{I}][\mathbf{A}^*\mathbf{A}(\mathbf{A}^*\mathbf{A})^-\mathbf{A}^*\mathbf{A} - \mathbf{A}^*\mathbf{A}], \tag{2.2.7}$$

which vanishes because the second factor in (2.2.7) is a zero matrix by definition. Hence the first part of (b) is established by using the result that $\mathbf{E}^*\mathbf{E} = \mathbf{0} \Rightarrow \mathbf{E} = \mathbf{0}$. The second part is proved similarly.

To prove (c) let us observe that if $R(\mathbf{A}^*\mathbf{VA}) = R(\mathbf{A})$ then

$$\mathbf{A}^*\mathbf{VA}\boldsymbol{\lambda} = \mathbf{0} \Rightarrow \mathbf{A}\boldsymbol{\lambda} = \mathbf{0}, \tag{2.2.8}$$

where $\boldsymbol{\lambda}$ is a vector. Since $\mathbf{A}^*\mathbf{V}[\mathbf{A}(\mathbf{A}^*\mathbf{VA})^-\mathbf{A}^*\mathbf{VA} - \mathbf{A}] = \mathbf{0}$, we have

$\mathbf{A}(\mathbf{A}^*\mathbf{V}\mathbf{A})^-\mathbf{A}^*\mathbf{V}\mathbf{A} - \mathbf{A} = 0$ by applying (2.2.8). This proves the first part of (c), and the second part of (c) is proved similarly.

To prove (d) observe that

$$\mathbf{A}^*\mathbf{V}[\mathbf{A}(\mathbf{A}^*\mathbf{V}\mathbf{A})_1^-\mathbf{A}^* - \mathbf{A}(\mathbf{A}^*\mathbf{V}\mathbf{A})_2^-\mathbf{A}^*] = \mathbf{0}, \tag{2.2.9}$$

where $(\mathbf{A}^*\mathbf{V}\mathbf{A})_i^-$ for $i = 1, 2$ are two choices of the g-inverse. Hence using (2.2.8), we have $\mathbf{A}(\mathbf{A}^*\mathbf{V}\mathbf{A})_1^-\mathbf{A}^* - \mathbf{A}(\mathbf{A}^*\mathbf{V}\mathbf{A})_2^-\mathbf{A}^* = \mathbf{0}$, which proves that $\mathbf{A}(\mathbf{A}^*\mathbf{V}\mathbf{A})^-\mathbf{A}^*$ is independent of the choice of the g-inverse. The other results in (d) are easily established. Proof of (e) is similar; (f) and (g) are easily proved.

COROLLARY 1: $\mathbf{A}(\mathbf{A}^*\mathbf{A})^-\mathbf{A}^*$ *is hermitian and idempotent.*

LEMMA 2.2.7 *If* $\mathbf{BC}$ *is idempotent and* $R(\mathbf{BC}) = R(\mathbf{B})$ *or* $R(\mathbf{C})$ *then* $\mathbf{CB}$ *is idempotent. In the former case* $\mathbf{C}$ *is a g-inverse of* $\mathbf{B}$, *and in the latter* $\mathbf{B}$ *is a g-inverse of* $\mathbf{C}$.

Proof: Let $R(\mathbf{BC}) = R(\mathbf{B})$. Then there exists a matrix $\mathbf{D}$ such that $\mathbf{B} = \mathbf{BCD}$ and

$$\mathbf{CBCB} = \mathbf{CBCBCD} = \mathbf{C}(\mathbf{BCBC})\mathbf{D} = \mathbf{C}(\mathbf{BC})\mathbf{D} = \mathbf{CBCD} = \mathbf{CB},$$

which shows that $\mathbf{CB}$ is idempotent. Further, $\mathbf{BCB} = \mathbf{BCBCD} = \mathbf{BCD} = \mathbf{B}$; that is, $\mathbf{C}$ is a g-inverse of $\mathbf{B}$.

The case $R(\mathbf{BC}) = R(\mathbf{C})$ can be considered similarly.

LEMMA 2.2.8 *For a square matrix* $\mathbf{H}$ *of order* n

$$R(\mathbf{I} - \mathbf{H}) = n - R(\mathbf{H}) \Rightarrow \mathbf{H}^2 = \mathbf{H}. \tag{2.2.10}$$

Proof: The proof is simple and is omitted.

THEOREM 2.2.1 *An n.s. condition that* $\mathbf{G}$ *of order* $n \times m$ *is a g-inverse of* $\mathbf{A}$ *of order* $m \times n$ *is that*

$$R(\mathbf{I} - \mathbf{GA}) = n - R(\mathbf{A}). \tag{2.2.11}$$

Proof: Necessity follows from the if part of Lemma 2.2.2(a). To prove sufficiency we observe that since $R(\mathbf{GA}) \leq R(\mathbf{A})$ the condition (2.2.11) implies that $R(\mathbf{I} - \mathbf{GA}) \leq n - R(\mathbf{GA})$. However, $R(\mathbf{I} - \mathbf{GA}) + R(\mathbf{GA}) \geq R(\mathbf{I}) = n$. Hence $R(\mathbf{I} - \mathbf{GA}) = n - R(\mathbf{GA})$. Then $R(\mathbf{GA}) = R(\mathbf{A})$ and the result of the theorem follows by applying Lemma 2.2.8 and using Definition 3 of a g-inverse.

2.3 SOLUTION OF CONSISTENT LINEAR EQUATIONS

THEOREM 2.3.1 *Let* $\mathbf{A}$ *be of order* $m \times n$ *and* $\mathbf{A}^-$ *be any g-inverse of* $\mathbf{A}$. *Further let* $\mathbf{H} = \mathbf{A}^-\mathbf{A}$. *Then the following hold;*

(a) *A general solution of the homogeneous equation* $\mathbf{Ax} = \mathbf{0}$ *is* $\mathbf{x} = (\mathbf{I} - \mathbf{H})\mathbf{z}$ *where* $\mathbf{z}$ *is an arbitrary vector.*

(b) *A general solution of a consistent nonhomogeneous equation* $\mathbf{Ax} = \mathbf{y}$ *is*

$$\mathbf{x} = \mathbf{A}^{-}\mathbf{y} + (\mathbf{I} - \mathbf{H})\mathbf{z}, \tag{2.3.1}$$

where $\mathbf{z}$ *is an arbitrary vector.*

(c) $\mathbf{Q}'\mathbf{x}$ *has a unique value for all solutions of* $\mathbf{Ax} = \mathbf{y}$, *iff*

$$\mathbf{H}'\mathbf{Q} = \mathbf{Q} \quad \text{or} \quad \mathbf{Q} \in \mathcal{M}(\mathbf{A}'). \tag{2.3.2}$$

(d) *A necessary and sufficient condition that* $\mathbf{Ax} = \mathbf{y}$ *is consistent is that*

$$\mathbf{AA}^{-}\mathbf{y} = \mathbf{y}. \tag{2.3.3}$$

Proof of (a): Note that this is equivalent to saying that $\mathcal{O}(\mathbf{A}^*) = \mathcal{M}(\mathbf{I} - \mathbf{H})$, which follows from the fact that $\mathbf{A}(\mathbf{I} - \mathbf{H}) = \mathbf{0}$ and $R(\mathbf{I} - \mathbf{H}) = n - R(\mathbf{A})$.

Proof of (b): (b) follows, since a general solution of $\mathbf{Ax} = \mathbf{y}$ is the sum of a particular solution of $\mathbf{Ax} = \mathbf{y}$ and a general solution of $\mathbf{Ax} = \mathbf{0}$.

Proof of (c): Observe that $\mathbf{Q}'[\mathbf{A}^{-}\mathbf{y} + (\mathbf{I} - \mathbf{H})\mathbf{z}] = \mathbf{Q}'\mathbf{A}^{-}\mathbf{y}, \forall\, \mathbf{z} \Leftrightarrow \mathbf{Q}'(\mathbf{I} - \mathbf{H})\mathbf{z} = \mathbf{0}, \forall\, \mathbf{z} \Leftrightarrow \mathbf{Q}' = \mathbf{Q}'\mathbf{H}$.

Proof of (d): Necessity follows from Definition 1 of a g-inverse, Sufficiency is trivial.

THEOREM 2.3.2 *A necessary and sufficient condition for the equation* $\mathbf{AXB} = \mathbf{C}$ *to have a solution is that*

$$\mathbf{AA}^{-}\mathbf{CB}^{-}\mathbf{B} = \mathbf{C}, \tag{2.3.4}$$

in which case the general solution is

$$\mathbf{X} = \mathbf{A}^{-}\mathbf{CB}^{-} + \mathbf{Z} - \mathbf{A}^{-}\mathbf{AZBB}^{-}, \tag{2.3.5}$$

where $\mathbf{Z}$ *is an arbitrary matrix.*

Proof: Necessity of (2.3.4) follows from the fact that if the equations are consistent there exists a matrix $\mathbf{X}$ such that $\mathbf{AXB} = \mathbf{C}$. Then $\mathbf{AA}^{-}\mathbf{CB}^{-}\mathbf{B} = \mathbf{AA}^{-}\mathbf{AXBB}^{-}\mathbf{B} = \mathbf{AXB} = \mathbf{C}$. Sufficiency is trivial since here $\mathbf{A}^{-}\mathbf{CB}^{-}$ is clearly a solution.

Observe that $\mathbf{X}$ defined by (2.3.5) satisfies the equation $\mathbf{AXB} = \mathbf{C}$. Also, any arbitrary solution $\mathbf{X}$ of this equation is obtainable through the formula (2.3.5) by a suitable choice of the matrix $\mathbf{Z}$; for example, $\mathbf{Z} = \mathbf{X} - \mathbf{A}^{-}\mathbf{CB}^{-}$ is one such choice. This shows that (2.3.5) provides the general solution.

The result (2.3.5) can be derived directly from (2.3.1) by observing that the equation $\mathbf{AXB} = \mathbf{C}$ can be written in the usual linear equation form,

$$(\mathbf{A} \otimes \mathbf{B}')\mathbf{x} = \mathbf{c},$$

where $(\mathbf{A} \otimes \mathbf{B}')$ is the Kronecker product defined in Section 1.5, $\mathbf{x}$ is the vector obtained by writing the columns of $\mathbf{X}$ one below the other [$(i + 1)$th following the ith], and $\mathbf{c}$ is the vector similarly obtained from $\mathbf{C}$.

By using (2.3.1) a general solution of $(\mathbf{A} \otimes \mathbf{B}')\mathbf{x} = \mathbf{c}$ is

$$\mathbf{x} = (\mathbf{A} \otimes \mathbf{B}')^{-}\mathbf{c} + [\mathbf{I} - (\mathbf{A} \otimes \mathbf{B}')^{-}(\mathbf{A} \otimes \mathbf{B}')]\mathbf{z},$$

where $\mathbf{z}$ is arbitrary. Observing that

$$(\mathbf{A} \otimes \mathbf{B}')^{-} = \mathbf{A}^{-} \otimes (\mathbf{B}')^{-}$$

$$(\mathbf{A} \otimes \mathbf{B}')^{-}(\mathbf{A} \otimes \mathbf{B}') = \mathbf{A}^{-}\mathbf{A} \otimes (\mathbf{B}')^{-}\mathbf{B}',$$

and using property (v) of Kronecker product in Section 1.5, the general solution is

$$\mathbf{x} = \mathbf{A}^{-} \otimes (\mathbf{B}')^{-}\mathbf{c} + [\mathbf{I} - \mathbf{A}^{-}\mathbf{A} \otimes (\mathbf{B}')^{-}\mathbf{B}']\mathbf{z}.$$

Writing $\mathbf{x}$ in matrix form,

$$\mathbf{X} = \mathbf{A}^{-}\mathbf{C}\mathbf{B}^{-} + \mathbf{Z} - \mathbf{A}^{-}\mathbf{A}\mathbf{Z}\mathbf{B}\mathbf{B}^{-},$$

as in (2.3.5).

COROLLARY 1: *A general solution of* $\mathbf{A}^*\mathbf{X}\mathbf{A} = \mathbf{0}$ *is* $\mathbf{X} = \mathbf{Z} - (\mathbf{A}\mathbf{A}^{-})^*\mathbf{Z}\mathbf{A}\mathbf{A}^{-}$ *where* $\mathbf{Z}$ *is arbitrary. The class of hermitian solutions is obtained by choosing* $\mathbf{Z}$ *as hermitian.*

We can also represent the solution as

$$\mathbf{Z} - \mathbf{P}_{\mathbf{A}}\mathbf{Z}\mathbf{P}_{\mathbf{A}}$$

where $\mathbf{P}_{\mathbf{A}}$ *is the projection operator onto* $\mathscr{M}(\mathbf{A})$, *with* $\mathbf{Z}$ *hermitian for the hermitian solutions.*

THEOREM 2.3.3 *Let* $\mathbf{A}(m \times n)$, $\mathbf{C}(m \times p)$, $\mathbf{B}(p \times q)$, $\mathbf{D}(n \times q)$ *be given matrices. A necessary and sufficient condition for the consistent equations* $\mathbf{A}\mathbf{X} = \mathbf{C}$, $\mathbf{X}\mathbf{B} = \mathbf{D}$ *to have a common solution is that*

$$\mathbf{A}\mathbf{D} = \mathbf{C}\mathbf{B}, \tag{2.3.6}$$

in which case the general expression for a common solution is

$$\mathbf{X} = \mathbf{A}^{-}\mathbf{C} + \mathbf{D}\mathbf{B}^{-} - \mathbf{A}^{-}\mathbf{A}\mathbf{D}\mathbf{B}^{-} + (\mathbf{I} - \mathbf{A}^{-}\mathbf{A})\mathbf{Z}(\mathbf{I} - \mathbf{B}\mathbf{B}^{-}), \tag{2.3.7}$$

where $\mathbf{Z}$ *is arbitrary.*

Proof: Suppose that a common solution $\mathbf{X}$ exists. Then $\mathbf{A}\mathbf{X}\mathbf{B} = \mathbf{C}\mathbf{B} = \mathbf{A}\mathbf{D}$, thus establishing necessity of (2.3.6).

If (2.3.6) holds, $\mathbf{X} = \mathbf{A}^{-}\mathbf{C} + \mathbf{D}\mathbf{B}^{-} - \mathbf{A}^{-}\mathbf{A}\mathbf{D}\mathbf{B}^{-}$ is a common solution. This proves sufficiency.

That (2.3.7) provides a general solution follows from Lemma 2.3.1.

LEMMA 2.3.1 $\mathbf{AY} = \mathbf{0}, \mathbf{YB} = \mathbf{0} \Leftrightarrow \mathbf{Y} = (\mathbf{I} - \mathbf{A}^-\mathbf{A})\mathbf{Z}(\mathbf{I} - \mathbf{BB}^-)$ *for some* $\mathbf{Z}$.

Proof: Let $\mathbf{Y}$ be a matrix of order $n \times p$ satisfying the equations $\mathbf{AY} = \mathbf{0}$, $\mathbf{YB} = \mathbf{0}$, where $\mathbf{A}$ and $\mathbf{B}$ are matrices of order $m \times n$ and $p \times q$, respectively. Consider a rank factorization of $\mathbf{Y}$: $\mathbf{Y} = \mathbf{UV}$, where $\mathbf{U}(n \times r)$ and $\mathbf{V}(r \times p)$ are matrices of rank r each. Observe

$$\mathbf{AY} = \mathbf{0} \Rightarrow \mathbf{AU} = \mathbf{0} \Rightarrow \mathbf{U} = (\mathbf{I} - \mathbf{A}^-\mathbf{A})\mathbf{D} \text{ for some } \mathbf{D}$$

$$\mathbf{YB} = \mathbf{0} \Rightarrow \mathbf{VB} = \mathbf{0} \Rightarrow \mathbf{V} = \mathbf{E}(\mathbf{I} - \mathbf{BB}^-) \text{ for some } \mathbf{E}.$$

Hence $\mathbf{Y} = (\mathbf{I} - \mathbf{A}^-\mathbf{A})\mathbf{DE}(\mathbf{I} - \mathbf{BB}^-)$. The '$\Leftarrow$' part is trivial.

COROLLARY 1: *A general solution of* $\mathbf{AX} = \mathbf{O}$ *is* $\mathbf{X} = (\mathbf{I} - \mathbf{A}^-\mathbf{A})\mathbf{Z}$ *or* $\mathbf{X} = \mathbf{Q}_{\mathbf{A}^*}\mathbf{Z}$ *where* $\mathbf{Q}_{\mathbf{A}^*}$ *is the projector onto* $\mathcal{O}(\mathbf{A}^*)$ *and* $\mathbf{Z}$ *is arbitrary. A general hermitian solution is* $(\mathbf{I} - \mathbf{A}^-)\mathbf{Z}(\mathbf{I} - \mathbf{A}^-\mathbf{A})^*$ *or* $\mathbf{Q}_{\mathbf{A}^*}\mathbf{Z}\mathbf{Q}_{\mathbf{A}^*}$ *where* $\mathbf{Z}$ *is hermitian.*

2.4 GENERAL REPRESENTATION OF g-INVERSE

We have seen in Theorem 2.3.1 that, given a g-inverse $\mathbf{A}^-$ of $\mathbf{A}$, all solutions of $\mathbf{Ax} = \mathbf{y}$ can be expressed as

$$\mathbf{A}^-\mathbf{y} + (\mathbf{I} - \mathbf{A}^-\mathbf{A})\mathbf{z}, \tag{2.4.1}$$

where $\mathbf{z}$ is an arbitrary vector. It is of some interest to obtain an explicit representation of all possible g-inverses as generated from any particular inverse.

THEOREM 2.4.1 *Two alternative representations of a general solution to g-inverse of a matrix* $\mathbf{A}$ *of order* $m \times n$ *are*

(a) $\mathbf{G} = \mathbf{A}^- + \mathbf{U} - \mathbf{A}^-\mathbf{AUAA}^-$ (2.4.2)

(b) $\mathbf{G} = \mathbf{A}^- + \mathbf{V}(\mathbf{I} - \mathbf{AA}^-) + (\mathbf{I} - \mathbf{A}^-\mathbf{A})\mathbf{W}$, (2.4.3)

where $\mathbf{A}^-$ *is a particular g-inverse and* $\mathbf{U}, \mathbf{V}, \mathbf{W}$ *are arbitrary matrices.*

The proof of (a) follows along the same lines as Theorem 2.3.2 by choosing $\mathbf{B} = \mathbf{C} = \mathbf{A}$.

To prove (b) first observe that $\mathbf{G}$ in (2.4.3) is a g-inverse of $\mathbf{A}$ for all $\mathbf{V}$ and $\mathbf{W}$. Conversely, if $\mathbf{G}$ is a g-inverse of $\mathbf{A}$, choose

$$\mathbf{V} = \mathbf{G} - \mathbf{A}^- \text{ and } \mathbf{W} = \mathbf{GAA}^- \tag{2.4.4}$$

and observe that $\mathbf{G}$ can be written in the form (2.4.3) with the above choices of $\mathbf{V}$ and $\mathbf{W}$.

To show that the two forms (a) and (b) are equivalent, put $\mathbf{U} = \mathbf{V}(\mathbf{I} - \mathbf{F}) + (\mathbf{I} - \mathbf{H})\mathbf{W}$ in form (a) where $\mathbf{F} = \mathbf{AA}^-$, $\mathbf{H} = \mathbf{A}^-\mathbf{A}$. Since $\mathbf{AUA} = \mathbf{0}$, here,

$$\mathbf{A}^- + \mathbf{U} - \mathbf{A}^-\mathbf{AUAA}^- = \mathbf{A}^- + \mathbf{U} = \mathbf{A}^- + \mathbf{V}(\mathbf{I} - \mathbf{F}) + (\mathbf{I} - \mathbf{H})\mathbf{W}.$$

This shows that an expression in form (b) can always be rewritten in form (a).

Conversely, observe

$$\mathbf{A}^- + \mathbf{U} - \mathbf{A}^-\mathbf{AUAA}^- = \mathbf{A}^- + \mathbf{U}(\mathbf{I} - \mathbf{AA}^-) + (\mathbf{I} - \mathbf{A}^-\mathbf{A})\mathbf{UAA}^-.$$

This shows an expression in form (a) can always be rewritten in form (b), choosing $\mathbf{V} = \mathbf{U}$, $\mathbf{W} = \mathbf{UAA}^-$.

COROLLARY 1: *The class of all solutions to a consistent nonhomogeneous linear equation* $\mathbf{Ax} = \mathbf{y}$ *is* $\mathbf{x} = \mathbf{Gy}$, $\mathbf{G} \in$ *class of all g-inverses of* $\mathbf{A}$.

Proof: Given arbitrary vectors $\mathbf{y}$ and $\mathbf{z}$, provided $\mathbf{y}$ is non-null, a matrix $\mathbf{W}$ can always be obtained such that $\mathbf{Wy} = \mathbf{z}$: for example, if the ith co-ordinate of $\mathbf{z}$ is z_i, the ith row of $\mathbf{W}$ can be taken to be $z_i\mathbf{y}^*/\mathbf{y}^*\mathbf{y}$. This choice of $\mathbf{W}$, together with $\mathbf{V} = \mathbf{0}$, in (2.4.3), leads to a g-inverse $\mathbf{G}$ of $\mathbf{A}$ for which

$$\mathbf{Gy} = \mathbf{A}^-\mathbf{y} + (\mathbf{I} - \mathbf{A}^-\mathbf{A})\mathbf{z}. \tag{2.4.5}$$

Thus Corollary 1 is proved.

THEOREM 2.4.2 *If every g-inverse of* $\mathbf{A}$ *is a g-inverse of* $\mathbf{B}$ *and vice versa, then* $\mathbf{A} = \mathbf{B}$ (*i.e., every matrix has a unique class of g-inverses*).

Proof. It is easily verified that $\mathbf{G} = \mathbf{A}^- + (\mathbf{I} - \mathbf{P}_{\mathbf{A}^*})\mathbf{B}^*$ is a g-inverse of $\mathbf{A}$. Then from the conditions $\mathbf{BGB} = \mathbf{B}$ and $\mathbf{BA}^-\mathbf{B} = \mathbf{B}$ we find $\mathbf{B} = \mathbf{BP}_{\mathbf{A}^*} \Rightarrow \mathbf{B} = \mathbf{EA}$ for some $\mathbf{E}$. Similarly, $\mathbf{B} = \mathbf{AD}$ for some $\mathbf{D}$. Now for any g-inverse $\mathbf{B}^-$

$$\mathbf{B} = \mathbf{BB}^-\mathbf{B} = \mathbf{EAB}^-\mathbf{AD} = \mathbf{EAD} = \mathbf{BD}.$$

Thus $\mathbf{B} = \mathbf{BD} = \mathbf{AD} = \mathbf{EA}$. Similarly, $\mathbf{A} = \mathbf{AF} = \mathbf{BF} = \mathbf{HB}$ for some $\mathbf{F}$ and $\mathbf{H}$. Now

$$\mathbf{B} - \mathbf{BD} = \mathbf{0} \Rightarrow \mathbf{D} = \mathbf{I} + \mathbf{C},$$

where

$$\mathbf{BC} = \mathbf{0} \Rightarrow \mathbf{HBC} = \mathbf{0} \Rightarrow \mathbf{AC} = \mathbf{0},$$

$$\mathbf{BD} = \mathbf{AD} \Rightarrow (\mathbf{B} - \mathbf{A})(\mathbf{I} + \mathbf{C}) = \mathbf{0} = (\mathbf{B} - \mathbf{A}) \quad \text{or} \quad \mathbf{B} = \mathbf{A}.$$

(Note that $\mathbf{P}_{\mathbf{A}^*}$ is the projector onto $\mathcal{M}(\mathbf{A}^*)$ as defined in (1.1.6)).

2.5 REFLEXIVE g-INVERSE

The true inverse of a nonsingular square matrix has the property that the inverse of the inverse is equal to the original matrix. This may not hold for any g-inverse as defined in Section 2.2. We shall show, however, that a sub-class of g-inverses possesses an analogous property. We give the following definition:

Definition 4: An $n \times m$ matrix $\mathbf{G}$ is said to be a reflexive g-inverse of an $m \times n$ matrix $\mathbf{A}$ if

$$\mathbf{AGA} = \mathbf{A} \text{ and } \mathbf{GAG} = \mathbf{G}. \tag{2.5.1}$$

We use the notation $\mathbf{A}_r^-$ to denote a reflexive g-inverse.

LEMMA 2.5.1 *A necessary and sufficient condition for a g-inverse* $\mathbf{G}$ *of* $\mathbf{A}$ *to be reflexive is that*

$$R(\mathbf{G}) = R(\mathbf{A}). \tag{2.5.2}$$

Proof. Necessity is easily established. Sufficiency follows from definition 3 since $R(\mathbf{GA}) = R(\mathbf{G})$.

Frame (1964) uses the term semi-inverse to denote a g-inverse that satisfies the rank condition (2.5.2). A semi-inverse is thus equivalent to a reflexive g-inverse. Lemma 2.5.1 by itself is not useful in obtaining a reflexive g-inverse, since it does not suggest how one could compute a g-inverse with rank equal to the rank of the original matrix. The characterizations of the class of reflexive g-inverse contained in Lemma 2.5.2 and 2.5.3 are useful in this respect.

LEMMA 2.5.2 *Let* $\mathbf{A} = \mathbf{MN}$ *be a rank factorization of* $\mathbf{A}(m \times n)$, *where* $\mathbf{M}(m \times r)$ *and* $\mathbf{N}(r \times n)$ *are matrices of rank* $r = R(\mathbf{A})$. *Then* $\mathbf{G}$ *is* $\mathbf{A}_r^-$ *iff it can be expressed as*

$$\mathbf{G} = \mathbf{N}_R^{-1}\mathbf{M}_L^{-1}, \tag{2.5.3}$$

where $\mathbf{N}_R^{-1}$ *is a right inverse of* $\mathbf{N}$ *and* $\mathbf{M}_L^{-1}$ *is a left inverse of* $\mathbf{M}$; *that is,* $\mathbf{NN}_R^{-1} = \mathbf{M}_L^{-1}\mathbf{M} = \mathbf{I}$.

Proof: The 'if' part is trivial. For the 'only if' part let $\mathbf{G}$ be $\mathbf{A}_r^-$.

$$\mathbf{MNGMN} = \mathbf{MN} \Rightarrow \mathbf{NGM} = \mathbf{I} \Rightarrow \mathbf{GM} = \mathbf{N}_R^{-1}, \mathbf{NG} = \mathbf{M}_L^{-1}. \tag{2.5.4}$$

But $\mathbf{G} = \mathbf{GMNG} \Rightarrow \mathbf{G} = \mathbf{N}_R^{-1}\mathbf{M}_L^{-1}$.

LEMMA 2.5.3 *A g-inverse* $\mathbf{G}$ *of* $\mathbf{A}$ *is* $\mathbf{A}_r^-$ *iff* $\mathbf{G}$ *can be expressed as*

$$\mathbf{G} = \mathbf{A}^-\mathbf{AA}^- \tag{2.5.5}$$

for some g-inverse $\mathbf{A}^-$ *of* $\mathbf{A}$.

Proof: Obviously $\mathbf{A}^-\mathbf{AA}^-$ is a g-inverse of $\mathbf{A}$ and $R(\mathbf{A}^-\mathbf{AA}^-) = R(\mathbf{A})$. Lemma 2.5.1 shows $\mathbf{A}^-\mathbf{AA}^-$ is $\mathbf{A}_r^-$. Conversely, let $\mathbf{G}$ be a reflexive g-inverse of $\mathbf{A}$. Then $\mathbf{G} = \mathbf{GAG}$. This shows that $\mathbf{G}$ can always be expressed in the form (2.5.5) choosing $\mathbf{A}^-$ to be $\mathbf{G}$ itself.

COROLLARY 1: *A g-inverse* $\mathbf{G}$ *of* $\mathbf{A}$ *is* $\mathbf{A}_r^-$ *iff* $\mathbf{G}$ *can be expressed as*

$$\mathbf{G} = \mathbf{G}_1\mathbf{AG}_2, \tag{2.5.6}$$

where $\mathbf{G}_1$ *and* $\mathbf{G}_2$ *are (possibly different) g-inverses of* $\mathbf{A}$.

For a different type of characterization of a reflexive inverse see Example 9 in the complements to Chapter 3.

We refer to $\mathbf{H} = \mathbf{A}^-\mathbf{A}$ and $\mathbf{F} = \mathbf{A}\mathbf{A}^-$ as the *right and left idempotents of* $\mathbf{A}^-$.

LEMMA 2.5.4 *For every g-inverse* $\mathbf{A}^-$ *of* $\mathbf{A}$, $\mathbf{A}^-\mathbf{A}\mathbf{A}^-$ *is unique reflexive g-inverse having right and left idempotents identical with those of* $\mathbf{A}^-$.

Proof: Let $\mathbf{A}_r^-$ be a reflexive inverse with the desired property. Since the right idempotents of $\mathbf{A}_r^-$ and $\mathbf{A}^-$ are identical, $\mathbf{A}_r^-\mathbf{A} = \mathbf{A}^-\mathbf{A} \Rightarrow \mathbf{A}_r^- = \mathbf{A}^-\mathbf{A}\mathbf{A}_r^-$. Similarly $\mathbf{A}\mathbf{A}_r^- = \mathbf{A}\mathbf{A}^- \Rightarrow \mathbf{A}_r^- = \mathbf{A}_r^-\mathbf{A}\mathbf{A}^-$. Hence $\mathbf{A}_r^- = \mathbf{A}^-\mathbf{A}\mathbf{A}_r^- = \mathbf{A}^-\mathbf{A}\mathbf{A}_r^-\mathbf{A}\mathbf{A}^- = \mathbf{A}^-\mathbf{A}\mathbf{A}^-$. The following result could be proved along the same lines.

LEMMA 2.5.5 *For every pair of g-inverses* $\mathbf{A}_{(1)}^-$ *and* $\mathbf{A}_{(2)}^-$ *of* $\mathbf{A}$, $\mathbf{A}_{(1)}^-\mathbf{A}\mathbf{A}_{(2)}^-$ *is the unique reflexive g-inverse having right idempotent same as that of* $\mathbf{A}_{(1)}^-$ *and left idempotent same as that of* $\mathbf{A}_{(2)}^-$.

The property of possessing identical left and right idempotents defines an equivalence relation among the generalized inverses. It is easily seen that two g-inverses belong to the same equivalence class iff they differ by a matrix of the type $(\mathbf{I} - \mathbf{A}^-\mathbf{A})\mathbf{U}(\mathbf{I} - \mathbf{A}\mathbf{A}^-)$, where $\mathbf{U}$ could be arbitrary or, in view of Lemma 2.5.4, we can state that two g-inverses $\mathbf{G}_1$ and $\mathbf{G}_2$ of $\mathbf{A}$ belong to the same equivalence class iff

$$\mathbf{G}_1\mathbf{A}\mathbf{G}_1 = \mathbf{G}_2\mathbf{A}\mathbf{G}_2, \tag{2.5.7}$$

since each equivalence class contains a unique reflexive inverse.

2.6 g-INVERSE FOR A BASIC SOLUTION OF Ax = y (CONSISTENT)

Definition 5: $\mathbf{x}_b$ is said to be a basic solution of the equation $\mathbf{A}\mathbf{x} = \mathbf{y}$ if (i) $\mathbf{A}\mathbf{x}_b = \mathbf{y}$ and (ii) $\mathbf{x}_b$ has utmost r nonzero components where $r = R(\mathbf{A})$.

Let $\mathbf{G}$ be a g-inverse which provides $\mathbf{x}_b$. Then $\mathbf{G}\mathbf{y}$ has $(n - r)$ components zero whenever $\mathbf{y} \in \mathcal{M}(\mathbf{A})$. We obtain $\mathbf{G}$ so that $\mathbf{G}\mathbf{A}$ has $n - r$ null rows. Suppose the rows $i_1, i_2, \ldots, i_r$ of $\mathbf{G}\mathbf{A}$ are non-null, and let $\mathbf{E}$ denote the matrix obtained from the unit matrix by bringing its i_1th, i_2th, ..., and i_rth rows to the first r positions. Then, if $\mathbf{E}\mathbf{G}$ be partitioned as

$$\mathbf{E}\mathbf{G} = \begin{pmatrix} \mathbf{G}_1 \\ \mathbf{G}_2 \end{pmatrix},$$

we have

$$\begin{pmatrix} \mathbf{G}_1 \\ \mathbf{G}_2 \end{pmatrix} \mathbf{A} = \begin{pmatrix} \mathbf{G}_1\mathbf{A} \\ \mathbf{0} \end{pmatrix}$$

where $\mathbf{G}_1$ is of order $r \times m$ and $\mathbf{G}_2$ of order $(n - r) \times m$. Also, since $\mathbf{E}\mathbf{E}' = \mathbf{I}$, if $\mathbf{G} = \mathbf{A}^-$, then $\mathbf{EG} = (\mathbf{AE}')^-$. Let $\mathbf{AE}'$ be partitioned as

$$\mathbf{AE}' = (\mathbf{A}_1 \vdots \mathbf{A}_2),$$

where $\mathbf{A}_1$ has r columns and $\mathbf{A}_2$ has $(n - r)$ columns.

$\mathbf{AE}'(\mathbf{EG})\mathbf{AE}' = \mathbf{AE}' \Rightarrow$

$$(\mathbf{A}_1 \vdots \mathbf{A}_2)\begin{pmatrix}\mathbf{G}_1\\ \mathbf{G}_2\end{pmatrix}\mathbf{AE}' = \mathbf{AE}',$$

$$\Rightarrow (\mathbf{A}_1 \vdots \mathbf{A}_2)\begin{pmatrix}\mathbf{G}_1\\ \mathbf{0}\end{pmatrix}(\mathbf{A}_1 \vdots \mathbf{A}_2) = (\mathbf{A}_1 \vdots \mathbf{A}_2),$$

$$\Rightarrow \mathbf{A}_1\mathbf{G}_1\mathbf{A}_1 = \mathbf{A}_1, \qquad \mathbf{A}_1\mathbf{G}_1\mathbf{A}_2 = \mathbf{A}_2 \Rightarrow \mathbf{G}_1 = \mathbf{A}_1^-,$$

the second equation being automatically satisfied, since $\mathscr{M}(\mathbf{A}_2) \subset \mathscr{M}(\mathbf{A}_1)$. To verify this, note that since $\mathbf{E}'$ is nonsingular $R(\mathbf{A}_1 \vdots \mathbf{A}_2) = R(\mathbf{A}) = r$. Also

$$\mathbf{A}_1\mathbf{G}_1(\mathbf{A}_1 \vdots \mathbf{A}_2) = (\mathbf{A}_1 \vdots \mathbf{A}_2)$$

$$\Rightarrow R(\mathbf{A}_1\mathbf{G}_1) \geq R(\mathbf{A}_1 \vdots \mathbf{A}_2) = r \Rightarrow R(\mathbf{A}_1) = R(\mathbf{G}_1) = r,$$

since $\mathbf{A}_1$ has only r columns and $\mathbf{G}_1$ has r rows. Since $\mathbf{A}_1$ is a matrix with full rank, clearly $\mathbf{G}_1 = (\mathbf{A}_1)_L^{-1}$. We have thus proved the 'necessity' part of Lemma 2.6.1 and the sufficiency part is easy to establish.

LEMMA 2.6.1 *Let* $\mathbf{A}(m \times n)$ *be a matrix of rank* r. $\mathbf{G}$ *is a* g-*inverse providing a basic solution of a consistent equation* $\mathbf{Ax} = \mathbf{y}$ *if there exists a permutation matrix*

$$\mathbf{E} = \begin{pmatrix}\mathbf{E}_1\\ \mathbf{E}_2\end{pmatrix} \tag{2.6.1}$$

such that $\mathbf{E}_1\mathbf{G} = (\mathbf{AE}_1')_L^{-1}$, $\mathbf{E}_2\mathbf{GA} = \mathbf{0}$, *where* $\mathbf{E}_1$ *is of order* $r \times n$ *and* $\mathbf{E}_2$ *of order* $(n - r) \times n$.

A g-inverse that provides a basic solution is denoted by $\mathbf{A}_b^-$. The need for and definition of such an inverse in linear programming computations were given by Rosen (1964).

A method of computing $\mathbf{A}_b^-$ is as follows. Since $R(\mathbf{A}) = r$, it is possible to rearrange the columns of $\mathbf{A}$ in such a way that the first r columns are independent. Let $\mathbf{B}_1$ be the matrix representing the r independent columns in the rearranged form which may be denoted by $\mathbf{B}$. Consider the matrix

$$\mathbf{H} = \begin{pmatrix}\mathbf{H}_1\\ \mathbf{H}_2\end{pmatrix} = \begin{pmatrix}(\mathbf{B}_1)_L^{-1}\\ \mathbf{H}_2\end{pmatrix}, \tag{2.6.2}$$

where $\mathbf{H}_2$ is any matrix with $(n - r)$ rows such that $\mathbf{H}_2\mathbf{B} = \mathbf{0}$. The matrix $\mathbf{H}$, as defined in (2.6.2), is a $\mathbf{B}_b^-$ for the matrix **B**. To obtain $\mathbf{A}_b^-$ we have only to carry out the corresponding rearrangement of *rows* in **H** as in the columns of **B** to obtain **A**.

Suppose that we wanted a solution of $\mathbf{Ax} = \mathbf{y}$ with utmost $s > r$ nonzero components, where r is the rank of **A**. As before, we obtain a matrix **B** by rearranging the columns of **A** in such a way that the first s columns of **A** contain r independent columns. Let $\mathbf{B}_1$ denote the matrix with first s columns of **B** and define

$$\mathbf{H} = \begin{pmatrix} \mathbf{B}_1^- \\ \mathbf{H}_2 \end{pmatrix},$$

where $\mathbf{B}_1^-$ is any g-inverse of $\mathbf{B}_1$ and $\mathbf{H}_2$ is any matrix of order $(n - s \times m)$ such that $\mathbf{H}_2\mathbf{B} = \mathbf{0}$. Let **G** be the matrix obtained by corresponding rearrangement of the rows of **H** as in the columns of **B** to obtain **A**. Then **G** is the required inverse. It is easy to see that **Gy** satisfies the equation $\mathbf{Ax} = \mathbf{y}$ and has at the utmost s nonzero components.

2.7 g-INVERSE OF A SPECIFIED RANK

In Lemma 2.2.3 it is shown that $R(\mathbf{A}^-) \geq R(\mathbf{A})$, and in Section 2.5 we have actually constructed an inverse with the property $R(\mathbf{A}^-) = R(\mathbf{A})$. Now we consider the construction of inverses with any rank s such that $R(\mathbf{A}) \leq s \leq \min(m, n)$. Theorem 2.7.1 provides a general representation of a g-inverse with a specified rank s.

First we shall prove a lemma which we need in proving Theorem 2.7.1.

LEMMA 2.7.1 *Let* **H** *be an idempotent matrix. Then there exists a positive definite matrix* **L** *such that* $\mathbf{L}^{-1}\mathbf{H}^*\mathbf{L} = \mathbf{H}$.

Proof: Let **H** be an idempotent matrix of order n and rank r. Observe that $\mathbf{I} - \mathbf{H}$ is also idempotent of order n and rank $n - r$. Consider the rank factorizations of **H** and $\mathbf{I} - \mathbf{H}$ as $\mathbf{H} = \mathbf{DE}$ and $\mathbf{I} - \mathbf{H} = \mathbf{D}_1\mathbf{E}_1$. Write

$$\mathbf{F}_1 = (\mathbf{D} \,\vdots\, \mathbf{D}_1) \quad \text{and} \quad \mathbf{F}_2' = (\mathbf{E}' \,\vdots\, \mathbf{E}_1')$$

and note that both $\mathbf{F}_1$ and $\mathbf{F}_2$ are square matrices of order n and $\mathbf{F}_1\mathbf{F}_2 = \mathbf{DE} + \mathbf{D}_1\mathbf{E}_1 = \mathbf{I}$.

Hence $\mathbf{F}_1 = \mathbf{F}_2^{-1} = \mathbf{L}^{-1}\mathbf{F}_2^*$, where $\mathbf{L} = \mathbf{F}_2^*\mathbf{F}_2$. Therefore $\mathbf{D} = \mathbf{L}^{-1}\mathbf{E}^*$ and $\mathbf{E} = \mathbf{D}^*\mathbf{L}$. Thus $\mathbf{H} = \mathbf{DE} = \mathbf{L}^{-1}\mathbf{E}^*\mathbf{D}^*\mathbf{L} = \mathbf{L}^{-1}\mathbf{H}^*\mathbf{L}$.

THEOREM 2.7.1 *Let* **A** *be a matrix of order* $m \times n$ *and rank* r, *and* s *an integer;* $r \leq s \leq \min(m, n)$. *Then* **G** *of order* $n \times m$ *is a g-inverse of* **A** *with rank* s, *iff it is of the form*

$$(\mathbf{A} + \mathbf{M}_2\mathbf{N}_2)_r^-, \tag{2.7.1}$$

where $\mathbf{M}_2$ *of order* $m \times (s-r)$ *and* $\mathbf{N}_2$ *of order* $(s-r) \times n$ *are arbitrary matrices such that*

$$R(\mathbf{A} \,\vdots\, \mathbf{M}_2) = R\begin{pmatrix}\mathbf{A}\\ \mathbf{N}_2\end{pmatrix} = s. \tag{2.7.2}$$

Proof (the if part): Let $\mathbf{A} = \mathbf{M}_1\mathbf{N}_1$ be a rank factorization of $\mathbf{A}$, where $\mathbf{M}_1$ and $\mathbf{N}_1$ are matrices of order $m \times r$ and $r \times n$ respectively. Observe that in view of (2.7.2)

$$\mathbf{A} + \mathbf{M}_2\mathbf{N}_2 = (\mathbf{M}_1 \,\vdots\, \mathbf{M}_2)\begin{pmatrix}\mathbf{N}_1\\ \mathbf{N}_2\end{pmatrix}$$

is a rank factorization of $\mathbf{A} + \mathbf{M}_2\mathbf{N}_2$. It is known that $(\mathbf{A} + \mathbf{M}_2\mathbf{N}_2)_r^- = \mathbf{Q}_1\mathbf{P}_1 + \mathbf{Q}_2\mathbf{P}_2$, where

$$(\mathbf{Q}_1 \,\vdots\, \mathbf{Q}_2)$$

is a right inverse of $\begin{pmatrix}\mathbf{N}_1\\ \mathbf{N}_2\end{pmatrix}$ and $\begin{pmatrix}\mathbf{P}_1\\ \mathbf{P}_2\end{pmatrix}$ is a left inverse of $(\mathbf{M}_1 \,\vdots\, \mathbf{M}_2)$, $\mathbf{Q}_1$, $\mathbf{Q}_2$, $\mathbf{P}_1$, and $\mathbf{P}_2$ being matrices of order $n \times r$, $n \times (s-r)$, $r \times m$ and $(s-r) \times m$, respectively.

$$\begin{pmatrix}\mathbf{N}_1\\ \mathbf{N}_2\end{pmatrix}(\mathbf{Q}_1\mathbf{P}_1 + \mathbf{Q}_2\mathbf{P}_2)(\mathbf{M}_1 \,\vdots\, \mathbf{M}_2) = \mathbf{I}_s,$$

$$\Rightarrow \mathbf{N}_1(\mathbf{Q}_1\mathbf{P}_1 + \mathbf{Q}_2\mathbf{P}_2)\mathbf{M}_1 = \mathbf{I}_r,$$

$$\Rightarrow \mathbf{M}_1\mathbf{N}_1(\mathbf{Q}_1\mathbf{P}_1 + \mathbf{Q}_2\mathbf{P}_2)\mathbf{M}_1\mathbf{N}_1 = \mathbf{M}_1\mathbf{N}_1.$$

This shows that $(\mathbf{A} + \mathbf{M}_2\mathbf{N}_2)_r^- = \mathbf{Q}_1\mathbf{P}_1 + \mathbf{Q}_2\mathbf{P}_2$, which is clearly of rank s, is a g-inverse of $\mathbf{A} = \mathbf{M}_1\mathbf{N}_1$.

Proof (the only if part): Let $\mathbf{G}$ be a g-inverse of $\mathbf{A}$ of rank s. $\mathbf{H} = \mathbf{GA}$ is idempotent and of rank r. Using Lemma 2.7.1, we observe that there exists a p.d. matrix $\mathbf{L}$, such that $\mathbf{L}^{-1}\mathbf{H}^*\mathbf{L} = \mathbf{H}$.

But $\mathbf{L}^{-1}\mathbf{H}^*\mathbf{L}$ is the adjoint $\mathbf{H}^\#$ of $\mathbf{H}$ if the inner product relation is induced by the $\mathbf{L}$-metric; that is, if $\mathbf{y}^*\mathbf{L}\mathbf{x} = (\mathbf{x}, \mathbf{y})$ is taken to be the inner product of two complex n-tuples $\mathbf{x}$ and $\mathbf{y}$. This implies that if the inner product is so defined, $\mathbf{GA}$ is the orthogonal projector onto $\mathcal{M}(\mathbf{GA})$. Consider

$$\mathbf{P}_\mathbf{G} = \mathbf{G}(\mathbf{G}^*\mathbf{L}\mathbf{G})^-\mathbf{G}^*\mathbf{L},$$

the orthogonal projector onto $\mathcal{M}(\mathbf{G})$ under the same inner product definition. Write

$$\mathbf{Y} = (\mathbf{G}^*\mathbf{L}\mathbf{G})^-\mathbf{G}^*\mathbf{L} - \mathbf{A}$$

and observe that $\mathbf{GY} = \mathbf{P_G} - \mathbf{GA}$ is the orthogonal projector onto the orthogonal complement of $\mathcal{M}(\mathbf{GA})$ in $\mathcal{M}(\mathbf{G})$. Hence $\mathbf{GY}$ is idempotent and $R(\mathbf{GY}) = R(\mathbf{P_G}) - R(\mathbf{GA}) = s - r$. Let us now put $\mathbf{X} = \mathbf{YGY}$ and observe that $\mathbf{GX} = \mathbf{GY}$ is idempotent and $R(\mathbf{X}) = R(\mathbf{GY}) = s - r$. Further, $\mathbf{G}(\mathbf{A} + \mathbf{X}) = \mathbf{P_G}$ is idempotent and of rank equal to $R(\mathbf{G}) = s = R(\mathbf{A}) + R(\mathbf{X}) = R(\mathbf{A} + \mathbf{X})$. Hence by Lemma 2.2.2 and Lemma 2.5.1, $\mathbf{G} = (\mathbf{A} + \mathbf{X})_r^-$. This proves the 'only if' part.

A g-Inverse of Maximum Rank

The special case where $s = \min(m, n)$ deserves special treatment. Suppose $m \leq n$. Consider the square matrix obtained from $\mathbf{A}$ by adding $(n - m)$ null rows, and a nonsingular matrix $\mathbf{C}$ such that $\mathbf{C}\begin{pmatrix}\mathbf{A}\\ \mathbf{0}\end{pmatrix} = \mathbf{H}$ is in the Hermite canonical form. Let $\mathbf{C}$ be partitioned as $(\mathbf{C}_1 \vdots \mathbf{C}_2)$. Observe that $\mathbf{C}_1\mathbf{A} = \mathbf{H}$, $R(\mathbf{H}) = R(\mathbf{A}) = r$, and $\mathbf{H}$ is idempotent. Hence, by Lemma 2.2.2 and Definition 3, $\mathbf{C}_1$ is $\mathbf{A}^-$. In fact $\mathbf{C}_1$ is a g-inverse of maximum rank in the sense explained above. Further $\mathbf{H}$ has precisely $n - r$ null and r non-null rows, one could use the g-inverse $\mathbf{C}_1$ so constructed to provide a test for consistency of given linear equations, different from the one suggested in Theorem 2.3.1(d). This is contained in Lemma 2.7.2.

LEMMA 2.7.2 *The equation* $\mathbf{Ax} = \mathbf{y}$ *is consistent iff a particular component of* $\mathbf{C}_1\mathbf{y} = \mathbf{h}$ *is zero, whenever the corresponding row of* $\mathbf{H} = \mathbf{C}_1\mathbf{A}$ *is a null row.*

When $m > n$, a g-inverse with the same property can be computed, first by adding $m - n$ null columns to $\mathbf{A}$, obtaining the nonsingular matrix $\mathbf{C}$ that reduces this square matrix to Hermite canonical form and deleting the last $m - n$ rows from the matrix $\mathbf{C}$ so obtained.

We note in passing that the g-inverse so constructed also provides a basic solution in the sense explained in Section 2.6; that is, if the equation $\mathbf{Ax} = \mathbf{y}$ is consistent, $\mathbf{h} = \mathbf{C}_1\mathbf{y}$ is a basic solution and any g-inverse of maximum rank providing a basic solution shares with $\mathbf{C}_1$ its property as claimed in Lemma 2.7.2.

2.8 USEFUL DECOMPOSITION THEOREMS FOR MATRICES

THEOREM 2.8.1. *Let* $\mathbf{A}_i$ *be* $m \times p_i$ *matrix of rank* r_i, $i = 1, \ldots, k$. *If* $\Sigma r_i = m$, *then the statements*

(a) $\mathbf{A}_i^*\mathbf{A}_j = \mathbf{0}$ *for all* $i \neq j$ (2.8.1)

(b) $\mathbf{I} = \mathbf{A}_1(\mathbf{A}_1^*\mathbf{A}_1)^-\mathbf{A}_1^* + \cdots + \mathbf{A}_k(\mathbf{A}_k^*\mathbf{A}_k)^-\mathbf{A}_k^*$ (2.8.2)

are equivalent.

Let Λ be a p.d. matrix. If $\Sigma r_i = m$, then the statements

$$\text{(a)}\quad \mathbf{A}_i^* \Lambda \mathbf{A}_j = 0 \quad \textit{for all } i \neq j \tag{2.8.3}$$

$$\text{(b)}\quad \Lambda^{-1} = \Sigma \mathbf{A}_i(\mathbf{A}_i^* \Lambda \mathbf{A}_i)^- \mathbf{A}_i^* \tag{2.8.4}$$

are equivalent.

We prove a more general theorem from which the results of Theorem 2.8.1 follow as special cases.

Let Λ be a matrix of order $m \times n$ and $\mathbf{A}_i, \mathbf{B}_i$ be matrices of order $m \times p_i$, $n \times q_i$, respectively, $i = 1, 2, \ldots, k$. Write $\mathbf{A} = (\mathbf{A}_1 \vdots \mathbf{A}_2 \vdots \cdots \vdots \mathbf{A}_k)$ and $\mathbf{B} = (\mathbf{B}_1 \vdots \mathbf{B}_2 \vdots \cdots \vdots \mathbf{B}_k)$. Consider the following statements

$$\text{(a)}\quad \mathbf{A}_i^* \Lambda \mathbf{B}_j = \mathbf{0}\ \forall\ i \neq j \tag{2.8.5}$$

$$\text{(b)}\quad \mathbf{G} = \Sigma\, \mathbf{B}_i(\mathbf{A}_i^* \Lambda \mathbf{B}_i)^- \mathbf{A}_i^* \text{ is a g-inverse of } \Lambda, \tag{2.8.6}$$

where $(\mathbf{A}_i^* \Lambda \mathbf{B}_i)^-$ is any g-inverse of $\mathbf{A}_i^* \Lambda \mathbf{B}_i$.

THEOREM 2.8.2

$$\text{(a)} \Rightarrow \text{(b) } \textit{iff } R(\mathbf{A}^* \Lambda \mathbf{B}) = R(\Lambda). \tag{2.8.7}$$

Proof: Observe that

$$\text{(a)} \Rightarrow \mathbf{A}^* \Lambda \mathbf{B} = \text{diag}(\mathbf{A}_1^* \Lambda \mathbf{B}_1, \mathbf{A}_2^* \Lambda \mathbf{B}_2, \ldots \mathbf{A}_k^* \Lambda \mathbf{B}_k)$$

$$\Rightarrow \mathbf{D} = \text{diag}((\mathbf{A}_1^* \Lambda \mathbf{B}_1)^-, (\mathbf{A}_2^* \Lambda \mathbf{B}_2)^- \ldots, (\mathbf{A}_k^* \Lambda \mathbf{B}_k)^-)$$

is a g-inverse of $\mathbf{A}^* \Lambda \mathbf{B}$. Hence by Lemma 2.2.5(d) it is seen that $\mathbf{G} = \mathbf{B}\mathbf{D}\mathbf{A}^*$ is a g-inverse of Λ iff $R(\mathbf{A}^* \Lambda \mathbf{B}) = R(\Lambda)$.

THEOREM 2.8.3

$$\text{(b)} \Rightarrow \text{(a) iff } \Sigma R(\mathbf{A}_i^* \Lambda) = \Sigma R(\Lambda \mathbf{B}_i) = R(\Lambda). \tag{2.8.8}$$

Proof (*the if part*):

$$\text{(b)} \Rightarrow R(\Lambda) = tr\, \mathbf{G}\Lambda = \Sigma\, tr\, \mathbf{B}_i(\mathbf{A}_i^* \Lambda \mathbf{B}_i)^- \mathbf{A}_i^* \Lambda$$

$$= \Sigma\, tr(\mathbf{A}_i^* \Lambda \mathbf{B}_i)^- \mathbf{A}_i^* \Lambda \mathbf{B}_i = \Sigma R(\mathbf{A}_i^* \Lambda \mathbf{B}_i)$$

This together with (2.8.8)

$$\Rightarrow R(\mathbf{A}_i^* \Lambda \mathbf{B}_i) = R(\mathbf{A}_i^* \Lambda) = R(\Lambda \mathbf{B}_i) \tag{2.8.9}$$

Observe next that (b) $\Rightarrow$

$$\mathbf{B}\mathbf{D}\mathbf{A}^* = \Lambda^- \Rightarrow \mathbf{A}^* \Lambda \mathbf{B} = \mathbf{A}^* \Lambda \mathbf{B}\mathbf{D}\mathbf{A}^* \Lambda \mathbf{B} \Rightarrow \mathbf{D} = (\mathbf{A}^* \Lambda \mathbf{B})^- \tag{2.8.10}$$

where $\mathbf{D}$ is as defined in the proof of Theorem 2.8.2.

We notice now that if (2.8.8) holds and (2.8.6) is true for some choice of g-inverses $(\mathbf{A}_i^*\mathbf{\Lambda B}_i)^-$, then it is true for every choice. This is because

$$\mathbf{\Lambda} = \mathbf{\Lambda G \Lambda} = \Sigma \mathbf{\Lambda B}_i(\mathbf{A}_i^*\mathbf{\Lambda B}_i)^- \mathbf{A}_i^*\mathbf{\Lambda}$$

and under (2.8.9), by Lemma 2.2.6 (g) the matrix $\mathbf{\Lambda B}_i(\mathbf{A}_i^*\mathbf{\Lambda B}_i)^-\mathbf{A}_i^*\mathbf{\Lambda}$ is independent of the choice of the g-inverse $(\mathbf{A}_i^*\mathbf{\Lambda B}_i)^-$.

Let us now choose and fix each g-inverse $(\mathbf{A}_i^*\mathbf{\Lambda B}_i)^-$ in $\mathbf{D}$ to be $(\mathbf{A}_i^*\mathbf{\Lambda B}_i)_r^-$. By Lemma 2.5.1 $R[(\mathbf{A}_i^*\mathbf{\Lambda B}_i)_r^-] = R(\mathbf{A}_i^*\mathbf{\Lambda B}_i)$. Hence

$$R(\mathbf{D}) = \Sigma R(\mathbf{A}_i^*\mathbf{\Lambda B}_i)_r^- = \Sigma R(\mathbf{A}_i^*\mathbf{\Lambda B}_i) = R(\mathbf{\Lambda}) = R(\mathbf{A}^*\mathbf{\Lambda B}),$$

since by Lemma 2.2.5(d) $\mathbf{BDA}^*$ can be a g-inverse of $\mathbf{\Lambda}$ iff $R(\mathbf{A}^*\mathbf{\Lambda B}) = R(\mathbf{\Lambda})$. This together with (2.8.10) $\Rightarrow \mathbf{D} = (\mathbf{A}^*\mathbf{\Lambda B})_r^-$ in view of Lemma 2.5.1. That is

$$\mathbf{DA}^*\mathbf{\Lambda BD} = \mathbf{D}$$

$$\Rightarrow (\mathbf{A}_i^*\mathbf{\Lambda B}_i)^-\mathbf{A}_i^*\mathbf{\Lambda B}_j(\mathbf{A}_j^*\mathbf{\Lambda B}_j)^- = \mathbf{0}, \qquad \forall\, i \neq j \Rightarrow \mathbf{A}_i^*\mathbf{\Lambda B}_j = 0, \qquad \forall\, i \neq j$$

using Lemma 2.2.6(f).

Proof (the only if part): Assume that both (a) and (b) hold

$$\mathbf{A}_i^*\mathbf{\Lambda} = \mathbf{A}_i^*\mathbf{\Lambda\Lambda}^-\mathbf{\Lambda} = \mathbf{A}_i^*\mathbf{\Lambda}\Sigma\mathbf{B}_j(\mathbf{A}_j^*\mathbf{\Lambda B}_j)^-\mathbf{A}_j^*\mathbf{\Lambda} = \mathbf{A}_i^*\mathbf{\Lambda B}_i(\mathbf{A}_i^*\mathbf{\Lambda B}_i)^-\mathbf{A}_i^*\mathbf{\Lambda}$$

$$\Rightarrow \mathbf{B}_i(\mathbf{A}_i^*\mathbf{\Lambda B}_i)^- = (\mathbf{A}_i^*\mathbf{\Lambda})^- \Rightarrow \mathbf{G} = \Sigma\mathbf{B}_i(\mathbf{A}_i^*\mathbf{\Lambda B}_i)^-\mathbf{A}_i^* = \Sigma(\mathbf{A}_i^*\mathbf{\Lambda})^-\mathbf{A}_i^*$$

$$\Rightarrow R(\mathbf{\Lambda}) = tr(\mathbf{G\Lambda}) = \Sigma tr(\mathbf{A}_i^*\mathbf{\Lambda})^-\mathbf{A}_i^*\mathbf{\Lambda} = \Sigma R(\mathbf{A}_i^*\mathbf{\Lambda}).$$

We prove $R(\mathbf{\Lambda}) = \Sigma R(\mathbf{\Lambda B}_i)$ similarly.

2.9 PRINCIPAL IDEMPOTENTS OF A SQUARE MATRIX

For every complex square matrix $\mathbf{A}$ (of order n) and each complex number λ, there exist square matrices $\mathbf{K}_\lambda$ (also of order n) such that

(i) $\mathbf{K}_\lambda\mathbf{K}_\mu = \delta_{\lambda\mu}\mathbf{K}_\lambda$ where $\delta_{\lambda\mu}$ is the Kronecker symbol,
(ii) $\sum_{\text{all }\lambda} \mathbf{K}_\lambda = \mathbf{I}$,
(iii) $\mathbf{K}_\lambda\mathbf{A} = \mathbf{AK}_\lambda$, and
(iv) $(\mathbf{A} - \lambda\mathbf{I})\mathbf{K}_\lambda$ is nilpotent.

Such matrices $\mathbf{K}_\lambda$ are called the principal idempotents of $\mathbf{A}$.

THEOREM 2.9.1 (explicit form for the principal idempotents) *The principal idempotents $\mathbf{K}_\lambda$ of $\mathbf{A}$ are given by*

$$\mathbf{K}_\lambda = \mathbf{F}_\lambda(\mathbf{E}_\lambda^*\mathbf{F}_\lambda)^-\mathbf{E}_\lambda^* \tag{2.9.1}$$

where

$$\mathbf{E}_\lambda^* = \mathbf{I} - (\mathbf{A} - \lambda\mathbf{I})^m\{(\mathbf{A} - \lambda\mathbf{I})^m\}^-$$

$$\mathbf{F}_\lambda = \mathbf{I} - \{(\mathbf{A} - \lambda\mathbf{I})^m\}^-\{(\mathbf{A} - \lambda\mathbf{I})^m\}$$

and m is a positive integer sufficiently large. The theorem remains true if $m = n$. The least value of m for which the theorem is true depends on **A** *and $m = 1$ iff* **A** *is semi-simple.*

Proof: Let $\lambda_1 \ldots, \lambda_u$ be the distinct eigen values of **A**. Consider the Jordan canonical representation of **A**

$$\mathbf{A} = \sum_{i=1}^{u} \mathbf{L}_i\mathbf{D}_i\mathbf{R}_i^* \tag{2.9.2}$$

where $\mathbf{D}_i = \text{diag}(\mathbf{J}_1(\lambda_1), \ldots, \mathbf{J}_{v_i}(\lambda_i))$ is a matrix of order r_i, $\mathbf{J}_j(\lambda_i)$ being a lower Jordan matrix with λ_i at the diagonal positions, 1 at the immediate subdiagonal positions and 0 elsewhere. Further if $\mathbf{L} = (\mathbf{L}_1 \vdots \mathbf{L}_2 \vdots \cdots \vdots \mathbf{L}_u)$ and $\mathbf{R} = (\mathbf{R}_1 \vdots \mathbf{R}_2 \vdots \cdots \vdots \mathbf{R}_u)$, then $\mathbf{LR}^* = \mathbf{R}^*\mathbf{L} = \mathbf{I}$. Hence

$$\mathbf{A} - \lambda\mathbf{I} = \sum_{i=1}^{u} \mathbf{L}_i(\mathbf{D}_i - \lambda\mathbf{I})\mathbf{R}_i^*,$$

where $\mathbf{D}_i - \lambda\mathbf{I}$ is nonsingular if $\lambda \neq \lambda_i$ and $\mathbf{D}_i - \lambda_i\mathbf{I}$ is nilpotent, that is, there exists an integer m_i such that

$$(\mathbf{D}_i - \lambda_i\mathbf{I})^{m_i} = \mathbf{0}$$

We assume that m_i is the least positive integer satisfying this condition. Let $m = \max(m_1, m_2, \ldots, m_u)$. Check that

$$(\mathbf{A} - \lambda_i\mathbf{I})^m = \sum_{j \neq i} \mathbf{L}_j(\mathbf{D}_j - \lambda_i\mathbf{I})^m\mathbf{R}_j^*$$

$$\Rightarrow \mathscr{M}(\mathbf{A} - \lambda_i\mathbf{I})^m = \mathscr{M}(\mathbf{L}_1 \vdots \mathbf{L}_2 \vdots \cdots \vdots \mathbf{L}_{i-1} \vdots \mathbf{L}_{i+1} \vdots \cdots \vdots \mathbf{L}_u),$$

$$\mathscr{M}[\{(\mathbf{A} - \lambda_i\mathbf{I})^m\}^*] = \mathscr{M}(\mathbf{R}_1 \vdots \mathbf{R}_2 \vdots \cdots \vdots \mathbf{R}_{i-1} \vdots \mathbf{R}_{i+1} \vdots \cdots \vdots \mathbf{R}_u)$$

Since $\mathbf{R}^*\mathbf{L} = \mathbf{L}^*\mathbf{R} = \mathbf{I}$, $\mathbf{E}^*_{\lambda_i}(\mathbf{A} - \lambda_i\mathbf{I})^m = \mathbf{0}$ together with

$$R(\mathbf{E}^*_{\lambda_i}) = n - R[(\mathbf{A} - \lambda_i\mathbf{I})^m] = r_i$$

implies $\mathscr{M}(\mathbf{E}_{\lambda_i}) = \mathscr{M}(\mathbf{R}_i)$. Similarly $\mathscr{M}(\mathbf{F}_{\lambda_i}) = \mathscr{M}(\mathbf{L}_i)$.

$$\mathbf{R}_i^*\mathbf{L}_j = \mathbf{0}\ \forall\ i \neq j \Rightarrow \mathbf{E}_{\lambda_i}^*\mathbf{F}_{\lambda_i} = \mathbf{0}\ \forall\ i \neq j.$$

Also

$$R(\mathbf{E}_{\lambda_1} \vdots \mathbf{E}_{\lambda_2} \vdots \cdots \vdots \mathbf{E}_{\lambda_u}) = R(\mathbf{R}_1 \vdots \mathbf{R}_2 \vdots \cdots \vdots \mathbf{R}_u) = n$$

and

$$R(\mathbf{F}_{\lambda_1} \vdots \mathbf{F}_{\lambda_2} \vdots \cdots \vdots \mathbf{F}_{\lambda_u}) = R(\mathbf{L}_1 \vdots \mathbf{L}_2 \vdots \cdots \vdots \mathbf{L}_u) = n.$$

Hence from Theorem 2.8.2, it follows that

$$\sum_{i=1}^{u} \mathbf{K}_{\lambda_i} = \mathbf{I}, \quad \mathbf{K}_{\lambda_i}\mathbf{K}_{\lambda_j} = \delta_{ij}\mathbf{K}_{\lambda_i}.$$

Since both $\mathbf{E}_\lambda$ and $\mathbf{F}_\lambda$ are null matrices, when λ is not an eigenvalue of $\mathbf{A}$, conditions (i) and (ii) are thus established. Now

$$\left(\sum_{i=1}^{u} \mathbf{K}_{\lambda_i}\right)\mathbf{A} = \mathbf{A}\left(\sum_{i=1}^{u} \mathbf{K}_{\lambda_i}\right)$$

$$\Rightarrow \sum_{i=1}^{u} \mathbf{K}_{\lambda_i}\mathbf{L}_i\mathbf{D}_i\mathbf{R}_i^* = \sum_{i=1}^{u} \mathbf{L}_i\mathbf{D}_i\mathbf{R}_i^*\mathbf{K}_{\lambda_i}$$

$$\Rightarrow \mathbf{K}_{\lambda_i}\mathbf{L}_i\mathbf{D}_i\mathbf{R}_i^* = \mathbf{L}_i\mathbf{D}_i\mathbf{R}_i^*\mathbf{K}_{\lambda_i} \ \forall\, i$$

since $\mathbf{L}$ is nonsingular. Hence

$$\mathbf{K}_{\lambda_i}\mathbf{A} = \mathbf{A}\mathbf{K}_{\lambda_i} \ \forall\, i.$$

Since $\mathbf{K}_\lambda\mathbf{A}$ is trivially equal to $\mathbf{A}\mathbf{K}_\lambda$ if λ is not an eigenvalue of $\mathbf{A}$, (iii) is established.

$$\text{(iii)} \Rightarrow (\mathbf{A} - \lambda\mathbf{I})\mathbf{K}_\lambda = \mathbf{K}_\lambda(\mathbf{A} - \lambda\mathbf{I})$$

$$\Rightarrow [(\mathbf{A} - \lambda\mathbf{I})\mathbf{K}_\lambda]^m = (\mathbf{A} - \lambda\mathbf{I})^m\mathbf{K}_\lambda^m = \mathbf{0},$$

since $(\mathbf{A} - \lambda\mathbf{I})^m\mathbf{F}_\lambda = \mathbf{0}, \forall\lambda$. Thus (iv) is also established.

If $\mathbf{A}$ is semisimple, each Jordan cell in the canonical representation of $\mathbf{A}$ is of order 1. Hence $m_i = 1 \ \forall\, i$ implying $m = 1$. Conversely if $m = 1$,

$$(\mathbf{A} - \lambda_i\mathbf{I})\mathbf{K}_{\lambda_i} = \mathbf{0} \ \forall\, i,$$

since $(\mathbf{A} - \lambda_i\mathbf{I})\mathbf{F}_{\lambda_i} = \mathbf{0} \ \forall\, i$. This implies that the nonnull columns of $\mathbf{K}_{\lambda_i}$ are eigenvectors of $\mathbf{A}$ corresponding to its eigenvalue λ_i. Since $\sum_{i=1}^{u} \mathbf{K}_{\lambda_i} = \mathbf{I}$, the eigenvectors of $\mathbf{A}$ span the whole space. Hence $\mathbf{A}$ is semisimple.

We now show that if the matrices $\mathbf{K}_\lambda$ satisfy (i)–(iv), they are always expressible as

$$\mathbf{K}_\lambda = \mathbf{F}_\lambda(\mathbf{E}_\lambda^*\mathbf{F}_\lambda)^-\mathbf{E}_\lambda^*$$

which in view of Lemma 2.2.6(g), incidentally, establishes the uniqueness of principal idempotents.

(iii) and (iv) $\Rightarrow (\mathbf{A} - \lambda\mathbf{I})^m\mathbf{K}_\lambda^m = \mathbf{0}$ for some integer m. Hence if λ is not an eigenvalue of $\mathbf{A}$, $(\mathbf{A} - \lambda\mathbf{I})^m$ is nonsingular. Hence $\mathbf{K}_\lambda^m = \mathbf{0}$. (i) implies $\mathbf{K}_\lambda^m = \mathbf{K}_\lambda$.

Thus $\mathbf{K}_\lambda = \mathbf{0}$ for any λ which is not an eigenvalue of $\mathbf{A}$. Let, as before, $\lambda_1, \lambda_2, \ldots, \lambda_u$ be the distinct eigenvalues of $\mathbf{A}$ and m_i be the least positive integer for which

$$(\mathbf{A} - \lambda_i\mathbf{I})^{m_i}\mathbf{K}_{\lambda_i}^{m_i} = (\mathbf{A} - \lambda_i\mathbf{I})^{m_i}\mathbf{K}_{\lambda_i} = \mathbf{0}.$$

If

$$m = \max(m_1, m_2, \ldots, m_u),$$

then

$$(\mathbf{A} - \lambda_i\mathbf{I})^m\mathbf{K}_{\lambda_i} = \mathbf{0}, \qquad \mathbf{K}_{\lambda_i}(\mathbf{A} - \lambda_i\mathbf{I})^m = \mathbf{0}$$

$\Rightarrow \mathbf{K}_{\lambda_i} = \mathbf{F}_{\lambda_i}\mathbf{W}_i\mathbf{E}_{\lambda_i}^*$ for some $\mathbf{W}_i$. Also

$$\Sigma\mathbf{K}_\lambda = \mathbf{I} \Rightarrow \sum_{i=1}^{u} \mathbf{K}_{\lambda_i} = \mathbf{I}$$

$$\mathbf{E}_{\lambda_i}^*\mathbf{F}_{\lambda_i} = \mathbf{E}_{\lambda_i}^*\left(\sum_{i=1}^{u} \mathbf{K}_{\lambda_i}\right)\mathbf{F}_{\lambda_i} = \mathbf{E}_{\lambda_i}^*\mathbf{K}_{\lambda_i}\mathbf{F}_{\lambda_i}$$

$$= \mathbf{E}_{\lambda_i}^*\mathbf{F}_{\lambda_i}\mathbf{W}_i\mathbf{E}_{\lambda_i}^*\mathbf{F}_{\lambda_i} \Rightarrow \mathbf{W}_i = (\mathbf{E}_{\lambda_i}^*\mathbf{F}_{\lambda_i})^-.$$

2.10 SPECTRAL DECOMPOSITION OF AN ARBITRARY MATRIX

THEOREM 2.10.1 *Any $m \times n$ matrix $\mathbf{A}$ can be written as*

$$\mathbf{A} = \sum \alpha_i\mathbf{U}_i, \tag{2.10.1}$$

where α_i^2, $i = 1, 2, \ldots, u$ are the distinct nonnull eigenvalues of $\mathbf{A}^*\mathbf{A}$ and the matrices

$$\mathbf{U}_i = \alpha_i^{-1}\mathbf{A}[\mathbf{I} - (\mathbf{A}^*\mathbf{A} - \alpha_i^2\mathbf{I})\{(\mathbf{A}^*\mathbf{A} - \alpha_i^2\mathbf{I})^2\}^-(\mathbf{A}^*\mathbf{A} - \alpha_i^2\mathbf{I})] \tag{2.10.2}$$

satisfy

$$\begin{aligned} &\mathbf{U}_i\mathbf{U}_i^*\mathbf{U}_i = \mathbf{U}_i \,\forall\, i \\ &\mathbf{U}_i\mathbf{U}_j^* = \mathbf{0}, \qquad \mathbf{U}_i^*\mathbf{U}_j = \mathbf{0} \,\forall\, i \neq j, \end{aligned} \tag{2.10.3}$$

(α_i is taken to be the positive square root of α_i^2 which is real and positive since $\mathbf{A}^*\mathbf{A}$ is hermitian and n.n.d.)

Proof: Let α_i^2 be an eigenvalue of $\mathbf{A}^*\mathbf{A}$ of multiplicity r_i. It is well known that the matrix $\mathbf{A}$ is diagonable in the form

$$\mathbf{B}^*\mathbf{AC} = \begin{pmatrix} \mathbf{\Lambda} & \mathbf{0} \\ \mathbf{0} & \mathbf{0} \end{pmatrix}$$

where **B** and **C** are unitary matrices of order m and n, respectively, and

$$\mathbf{\Lambda} = \text{diag}\,(\alpha_1\mathbf{I}_{r_1}, \alpha_2\mathbf{I}_{r_2}, \ldots, \alpha_u\mathbf{I}_{r_u}).$$

Let us partition

$$\mathbf{B} = (\mathbf{B}_1 \vdots \mathbf{B}_2 \vdots \cdots \vdots \mathbf{B}_u \vdots \mathbf{B}_{u+1})$$
$$\mathbf{C} = (\mathbf{C}_1 \vdots \mathbf{C}_2 \vdots \cdots \vdots \mathbf{C}_u \vdots \mathbf{C}_{u+1}),$$

where $\mathbf{B}_i$, $\mathbf{C}_i$ are matrices of order $m \times r_i$ $(i = 1, 2, \ldots, u)$. $\mathbf{B}_{u+1}$, $\mathbf{C}_{u+1}$ are of order $m \times (n - r)$ and $r = \Sigma_1^u r_i$. Observe

$$\mathbf{A} = \mathbf{B}\begin{pmatrix} \mathbf{\Lambda} & \mathbf{0} \\ \mathbf{0} & \mathbf{0} \end{pmatrix} \mathbf{C}^* = \Sigma\alpha_i\mathbf{B}_i\mathbf{C}_i^* = \Sigma\alpha_i\mathbf{U}_i,$$
$$\mathbf{U}_i = \mathbf{B}_i\mathbf{C}_i^*. \tag{2.10.4}$$

Since **B** and **C** are unitary matrices, the matrices $\mathbf{U}_i$ are easily seen to obey (2.10.3). From (2.10.4)

$$\mathbf{A}^*\mathbf{A} = \mathbf{CDC}^*,$$

where $\mathbf{D} = \text{diag}\,(\mathbf{D}_1, \mathbf{D}_2, \ldots, \mathbf{D}_{u+1})$, $\mathbf{D}_i = \alpha_i^2\mathbf{I}_{r_i}$, $i = 1, 2, \ldots, u$ and $\mathbf{D}_{u+1} = \mathbf{0}$. Hence

$$\mathbf{A}^*\mathbf{A} - \alpha_i^2\mathbf{I} = \mathbf{C}(\mathbf{D} - \alpha_i^2\mathbf{I})\mathbf{C}^*,$$

and

$$\mathbf{D} - \alpha_i^2\mathbf{I} = \text{diag}\,(\mathbf{D}_1 - \alpha_i^2\mathbf{I}_{r_1}, \mathbf{D}_2 - \alpha_i^2\mathbf{I}_{r_2}, \ldots, \mathbf{D}_u - \alpha_i^2\mathbf{I}_{r_u}, -\alpha_i^2\mathbf{I}_{r_{u+1}}),$$

where $r_{u+1} = n - r$.

Observe that the ith diagonal submatrix in $\mathbf{D} - \alpha_i^2\mathbf{I}$, is $\mathbf{D}_i - \alpha_i^2\mathbf{I}_{r_i}$ which is a null matrix, while the other diagonal submatrices are strictly nonsingular. Hence

$$\mathscr{M}(\mathbf{A}^*\mathbf{A} - \alpha_i^2\mathbf{I}) = \mathscr{M}(\mathbf{C}_1 \vdots \mathbf{C}_2 \vdots \cdots \vdots \mathbf{C}_{i-1} \vdots \mathbf{C}_{i+1} \vdots \cdots \vdots \mathbf{C}_{u+1})$$

and the projector

$$(\mathbf{A}^*\mathbf{A} - \alpha_i^2\mathbf{I})\{(\mathbf{A}^*\mathbf{A} - \alpha_i^2\mathbf{I})^2\}^-(\mathbf{A}^*\mathbf{A} - \alpha_i^2\mathbf{I}) = \sum_{j \neq i} \mathbf{C}_j\mathbf{C}_j^*.$$

Therefore

$$\mathbf{I} - (\mathbf{A}^*\mathbf{A} - \alpha_i^2\mathbf{I})\{(\mathbf{A}^*\mathbf{A} - \alpha_i^2\mathbf{I})^2\}^-(\mathbf{A}^*\mathbf{A} - \alpha_i^2\mathbf{I}) = \mathbf{C}_i\mathbf{C}_i^*$$

since $\Sigma\mathbf{C}_j\mathbf{C}_j^* = \mathbf{I}$ and

$$\alpha_i^{-1}\mathbf{A}[\mathbf{I} - (\mathbf{A}^*\mathbf{A} - \alpha_i^2\mathbf{I})\{(\mathbf{A}^*\mathbf{A} - \alpha_i^2\mathbf{I})^2\}^-(\mathbf{A}^*\mathbf{A} - \alpha_i^2\mathbf{I})]$$
$$= \alpha_i^{-1}(\Sigma\alpha_j\mathbf{B}_j\mathbf{C}_j^*)\mathbf{C}_i\mathbf{C}_i^* = \alpha_i^{-1}\alpha_i\mathbf{B}_i\mathbf{C}_i^*\mathbf{C}_i\mathbf{C}_i^* = \mathbf{B}_i\mathbf{C}_i^* = \mathbf{U}_i$$

as defined in (2.10.4).

COMPLEMENTS

1. Let $\mathbf{I}$ be the unit matrix of order n and $\mathbf{J}(n \times n)$ be a matrix whose elements are each equal to 1. If $a + (n-1)b = 0$, then $(a-b)^{-1}\mathbf{I}$ is a g-inverse of $(a-b)\mathbf{I} + b\mathbf{J}$.

2(a). Let $\mathbf{A}$ be any square matrix. It is well known that there exists a nonsingular matrix $\mathbf{C}$ such $\mathbf{CA} = \mathbf{H}$ is in the Hermite canonical form. Then $\mathbf{C}$ is a g-inverse of $\mathbf{A}$.

2(b). Let $\mathbf{A}$ be $m \times n$ $(m > n)$ matrix. Consider $\mathbf{A}_0 = (\mathbf{A} \vdots \mathbf{0})$, where $\mathbf{0}$ is $m \times (m-n)$ zero matrix. Let $\mathbf{B}_0\mathbf{A}_0 = \mathbf{H}$ (Hermite) with $|\mathbf{B}_0| \neq 0$. Consider the partition $\mathbf{B}_0' = (\mathbf{B}' \vdots \mathbf{B}_1')$. Then $\mathbf{B} = \mathbf{A}^-$.

3(a). Let $\mathbf{A}$ be a matrix of order $m \times n$, and $\mathbf{u}, \mathbf{v}$ be column vectors of order $m \times 1$ and $n \times 1$, respectively. If either $\mathbf{u} \in \mathcal{M}(\mathbf{A})$ or $\mathbf{v} \in \mathcal{M}(\mathbf{A}')$

$$\mathbf{A}^- - \frac{(\mathbf{A}^-\mathbf{u})(\mathbf{v}'\mathbf{A}^-)}{1 + \mathbf{v}'\mathbf{A}^-\mathbf{u}}$$

is a g-inverse of $(\mathbf{A} + \mathbf{uv}')$, provided $\mathbf{v}'\mathbf{A}^-\mathbf{u} \neq -1$.

3(b). Further

$$(\mathbf{A} + \mathbf{uv}')^- + \frac{(\mathbf{A} + \mathbf{uv}')^-\mathbf{uv}'(\mathbf{A}' + \mathbf{vu}')^-}{1 - \mathbf{v}'(\mathbf{A} + \mathbf{uv}')^-\mathbf{u}}$$

is a g-inverse of $\mathbf{A}$, provided $\mathbf{u} \in \mathcal{M}(\mathbf{A} + \mathbf{uv}')$ or $\mathbf{v} \in \mathcal{M}(\mathbf{A}' + \mathbf{vu}')$, and $\mathbf{v}'(\mathbf{A} + \mathbf{uv}')^-\mathbf{u} \neq 1$.

4. If $\vec{\mathbf{a}} = (a_1, a_2, a_3)$ and $\vec{\mathbf{b}} = (b_1, b_2, b_3)$ are two vectors in the Euclidean three-dimensional space, the vector cross product $\vec{\mathbf{a}} \times \vec{\mathbf{b}}$ is defined by

$$\vec{\mathbf{a}} \times \vec{\mathbf{b}} = (a_2b_3 - a_3b_2, a_3b_1 - a_1b_3, a_1b_2 - a_2b_1).$$

(a) The operator $(\vec{\mathbf{a}} \times \cdot)$ is a linear operator which corresponds to the 3×3 skew symmetric matrix

$$\mathbf{T} = \begin{pmatrix} 0 & -a_3 & a_2 \\ a_3 & 0 & -a_1 \\ -a_2 & a_1 & 0 \end{pmatrix}$$

(b) $-(a_1^2 + a_2^2 + a_3^2)^{-1}\mathbf{T}$ is a g-inverse of $\mathbf{T}$.

5. Let $\mathbf{A}$ be $n \times m$ matrix of rank r and $\mathbf{B}$ a $s \times m$ matrix of rank $m - r$ such that $\mathcal{M}(\mathbf{A}^*) \cap \mathcal{M}(\mathbf{B}^*)$ is a set containing only the null vector. Then

(a) $\mathbf{A}^*\mathbf{A} + \mathbf{B}^*\mathbf{B}$ is nonsingular

(b) $(\mathbf{A}^*\mathbf{A} + \mathbf{B}^*\mathbf{B})^{-1}$ is a g-inverse of $\mathbf{A}^*\mathbf{A}$

(c) if $s = m - r$, then

$$\begin{pmatrix} \mathbf{A}^*\mathbf{A} & \mathbf{B}^* \\ \mathbf{B} & \mathbf{0} \end{pmatrix} \text{ is nonsingular.}$$

(d) If

$$\begin{pmatrix} \mathbf{A}^*\mathbf{A} & \mathbf{B}^* \\ \mathbf{B} & \mathbf{0} \end{pmatrix}^{-1} = \begin{pmatrix} \mathbf{C}_1 & \mathbf{C}_2^* \\ \mathbf{C}_2 & \mathbf{C}_3 \end{pmatrix}, \text{ then}$$

(i) $\mathbf{C}_2^*$ is a right inverse of $\mathbf{B}$ and $\mathbf{AC}_2^* = \mathbf{0}$.
(ii) $\mathbf{C}_1$ is g-inverse of $\mathbf{A^*A}$.

6. Let

$$\mathbf{M} = \begin{pmatrix} \mathbf{A} & \mathbf{C} \\ \mathbf{C}^* & \mathbf{B} \end{pmatrix},$$

where $\mathbf{A} = \mathbf{X}_1^*\mathbf{X}_1, \mathbf{B} = \mathbf{X}_2^*\mathbf{X}_2, \mathbf{C} = \mathbf{X}_1^*\mathbf{X}_2$.
If $\mathbf{D} = \mathbf{B} - \mathbf{C}^*\mathbf{A}^-\mathbf{C}$, then

$$\mathbf{G} = \begin{pmatrix} \mathbf{A}^- + \mathbf{A}^-\mathbf{C}\mathbf{D}^-\mathbf{C}^*\mathbf{A}^- & -\mathbf{A}^-\mathbf{C}\mathbf{D}^- \\ -\mathbf{D}^-\mathbf{C}^*\mathbf{A}^- & \mathbf{D}^- \end{pmatrix}$$

is a g-inverse of $\mathbf{M}$. (Rohde, 1965)

7. A hermitian matrix has a hermitian g-inverse and n.n.d. matrix has a n.n.d. g-inverse.

8. A hermitian matrix $\mathbf{A}$ has a n.n.d. g-inverse if and only if $\mathbf{A}$ is n.n.d.

9. Let $\mathbf{A}(n \times m)$ and $\mathbf{R}(s \times m)$ be given complex matrices and $\mathbf{N}(m \times t)$ be any matrix of rank $= m -$ Rank $(\mathbf{R})$ such that $\mathbf{RN} = \mathbf{0}$. Write

$$\mathbf{D} = \mathbf{N}(\mathbf{N}^*\mathbf{A}^*\mathbf{A}\mathbf{N})^-\mathbf{N}^*\mathbf{A}^*, \text{ and } \mathbf{E} = \mathbf{A}^* - \mathbf{A}^*\mathbf{A}\mathbf{D}.$$

Then

(a) $\mathscr{M}(\mathbf{E}) \subset \mathscr{M}(\mathbf{R}^*)$
(b) The equations

$$\begin{pmatrix} \mathbf{A}^*\mathbf{A} & \mathbf{R}^* \\ \mathbf{R} & \mathbf{0} \end{pmatrix} \begin{pmatrix} \mathbf{X} \\ \mathbf{Y} \end{pmatrix} = \begin{pmatrix} \mathbf{A}^* \\ \mathbf{0} \end{pmatrix}$$

are consistent.
(Verify that $\mathbf{X} = \mathbf{D}, \mathbf{Y} = (\mathbf{R}^*)^-\mathbf{E}$ is a solution)
(c) If

$$\begin{pmatrix} \mathbf{A}^*\mathbf{A} & \mathbf{R}^* \\ \mathbf{R} & \mathbf{0} \end{pmatrix}^- = \begin{pmatrix} \mathbf{C}_1 & \mathbf{C}_2^* \\ \mathbf{C}_2 & \mathbf{C}_3 \end{pmatrix}$$

then

(i) $\mathbf{C}_2^*$ is a g-inverse of $\mathbf{R}$.
(ii) $\mathbf{RC}_1\mathbf{R}^* = \mathbf{0}$.
(iii) $\mathbf{AC}_1\mathbf{A}^*$ is idempotent.
(iv) $\mathbf{C}_1$ is a g-inverse of $\mathbf{A^*A}$ if and only if $\mathscr{M}(\mathbf{A}^*)$ and $\mathscr{M}(\mathbf{R}^*)$ are virtually disjoint, in which case $\mathbf{C}_1\mathbf{A}^*$ is a g-inverse of $\mathbf{A}$. (Khatri, 1968.)

10. For every matrix $\mathbf{A}(m \times n)$, there exist nonsingular matrices $\mathbf{B}$, $\mathbf{C}$ such that $\mathbf{BAC} = \mathbf{\Lambda} = (\lambda_{ij})$ is diagonal in the sense that, excluding possibly λ_{ii}, $i = 1, 2, \ldots,$ min (m, n), all the other λ_{ij}'s are zero. Let $\mathbf{\Lambda}^-$ be identical with the transpose of $\mathbf{\Lambda}$, except that the i-th diagonal element in $\mathbf{\Lambda}^-$ is $1/\lambda_{ii}$, whenever $\lambda_{ii} \neq 0$. Then $\mathbf{C\Lambda^-B}$ is $\mathbf{A}_r^-$.

11. A matrix **A** is a g-inverse of itself iff it is of the form

$$\mathbf{A} = \mathbf{L}\mathbf{D}\mathbf{L}^{-1},$$

where **D** is a diagonal matrix with diagonal elements equal to 0, +1, or −1.

12. *Numerical computation of* $\mathbf{A}^-$. We append a unit matrix to **A** and carry out pivotal condensation by sweep-out method.

A			I				Row No.	Row Operation
2	4	6	1	.	.	.	(1)	
3	6	12	.	1	.	.	(2)	
4	8	12	.	.	1	.	(3)	
5	10	25	.	.	.	1	(4)	
1	2	3	1/2	.	.	.	(5)	= (1) ÷ 2
.	.	3	−3/2	1	.	.	(6)	= (2) − 3 × (5)
.	.	.	−2	.	1	.	(7)	= (3) − 4 × (5)
.	.	10	−5/2	.	.	1	(8)	= (4) − 5 × (5)
1	2	0	5/4	.	.	−3/10	(9)	= 5 − 3 × (12)
.	.	.	−3/4	1	.	−3/10	(10)	= (6) − 3 × (12)
.	.	.	−2	.	1	.	(11)	= (7)
.	.	1	−1/4	.	.	1/10	(12)	= (8) ÷ 10

Interchanging rows (11) and (12) to bring the matrix under **A** in the third block to Hermite canonical form (see Section 1.2 for definition), we find the corresponding matrix under **I** [in rows (9), (10), (12)]

5/4	.	.	−3/10	(9)
−3/4	1	.	−3/10	(10)
−1/4	.	.	1/10	(12),

which is $\mathbf{A}^-$ with full rank, equal to 3. Actually the rank of **A** is 2, and $\mathbf{A}^-$ with rank 2 is obtained by replacing the row (10), which has a zero pivot in the Hermite canonical form, by a null row.

13. A matrix **S** is semisimple with real eigenvalues if and only if there exists a p.d. matrix **L**, such that $\mathbf{L}^{-1}\mathbf{S}^*\mathbf{L} = \mathbf{S}$. (Since **S** is semisimple, there exists a nonsingular matrix **P**, such that

$$\mathbf{S} = \mathbf{P}\mathbf{D}\mathbf{P}^{-1},$$

where **D** is a diagonal matrix of eigenvalues of **S**. $\mathbf{D}^* = \mathbf{D} \Rightarrow \mathbf{S}^* = (\mathbf{P}^*)^{-1}\mathbf{D}\mathbf{P}^* = (\mathbf{P}^*)^{-1}\mathbf{P}^{-1}\mathbf{S}\mathbf{P}\mathbf{P}^*$. The other part is easy.)

This incidentally gives an alternative proof of Lemma 2.7.1, since an idempotent matrix is necessarily semisimple, and its only eigenvalues are 0 and 1. (Bhimasankaram)

14. The conditions given in Lemma 2.2.4(iii) are both necessary and sufficient for $\mathbf{BA^-C}$ to be invariant under any choice of $\mathbf{A}^-$.

15. (The generalized eigenvalue equation $\mathbf{Ax} = \lambda\mathbf{Bx}$)

Methods for solving this equation are well known for the case where $\mathbf{B}$ is nonsingular. To avoid trivialities, therefore, we assume $\mathbf{B}$ is singular.

(a) Let $\mathbf{x}$ be an eigenvector of $\mathbf{B^-A}$ corresponding to its eigenvalue λ, then $\mathbf{x}$, λ satisfy the equation $\mathbf{Ax} = \lambda\mathbf{Bx}$, if $\mathbf{Ax}$ is an eigenvector of $\mathbf{BB^-}$ for its eigenvalue 1.

(b) Let $\mathbf{x}$, λ satisfy the equation $\mathbf{Ax} = \lambda\mathbf{Bx}$. Then $\mathbf{x}$ is an eigenvector of $\mathbf{B^-A}$ corresponding to its eigenvalue λ, if $\mathbf{x}$ is also an eigenvector of $\mathbf{B^-B}$ for its eigenvalue 1.

16. Let $\mathbf{G}$, $\mathbf{A}$ and $\mathbf{H}$ be matrices of order $n \times m$, $m \times n$ and $n \times p$ respectively. If $R(\mathbf{A}) = R(\mathbf{AH})$ then $\mathbf{GAH} = \mathbf{H} \Rightarrow \mathbf{G} = \mathbf{A}^-$.

CHAPTER 3

Three Basic Types of g-Inverses

3.1 g-INVERSE FOR A MINIMUM NORM SOLUTION OF $\mathbf{Ax} = \mathbf{y}$ (CONSISTENT)

In Chapter 2 we obtained the complete class of solutions of the equation $\mathbf{Ax} = \mathbf{y}$ and showed that every solution can be expressed in the form $\mathbf{x} = \mathbf{Gy}$, where $\mathbf{G}$ is a g-inverse. We now inquire whether there exists a particular choice of g-inverse independently of $\mathbf{y}$, such that $\mathbf{Gy}$ has the smallest norm in the class of all solutions of $\mathbf{Ax} = \mathbf{y}$, that is, we wish to find $\mathbf{G}$, if it exists, such that

$$\min_{\mathbf{Ax}=\mathbf{y}} \|\mathbf{x}\| = \|\mathbf{Gy}\| .$$

We denote the inner product of two vectors $\mathbf{x}, \mathbf{y} \in \mathscr{E}^s$ by $(\mathbf{x}, \mathbf{y})_s$. The adjoint of a $m \times n$ matrix $\mathbf{A}$, denoted by $\mathbf{A}^{\#}$, is defined by the relation

$$(\mathbf{Ax}, \mathbf{y})_m = (\mathbf{x}, \mathbf{A}^{\#}\mathbf{y})_n .$$

The norm of $\mathbf{x}$ is denoted by $\|\mathbf{x}\| = \sqrt{(\mathbf{x}, \mathbf{x})}$. It may be verified that

$$(\mathbf{A}^{\#})^{\#} = \mathbf{A} \quad (\mathbf{AB})^{\#} = \mathbf{B}^{\#}\mathbf{A}^{\#} .$$

THEOREM 3.1.1 *Let* $\mathbf{G}$ *be a g-inverse of* $\mathbf{A}$ *such that* $\mathbf{Gy}$ *is a minimum norm solution of* $\mathbf{Ax} = \mathbf{y}$ *for any* $\mathbf{y} \in \mathscr{M}(\mathbf{A})$. *Then it is n.s. that*

$$\mathbf{AGA} = \mathbf{A}, \qquad (\mathbf{GA})^{\#} = \mathbf{GA} . \tag{3.1.1}$$

Proof: Using Theorem 2.3.1 a general solution of $\mathbf{Ax} = \mathbf{y}$ is $\mathbf{Gy} + (\mathbf{I} - \mathbf{GA})\mathbf{z}$ where $\mathbf{z}$ is arbitrary. If $\mathbf{Gy}$ has minimum norm, then

$$\|\mathbf{Gy}\| \leq \|\mathbf{Gy} + (\mathbf{I} - \mathbf{GA})\mathbf{z}\| \quad \forall\ \mathbf{y} \in \mathscr{M}(\mathbf{A}), \mathbf{z} \in \mathscr{E}^n$$

or

$$\|\mathbf{GAb}\| \le \|\mathbf{GAb} + (\mathbf{I} - \mathbf{GA})\mathbf{z}\| \quad \forall\, \mathbf{b}, \mathbf{z}$$
$$\Leftrightarrow (\mathbf{GAb}, (\mathbf{I} - \mathbf{GA})\mathbf{z}) = 0 \quad \forall\, \mathbf{b}, \mathbf{z}$$
$$\Leftrightarrow (\mathbf{GA})^{\#}(\mathbf{I} - \mathbf{GA}) = \mathbf{0} \Leftrightarrow (\mathbf{GA})^{\#} = \mathbf{GA}.$$

The condition (3.1.1) is established.

Note 1. The following conditions are equivalent:

(i) $\mathbf{AGA} = \mathbf{A}, \quad (\mathbf{GA})^{\#} = \mathbf{GA}$
(ii) $\mathbf{GAA}^{\#} = \mathbf{A}^{\#}$
(iii) $\mathbf{GA} = \mathbf{P}_{\mathbf{A}^{\#}}$, where $\mathbf{P}$ is the projection operator.

The equivalence follows from the definitions of adjoint and projection operators.

Note 2. If $\|\mathbf{x}\| = (\mathbf{x}^*\mathbf{N}\mathbf{x})^{1/2}$ [or $(\mathbf{x}, \mathbf{y}) = \mathbf{y}^*\mathbf{N}\mathbf{x}$], where $\mathbf{N}$ is p.d., then the conditions in *Note* 1 can be written:

(i) $\mathbf{AGA} = \mathbf{A} \quad (\mathbf{GA})^*\mathbf{N} = \mathbf{NGA}$
(ii) $\mathbf{GAN}^{-1}\mathbf{A}^* = \mathbf{N}^{-1}\mathbf{A}^*$
(iii) $\mathbf{GA} = \mathbf{P}_{\mathbf{N}^{-1}\mathbf{A}^*}$ (or $(\mathbf{GA})^* = \mathbf{P}_{\mathbf{A}^*}$ under the inner product $(\mathbf{x}, \mathbf{y}) = \mathbf{y}^*\mathbf{N}^{-1}\mathbf{x}$).

Note 3. In the special case $\mathbf{N} = \mathbf{I}$ the conditions reduce to

(i) $\mathbf{AGA} = \mathbf{A} \quad (\mathbf{GA})^* = \mathbf{GA}$
(ii) $\mathbf{GAA}^* = \mathbf{A}^*$
(iii) $\mathbf{GA} = \mathbf{P}_{\mathbf{A}^*}$.

Note 4. The minimum norm solution is unique, although minimum norm g-inverse may not be.

Let $\mathbf{G}_1$ and $\mathbf{G}_2$ be two such inverses. Then using condition (ii) of *Note* 1, $(\mathbf{G}_1 - \mathbf{G}_2)\mathbf{AA}^{\#} = \mathbf{0} \Leftrightarrow (\mathbf{G}_1 - \mathbf{G}_2)\mathbf{A} = \mathbf{0}$ or $\mathbf{GA}$ is unique and hence $\mathbf{Gy} = \mathbf{GAb}$ is unique.

Note 5. A matrix $\mathbf{G}$ which provides the minimum norm solution of $\mathbf{Ax} = \mathbf{y}$ is denoted by $\mathbf{A}_m^-$, or more explicitly by $\mathbf{A}_{m(\mathbf{N})}^-$, indicating the particular norm used as in *Note* 2 and referred to as minimum $\mathbf{N}$-norm g-inverse.

Note 6. It is easily verified that one choice of $\mathbf{A}_{m(\mathbf{N})}^-$ is $\mathbf{N}^{-1}\mathbf{A}^*(\mathbf{AN}^{-1}\mathbf{A}^*)^-$, and by applying Theorem 2.3.1(b), the general expression for $\mathbf{A}_{m(\mathbf{N})}^-$ is

$$\mathbf{N}^{-1}\mathbf{A}^*(\mathbf{AN}^{-1}\mathbf{A}^*)^- + \mathbf{U}[\mathbf{I} - \mathbf{AN}^{-1}\mathbf{A}^*(\mathbf{AN}^{-1}\mathbf{A}^*)^-], \qquad (3.1.2)$$

where $\mathbf{U}$ is arbitrary.

A reflexive g-inverse providing a minimum norm solution is denoted by $\mathbf{A}_{mr}^-$.

THEOREM 3.1.2 *Let* $\|\mathbf{x}\| = (\mathbf{x}^*\mathbf{x})^{1/2}$. *Then*

(i) $\mathbf{A}_{mr}^- = \mathbf{A}^*(\mathbf{A}\mathbf{A}^*)^-$ (3.1.3)

(ii) *If* $\mathbf{A}$ *is of order* $m \times n$ *and rank* n, *every g-inverse of* $\mathbf{A}$ *is* $\mathbf{A}_{mr}^-$.

Proof: Clearly, $\mathbf{G} = \mathbf{A}^*(\mathbf{A}\mathbf{A}^*)^- \Rightarrow R(\mathbf{G}) = R(\mathbf{A}) \Rightarrow \mathbf{G}$ is reflexive by Lemma 2.5.1.

Conversely, let $\mathbf{G}$ be $\mathbf{A}_{mr}^-$. Then $\mathbf{GAA}^* = \mathbf{A}^*$ and $R(\mathbf{G}) = R(\mathbf{A})$. Hence, $\mathcal{M}(\mathbf{G}) = \mathcal{M}(\mathbf{A}^*) \Rightarrow \mathbf{G} = \mathbf{A}^*\mathbf{D}$ for some $\mathbf{D} \Rightarrow \mathbf{A} = \mathbf{AGA} = \mathbf{AA}^*\mathbf{DA} \Rightarrow \mathbf{AA}^* = \mathbf{AA}^*\mathbf{DAA}^* \Rightarrow \mathbf{D} = (\mathbf{AA}^*)^- \Rightarrow \mathbf{G} = \mathbf{A}^*\mathbf{D} = \mathbf{A}^*(\mathbf{AA}^*)^-$. This completes the proof of (i).

If $R(\mathbf{A}) = n$, $\mathcal{M}(\mathbf{A}^*) = \mathscr{E}^n$. Hence every matrix with n rows, including a g-inverse $\mathbf{G}$ of $\mathbf{A}$, is expressible as $\mathbf{A}^*\mathbf{D}$ for a suitable choice of $\mathbf{D}$. Rest of the proof of (ii) follows as in the proof of (i), above.

THEOREM 3.1.3 *Let* $\|\mathbf{x}\| = (\mathbf{x}^*\mathbf{N}\mathbf{x})^{1/2}$, *where* $\mathbf{N}$ *is a p.d. matrix. Then*

(i) $\mathbf{A}_{mr}^- = \mathbf{N}^{-1}\mathbf{A}^*(\mathbf{A}\mathbf{N}^{-1}\mathbf{A}^*)^-$ (3.1.4)

(ii) *If* $\mathbf{A}$ *is of order* $m \times n$ *and rank* n, *every g-inverse of* $\mathbf{A}$ *is* $\mathbf{A}_{mr}^-$.

The proof of Theorem 3.1.3 follows along the same lines as in Theorem 3.1.2.

We shall now consider the problem of finding a minimum seminorm solution.

THEOREM 3.1.4 *Let the seminorm of* $\mathbf{x}$ *be* $(\mathbf{x}^*\mathbf{N}\mathbf{x})^{1/2}$, *where* $\mathbf{N}$ *is p.s.d. matrix and let* $\mathbf{G}$ *be a g-inverse of* $\mathbf{A}$, *such that* $\mathbf{Gy}$ *is a minimum seminorm solution of* $\mathbf{Ax} = \mathbf{y}$, $\mathbf{y} \in \mathcal{M}(\mathbf{A})$. *Then it is n.s. that*

$$\mathbf{AGA} = \mathbf{A} \quad (\mathbf{GA})^*\mathbf{N} = \mathbf{NGA}. \tag{3.1.5}$$

The proof follows on the same lines as that of Theorem 3.1.1. We use the same notation $\mathbf{A}_{m(\mathbf{N})}^-$ to indicate a matrix satisfying the conditions (3.1.5) and refer to it as minimum $\mathbf{N}$-seminorm g-inverse. Theorem 3.1.5 provides explicit expressions for minimum seminorm inverses.

THEOREM 3.1.5 *Let the seminorm of* $\mathbf{x}$ *be* $(\mathbf{x}^*\mathbf{N}\mathbf{x})^{1/2}$ *where* $\mathbf{N}$ *is p.s.d. Then* $\mathbf{G}$, *as defined in Theorem 3.1.4, exists and one choice of* $\mathbf{G}$ *is*

$$(\mathbf{N} + \mathbf{A}^*\mathbf{A})^-\mathbf{A}^*[\mathbf{A}(\mathbf{N} + \mathbf{A}^*\mathbf{A})^-\mathbf{A}^*]^- \tag{3.1.6}$$

and in particular when $\mathcal{M}(\mathbf{A}^*) \subset \mathcal{M}(\mathbf{N})$ *it has the simpler form*

$$\mathbf{N}^-\mathbf{A}^*(\mathbf{A}\mathbf{N}^-\mathbf{A}^*)^-. \tag{3.1.7}$$

Proof: The problem is to minimize $\mathbf{x}^*\mathbf{N}\mathbf{x}$ subject to the condition $\mathbf{Ax} = \mathbf{y}$. Using the method of Lagrangian multipliers, the minimizing equations are

(see Rao, 1965, for matrix derivatives)

$$\begin{aligned} \mathbf{N}\mathbf{x} + \mathbf{A}^*\boldsymbol{\mu} &= \mathbf{0} \\ \mathbf{A}\mathbf{x} &= \mathbf{y}, \end{aligned} \tag{3.1.8}$$

where $\boldsymbol{\mu}$ is the vector of Lagrangian multipliers. We observe that equations (3.1.8) are soluble. Let $\hat{\mathbf{x}}$ and $\hat{\boldsymbol{\mu}}$ be one solution and $\mathbf{x}$ any vector satisfying $\mathbf{A}\mathbf{x} = \mathbf{y}$. Then

$$\begin{aligned} \mathbf{x}^*\mathbf{N}\mathbf{x} &= [(\mathbf{x} - \hat{\mathbf{x}}) + \hat{\mathbf{x}}]^*\mathbf{N}[(\mathbf{x} - \hat{\mathbf{x}}) + \hat{\mathbf{x}}] \\ &= (\mathbf{x} - \hat{\mathbf{x}})^*\mathbf{N}(\mathbf{x} - \hat{\mathbf{x}}) + \hat{\mathbf{x}}^*\mathbf{N}\hat{\mathbf{x}}, \end{aligned} \tag{3.1.9}$$

since terms like (using the relation $\mathbf{N}\hat{\mathbf{x}} = -\mathbf{A}^*\hat{\boldsymbol{\mu}}$)

$$\begin{aligned} (\mathbf{x} - \hat{\mathbf{x}})^*\mathbf{N}\hat{\mathbf{x}} &= -(\mathbf{x} - \hat{\mathbf{x}})^*\mathbf{A}^*\hat{\boldsymbol{\mu}} \\ &= -[\mathbf{A}\mathbf{x} - \mathbf{A}\hat{\mathbf{x}}]^*\hat{\boldsymbol{\mu}} = -(\mathbf{y} - \mathbf{y})^*\hat{\boldsymbol{\mu}} = 0. \end{aligned}$$

Equation (3.1.9) shows that $\hat{\mathbf{x}}^*\mathbf{N}\hat{\mathbf{x}}$ is a minimum for any solution of (3.1.8).

To solve Equation (3.1.8) some care is needed, as the matrices involved do not have full rank. Equation (3.1.8) is equivalent to

$$\begin{aligned} (\mathbf{N} + \mathbf{A}^*\mathbf{A})\mathbf{x} + \mathbf{A}^*\boldsymbol{\mu} &= \mathbf{A}^*\mathbf{y} \\ \mathbf{A}\mathbf{x} &= \mathbf{y}. \end{aligned} \tag{3.1.10}$$

From the first equation $\mathbf{x} = (\mathbf{N} + \mathbf{A}^*\mathbf{A})^-\mathbf{A}^*(\mathbf{y} - \boldsymbol{\mu})$, and from the second $\mathbf{A}(\mathbf{N} + \mathbf{A}^*\mathbf{A})^-\mathbf{A}^*(\mathbf{y} - \boldsymbol{\mu}) = \mathbf{y} \Rightarrow \mathbf{y} - \boldsymbol{\mu} = [\mathbf{A}(\mathbf{N} + \mathbf{A}^*\mathbf{A})^-\mathbf{A}^*]^-\mathbf{y}$. Then a solution for $\mathbf{x}$ is

$$\mathbf{x} = (\mathbf{N} + \mathbf{A}^*\mathbf{A})^-\mathbf{A}^*[\mathbf{A}(\mathbf{N} + \mathbf{A}^*\mathbf{A})^-\mathbf{A}^*]^-\mathbf{y}, \tag{3.1.11}$$

which proves that (3.1.6) is the required g-inverse.

In the case $\mathcal{M}(\mathbf{A}^*) \subset \mathcal{M}(\mathbf{N})$, equations (3.1.8) are directly solved without converting them into the equivalent form (3.1.10) to obtain (3.1.7).

COROLLARY 1 *Let*

$$\begin{pmatrix} \mathbf{N} & \mathbf{A}^* \\ \mathbf{A} & \mathbf{0} \end{pmatrix}^- = \begin{pmatrix} \mathbf{C}_1 & \mathbf{C}_2 \\ \mathbf{C}_3 & \mathbf{C}_4 \end{pmatrix} \tag{3.1.12}$$

for any choice of the g-inverse. Then $\mathbf{C}_2$ *is a g-inverse of* $\mathbf{A}$ *which provides the minimum seminorm solution of* $\mathbf{A}\mathbf{x} = \mathbf{y}$.

In many practical problems it may be easier to obtain a g-inverse of the partitioned matrix on the left-hand side of (3.1.12) and then choose $\mathbf{C}_2$ instead of using the explicit forms (3.1.6) and (3.1.7), given in Theorem 3.1.5.

3.2 g-INVERSE FOR A LEAST-SQUARES SOLUTION $\mathbf{Ax} = \mathbf{y}$ (INCONSISTENT)

Let us consider an inconsistent equation $\mathbf{Ax} = \mathbf{y}$. We say that $\hat{\mathbf{x}}$ is a least-squares solution (l.s.s.) if

$$\|\mathbf{A}\hat{\mathbf{x}} - \mathbf{y}\| = \inf_{\mathbf{x}} \|\mathbf{Ax} - \mathbf{y}\|. \tag{3.2.1}$$

The study of least-squares inverses is important in the context of the theory of linear estimation in the Gauss–Markoff model (see Chapter 7), smoothing of time series, etc.

THEOREM 3.2.1 *Let* $\mathbf{G}$ *be a matrix (not necessarily a g-inverse) such that* $\mathbf{Gy}$ *is a l.s.s. of* $\mathbf{Ax} = \mathbf{y}$ *for any* $\mathbf{y} \in \mathscr{E}^m$. *Then it is n.s. that*

$$\mathbf{AGA} = \mathbf{A} \quad (\mathbf{AG})^{\#} = \mathbf{AG}. \tag{3.2.2}$$

Proof: By hypothesis

$$\begin{aligned} \|\mathbf{AGy} - \mathbf{y}\| &\le \|\mathbf{Ax} - \mathbf{y}\| \ \forall\, \mathbf{x}, \mathbf{y} \\ &\le \|\mathbf{AGy} - \mathbf{y} + \mathbf{Aw}\| \ \forall\, \mathbf{y}, \mathbf{w} = \mathbf{x} - \mathbf{Gy} \\ &\Leftrightarrow (\mathbf{Aw}, (\mathbf{AG} - \mathbf{I})\mathbf{y}) = 0 \ \forall\, \mathbf{y}, \mathbf{w} \\ &\Leftrightarrow \mathbf{A}^{\#}\mathbf{AG} = \mathbf{A}^{\#}, \end{aligned}$$

which is equivalent to the two conditions in (3.2.2).

Note 1. The following conditions are equivalent:

(i) $\mathbf{AGA} = \mathbf{A}, \quad (\mathbf{AG})^{\#} = \mathbf{AG}$,
(ii) $\mathbf{A}^{\#}\mathbf{AG} = \mathbf{A}^{\#}$,
(iii) $\mathbf{AG} = \mathbf{P_A}$.

Note 2. If $\|\mathbf{x}\| = \mathbf{x}^*\mathbf{Mx}$ where $\mathbf{M}$ is p.d., then the conditions of *Note 1* can be written:

(i) $\mathbf{AGA} = \mathbf{A}, \quad (\mathbf{AG})^*\mathbf{M} = \mathbf{MAG}$,
(ii) $\mathbf{A}^*\mathbf{MAG} = \mathbf{A}^*\mathbf{M}$,
(iii) $\mathbf{AG} = \mathbf{P_A}$.

Note 3. When $\mathbf{M} = \mathbf{I}$, the conditions reduce to:

(i) $\mathbf{AGA} = \mathbf{A}, \quad (\mathbf{AG})^* = \mathbf{AG}$,
(ii) $\mathbf{A}^*\mathbf{AG} = \mathbf{A}^*$,
(iii) $\mathbf{AG} = \mathbf{P_A}$.

Note 4. An l.s.s. may not be unique, but $\min \|\mathbf{Ax} - \mathbf{y}\|$ is unique. Indeed, for any l.s.s. $\mathbf{x}$, the vectors $\mathbf{Ax}$ and $\mathbf{Ax} - \mathbf{y}$ are unique. The result follows since

$\mathbf{P_A} = \mathbf{AG}$ must be unique for any l.s. g-inverse $\mathbf{G}$. Further, if $\mathbf{Gy}$ is an l.s.s., then the class of l.s. solutions is

$$\mathbf{Gy} + (\mathbf{I} - \mathbf{GA})\mathbf{z}, \quad \mathbf{z} \text{ arbitrary.} \tag{3.2.3}$$

Note 5. Let $\mathbf{M}$ be p.s.d. and $\mathbf{G}$ be a matrix such that $\|\mathbf{AGy} - \mathbf{y}\| \leq \|\mathbf{Ax} - \mathbf{y}\|$ for all $\mathbf{x}, \mathbf{y}$ where $\|\mathbf{w}\| = \sqrt{(\mathbf{w}^*\mathbf{Mw})}$ is a seminorm. Then it is n.s. that

$$\mathbf{A}^*\mathbf{MAG} = \mathbf{A}^*\mathbf{M},$$

which is equivalent to the two conditions

$$\mathbf{MAGA} = \mathbf{MA}, \qquad (\mathbf{AG})^*\mathbf{M} = \mathbf{MAG}. \tag{3.2.4}$$

It is seen that the conditions (3.2.4) are different from those when $\mathbf{M}$ is p.d. [see condition (i) in *Note* 2]. When $\mathbf{M}$ is not p.d., $\mathbf{G}$ need not be a g-inverse.

Note 6. A g-inverse which provides a l.s.s. of $\mathbf{Ax} = \mathbf{y}$ is denoted by $\mathbf{A}_l^-$, or, more explicitly, by $\mathbf{A}_{l(\mathbf{M})}^-$ indicating the norm involved as in *Note 2* and *Note 3* and referred to as $\mathbf{M}$-least squares g-inverse.

Note 7. Let $\mathbf{M}$ be p.d. or p.s.d. Then one choice of $\mathbf{A}_{l(\mathbf{M})}^-$ is $(\mathbf{A}^*\mathbf{MA})^-\mathbf{A}^*\mathbf{M}$ and a general solution, applying Theorem 2.3.1(*b*), is

$$(\mathbf{A}^*\mathbf{MA})^-\mathbf{A}^*\mathbf{M} + [\mathbf{I} - (\mathbf{A}^*\mathbf{MA})^-\mathbf{A}^*\mathbf{MA}]\mathbf{U}, \tag{3.2.5}$$

where $\mathbf{U}$ is arbitrary.

A reflexive g-inverse providing an l.s.s. solution is denoted by $\mathbf{A}_{lr}^-$.

THEOREM 3.2.2 *Let* $\|\mathbf{y}\| = (\mathbf{y}^*\mathbf{My})^{1/2}$, *where* $\mathbf{M}$ *is p.d. Then*

(i) $\mathbf{A}_{lr}^- = (\mathbf{A}^*\mathbf{MA})^-\mathbf{A}^*\mathbf{M}$. (3.2.6)

(ii) *If* $\mathbf{A}$ *is a matrix of order* $m \times n$ *and rank* m, *then every g-inverse of* $\mathbf{A}$ *is* $\mathbf{A}_{lr}^-$.

The proof of Theorem 3.2.2 is similar to that of Theorem 3.1.3.

Let the spaces $\mathscr{E}^m$ and $\mathscr{E}^n$ be furnished with inner products through p.d. matrices $\mathbf{M}$ and $\mathbf{N}$, respectively, so that for a matrix $\mathbf{A}$ of order $m \times n$, representing a transformation from $\mathscr{E}^n$ to $\mathscr{E}^m$, the adjoint matrix $\mathbf{A}^\#$ is given by $\mathbf{N}^{-1}\mathbf{A}^*\mathbf{M}$ and for a matrix $\mathbf{G}$ of order $n \times m$, representing a transformation from $\mathscr{E}^m$ to $\mathscr{E}^n$, the adjoint matrix $\mathbf{G}^\#$ is given by $\mathbf{M}^{-1}\mathbf{G}^*\mathbf{N}$. The following theorem gives an interesting characterization of the g-inverses $\mathbf{A}_{l(\mathbf{M})}^-$.

THEOREM 3.2.3 *If* $\mathbf{G}$ *is a g-inverse of* $\mathbf{A}$, $\mathbf{GG}^\#$ *is a g-inverse of* $\mathbf{A}^\#\mathbf{A}$ *iff* $\mathbf{G}$ *is* $\mathbf{A}_{l(\mathbf{M})}^-$, *that is,*

$$\mathbf{A} = \mathbf{AGA}, \qquad \mathbf{A}^\#\mathbf{A} = \mathbf{A}^\#\mathbf{AGG}^\#\mathbf{A}^\#\mathbf{A} \Leftrightarrow \mathbf{A}^\# = \mathbf{A}^\#\mathbf{AG}. \tag{3.2.7}$$

Proof: The '$\Leftarrow$' part follows from *Note 1* to Theorem 3.2.1. For the '$\Rightarrow$' part note that

$$\begin{aligned}\mathbf{A}^{\#}\mathbf{A}\mathbf{G}\mathbf{G}^{\#}\mathbf{A}^{\#}\mathbf{A} = \mathbf{A}^{\#}\mathbf{A} &\Rightarrow \mathbf{A}^{\#}\mathbf{A} + \mathbf{A}^{\#}\mathbf{A}\mathbf{G}\mathbf{G}^{\#}\mathbf{A}^{\#}\mathbf{A} - 2\mathbf{A}^{\#}\mathbf{A} = \mathbf{0}\\ &\Rightarrow \mathbf{A}^{\#}(\mathbf{I} + \mathbf{A}\mathbf{G}\mathbf{G}^{\#}\mathbf{A}^{\#} - \mathbf{G}^{\#}\mathbf{A}^{\#} - \mathbf{A}\mathbf{G})\mathbf{A} = \mathbf{0}\\ &\Rightarrow \mathbf{A}^{\#}(\mathbf{I} - \mathbf{A}\mathbf{G})(\mathbf{I} - \mathbf{G}^{\#}\mathbf{A}^{\#})\mathbf{A} = \mathbf{0}\\ &\Rightarrow \mathbf{A}^{\#} = \mathbf{A}^{\#}\mathbf{A}\mathbf{G}.\end{aligned}$$

An extremely important theorem which provides a duality relationship between minimum norm and least-squares inverses and which has statistical applications (estimation from linear models considered in Chapter 7) is as follows.

THEOREM 3.2.4 *Let* $\mathbf{M}$ *be p.d. Then*

$$(\mathbf{A}^*)^-_{m(\mathbf{M})} = [\mathbf{A}^-_{l(\mathbf{M}^{-1})}]^*. \tag{3.2.8}$$

Proof: The result follows from definitions. Let $\mathbf{G}$ denote left-hand side of (3.2.8). Then by definition

$$\mathbf{A}^*\mathbf{G}\mathbf{A}^* = \mathbf{A}^*, (\mathbf{G}\mathbf{A}^*)^*\mathbf{M} = \mathbf{M}(\mathbf{G}\mathbf{A}^*)$$

which can be rewritten as

$$\mathbf{A}\mathbf{G}^*\mathbf{A} = \mathbf{A}, (\mathbf{A}\mathbf{G}^*)^*\mathbf{M}^{-1} = \mathbf{M}^{-1}\mathbf{A}\mathbf{G}^*,$$

which means that $\mathbf{G}^*$ is $\mathbf{A}^-_{l(\mathbf{M}^{-1})}$. The theorem is proved.

The relationship (3.2.8) is significant in two ways. First, it shows that the computation of a least-squares inverse can be made to depend on that of a minimum norm inverse and vice versa. Second, it provides a direct demonstration of the minimum variance property of estimators obtained by least-squares theory and shows how the method of least-squares comes in a natural way in computing minimum variance estimators (see Section 7.2 in Chapter 7).

3.3 g-INVERSE FOR MINIMUM NORM LEAST-SQUARES SOLUTION OF $\mathbf{Ax} = \mathbf{y}$ (INCONSISTENT)

3.3.1 Moore–Penrose Inverse

The concept of a generalized inverse of an arbitrary $m \times n$ complex matrix $\mathbf{A}$ is originally due to Moore (1920). He developed it in the context of linear transformations from n-dimensional to m-dimensional vector space over a complex field with the usual Euclidean norm. Moore's definition was essentially as follows.

Definition 1 (***Moore***)***:*** A matrix $\mathbf{G}$ of order $(n \times m)$ is the generalized inverse of $\mathbf{A}$ of order $m \times n$, if

$$\mathbf{AG} = \mathbf{P_A} \quad \mathbf{GA} = \mathbf{P_G}, \tag{3.3.1}$$

where $\mathbf{P_A}$ is the operator (matrix), projecting vectors in $\mathscr{E}^m$ onto $\mathscr{M}(\mathbf{A})$, and $\mathbf{P_G}$ is the operator (matrix), projecting vectors in $\mathscr{E}^n$ onto $\mathscr{M}(\mathbf{G})$.

Unaware of Moore's work, Penrose (1955) defined a generalized inverse, which is the same as the Moore inverse when the associated norms in $\mathscr{E}^m$ and $\mathscr{E}^n$ are of the simple type $\|\mathbf{x}\| = (\mathbf{x}^*\mathbf{x})^{1/2}$.

Definition 2 (***Penrose***)***:*** $\mathbf{G}$ is the generalized inverse of $\mathbf{A}$ if

$$\mathbf{AGA} = \mathbf{A}, \qquad (\mathbf{AG})^* = \mathbf{AG}, \qquad \mathbf{GAG} = \mathbf{G}, \qquad (\mathbf{GA})^* = \mathbf{GA}. \tag{3.3.2}$$

3.3.2 g-Inverse for Minimum Norm Least-Squares Solution

In Section 3.1 we determined a g-inverse which provides a minimum norm solution of a consistent equation $\mathbf{Ax} = \mathbf{y}$ and in Section 3.2, a g-inverse for a least-squares solution (l.s.s.) of an inconsistent equation $\mathbf{Ax} = \mathbf{y}$. An l.s.s. is not, however, unique. We shall investigate the problem by choosing from among the l.s. solutions one with a minimum norm by a suitable choice of g-inverse.

Definition 3: $\mathbf{G}$ is said to be a minimum norm (or seminorm) least-squares g-inverse of $\mathbf{A}$ if $\mathbf{G}$ is $\mathbf{A}_l^-$ and for any $\mathbf{y} \in \mathscr{E}^m$

$$\|\mathbf{Gy}\|_n \leq \|\mathbf{x}\|_n \,\forall\, \mathbf{x} \in \{\mathbf{x}: \|\mathbf{Ax} - \mathbf{y}\|_m \leq \|\mathbf{Az} - \mathbf{y}\|_m \,\forall\, \mathbf{z} \in \mathscr{E}^n\}, \tag{3.3.3}$$

where $\|\cdot\|_n$ and $\|\cdot\|_m$ are norms (or seminorms) in $\mathscr{E}^n$ and $\mathscr{E}^m$, respectively.

THEOREM 3.3.1 *Let there exist a matrix* $\mathbf{G}$ *such that* $\mathbf{Gy}$ *is a minimum norm least-squares solution of* $\mathbf{Ax} = \mathbf{y}$. *Then it is n.s. that*

$$\mathbf{AGA} = \mathbf{A}, \qquad (\mathbf{AG})^\# = \mathbf{AG}, \qquad \mathbf{GAG} = \mathbf{G}, \qquad (\mathbf{GA})^\# = \mathbf{GA}. \tag{3.3.4}$$

Proof: Since $\mathbf{Gy}$ is an l.s.s., the first two conditions of (3.3.4) follow from Theorem 3.2.1.

It is shown in (3.2.3) that a general l.s.s. is $\mathbf{Gy} + (\mathbf{I} - \mathbf{GA})\mathbf{z}$. Then

$$\|\mathbf{Gy}\| \leq \|\mathbf{Gy} + (\mathbf{I} - \mathbf{GA})\mathbf{z}\| \;\forall\, \mathbf{y}, \mathbf{z},$$

$$\Leftrightarrow (\mathbf{Gy}, (\mathbf{I} - \mathbf{GA})\mathbf{z}) = 0 \;\forall\, \mathbf{y}, \mathbf{z} \Leftrightarrow \mathbf{G}^\#(\mathbf{I} - \mathbf{GA}) = \mathbf{0}.$$

Then

$$\mathbf{G}^\# = \mathbf{G}^\#\mathbf{GA} \Leftrightarrow \mathbf{GAG} = \mathbf{G}, \qquad (\mathbf{GA})^\# = \mathbf{GA}.$$

Note 1. The minimum norm least-squares inverse of $\mathbf{G}$ of Theorem 3.3.1 is unique. This result is a simple consequence of the fact that a minimum norm solution is unique.

Note 2. The following conditions are equivalent:

(i) $\mathbf{AGA} = \mathbf{A}$, $\quad \mathbf{GAG} = \mathbf{G}$, $\quad (\mathbf{AG})^{\#} = \mathbf{AG}$, $\quad (\mathbf{GA})^{\#} = \mathbf{GA}$,
(ii) $\mathbf{A}^{\#}\mathbf{AG} = \mathbf{A}$, $\quad \mathbf{G}^{\#}\mathbf{GA} = \mathbf{G}$,
(iii) $\mathbf{AG} = \mathbf{P_A}$, $\quad \mathbf{GA} = \mathbf{P_G}$.

Note 3. If $\|\mathbf{y}\|_m = (\mathbf{y}^*\mathbf{My})^{1/2}$ and $\|\mathbf{x}\|_n = (\mathbf{x}^*\mathbf{Nx})^{1/2}$, where $\mathbf{M}$ and $\mathbf{N}$ are p.d., then the conditions in *Note 1* can be written

(i) $\mathbf{AGA} = \mathbf{A}$, $\quad \mathbf{GAG} = \mathbf{G}$, $\quad (\mathbf{AG})^*\mathbf{M} = \mathbf{MAG}$, $\quad (\mathbf{GA})^*\mathbf{N} = \mathbf{NGA}$,
(ii) $\mathbf{A}^*\mathbf{MAG} = \mathbf{A}^*\mathbf{M}$, $\quad \mathbf{G}^*\mathbf{NGA} = \mathbf{G}^*\mathbf{N}$,
(iii) $\mathbf{AG} = \mathbf{P_A}$, $\quad \mathbf{GA} = \mathbf{P_G}$.

Note 4. If $\mathbf{M}$ and $\mathbf{N}$ are identity matrices the conditions reduce to

(i) $\mathbf{AGA} = \mathbf{A}$, $\quad \mathbf{GAG} = \mathbf{G}$, $\quad (\mathbf{AG})^* = \mathbf{AG}$, $\quad (\mathbf{GA})^* = \mathbf{GA}$,
(ii) $\mathbf{A}^*\mathbf{AG} = \mathbf{A}^*$, $\quad \mathbf{G}^*\mathbf{GA} = \mathbf{G}^*$,
(iii) $\mathbf{AG} = \mathbf{P_A}$, $\quad \mathbf{GA} = \mathbf{P_G}$.

The conditions in (i) of *Note 2* is a general formulation of the conditions in (i) of *Note 4*, originally given by Penrose.

Note 5. Let $\|\mathbf{y}\|_m = (\mathbf{y}^*\mathbf{My})^{1/2}$ and $\|\mathbf{x}\|_n = (\mathbf{x}^*\mathbf{Nx})^{1/2}$, where $\mathbf{M}$ and $\mathbf{N}$ may be p.s.d. If there exists a $\mathbf{G}$ satisfying definition 3, then it is n.s. that

$$\mathbf{MAGA} = \mathbf{MA}, \quad \mathbf{NGAG} = \mathbf{NG}, \quad (\mathbf{AG})^*\mathbf{M} = \mathbf{MAG}, \quad (\mathbf{GA})^*\mathbf{N} = \mathbf{NGA} \tag{3.3.5}$$

or equivalently

$$\mathbf{AG} = \mathbf{P_A}, \qquad \mathbf{GA} = \mathbf{P_G}, \tag{3.3.6}$$

where $\mathbf{P_A}$ and $\mathbf{P_G}$ are projection operators in the extended sense of Section 1.1 of Chapter 1.

A g-inverse which satisfies the conditions in *Note 3* or *Note 5* is denoted by $\mathbf{A}^+_{\mathbf{MN}}$ and referred to as minimum $\mathbf{N}$-norm (or seminorm) $\mathbf{M}$-least-squares g-inverse of $\mathbf{A}$. When $\mathbf{M}$ and $\mathbf{N}$ are identity matrices, we use the notation $\mathbf{A}^+$ dropping the suffixes.

Note 6. $\mathbf{A}^+_{\mathbf{MN}}$ is unique if $\mathbf{N}$ is p.d. When both $\mathbf{M}$ and $\mathbf{N}$ are p.s.d., $\mathbf{A}^+_{\mathbf{MN}}$ may not be unique.

Note 7. If $\mathbf{M}$ and $\mathbf{N}$ are p.d., then

(a) $(\mathbf{A}^+_{\mathbf{MN}})^+_{\mathbf{NM}} = \mathbf{A}$, (3.3.7)
(b) $(\mathbf{A}^+_{\mathbf{MN}})^* = (\mathbf{A}^*)^+_{\mathbf{N}^{-1}\mathbf{M}^{-1}}$. (3.3.8)

Note 8. For any matrix $\mathbf{A}$, $\mathbf{A}^+ = \mathbf{P}_{\mathbf{A}^*}\mathbf{A}^-\mathbf{P_A}$, where $\mathbf{P}_{\mathbf{A}^*}$ and $\mathbf{P_A}$ are projection operators.

THEOREM 3.3.2 $\mathbf{G} = \mathbf{A}^+_{\mathbf{MN}}$ *iff*

$$\mathbf{G} = \mathbf{A}_r^-, \text{ and } \mathbf{G}^\#\mathbf{G}\mathbf{G}^\# = (\mathbf{A}^\#\mathbf{A}\mathbf{A}^\#)^-,$$

where the adjoint matrices $\mathbf{A}^\#$ *and* $\mathbf{G}^\#$ *are given by* $\mathbf{A}^\# = \mathbf{N}^{-1}\mathbf{A}^*\mathbf{M}$ *and* $\mathbf{G}^\# = \mathbf{M}^{-1}\mathbf{G}^*\mathbf{N}$, *respectively.*

Proof: 'Only if' part is trivial. To prove the 'if' part, let

$$\mathbf{A} = \mathbf{U}\begin{pmatrix} \mathbf{D} & \mathbf{0} \\ \mathbf{0} & \mathbf{0} \end{pmatrix}\mathbf{V}^*$$

be the singular value decomposition of $\mathbf{A}$, where $\mathbf{U}^*\mathbf{M}\mathbf{U} = \mathbf{I}_m$, $\mathbf{V}^*\mathbf{N}^{-1}\mathbf{V} = \mathbf{I}_n$ and $\mathbf{D}$ is a diagonal matrix with strictly positive diagonal elements, which are positive square roots of the nonnull eigenvalues of $\mathbf{N}^{-1}\mathbf{A}^*\mathbf{M}\mathbf{A} = \mathbf{A}^\#\mathbf{A}$

$$\mathbf{G} = \mathbf{A}^- \Rightarrow \mathbf{G} = \mathbf{N}^{-1}\mathbf{V}\begin{pmatrix} \mathbf{D}^{-1} & \mathbf{J} \\ \mathbf{F} & \mathbf{H} \end{pmatrix}\mathbf{U}^*\mathbf{M},$$

where $\mathbf{J}$, $\mathbf{F}$, and $\mathbf{H}$ are arbitrary. Further,

$$\mathbf{G} = \mathbf{G}\mathbf{A}\mathbf{G} \Rightarrow \mathbf{H} = \mathbf{F}\mathbf{D}\mathbf{J}.$$

Also

$$\begin{aligned} &\mathbf{G}^\#\mathbf{G}\mathbf{G}^\# = (\mathbf{A}^\#\mathbf{A}\mathbf{A}^\#)^- \\ &\Rightarrow \mathbf{F}^*\mathbf{F}\mathbf{D}^{-1} + \mathbf{D}^{-1}\mathbf{J}\mathbf{J}^* + \mathbf{F}^*\mathbf{F}\mathbf{D}\mathbf{J}\mathbf{J}^* = \mathbf{0} \\ &\Leftrightarrow \mathbf{D}\mathbf{X}\mathbf{D} + \mathbf{D}\mathbf{Y}\mathbf{D} + \mathbf{D}\mathbf{X}\mathbf{D}^2\mathbf{Y}\mathbf{D} = \mathbf{0} \\ &\Leftrightarrow (\mathbf{I} + \mathbf{D}\mathbf{X}\mathbf{D})^{-1} = \mathbf{I} + \mathbf{D}\mathbf{Y}\mathbf{D}, \end{aligned} \tag{3.3.9}$$

where $\mathbf{X} = \mathbf{D}^{1/2}\mathbf{F}^*\mathbf{F}\mathbf{D}^{-1/2}$ and $\mathbf{Y} = \mathbf{D}^{-1/2}\mathbf{J}\mathbf{J}^*\mathbf{D}^{1/2}$. Observe that the eigenvalues of $\mathbf{D}\mathbf{X}\mathbf{D}$ and $\mathbf{D}\mathbf{Y}\mathbf{D}$ are strictly nonnegative. This together with (3.3.9) implies that

$$\mathbf{D}\mathbf{X}\mathbf{D} = \mathbf{D}\mathbf{Y}\mathbf{D} = \mathbf{0} \Rightarrow \mathbf{F} = \mathbf{0}, \qquad \mathbf{J} = \mathbf{0} \Rightarrow \mathbf{H} = \mathbf{0}$$

$$\Rightarrow \mathbf{G} = \mathbf{N}^{-1}\mathbf{V}\begin{pmatrix} \mathbf{D}^{-1} & \mathbf{0} \\ \mathbf{0} & \mathbf{0} \end{pmatrix}\mathbf{U}^*\mathbf{M} = \mathbf{A}^+_{\mathbf{MN}}.$$

This completes the proof of Theorem 3.3.2.

3.3.3 Expressions for $\mathbf{A}^+_{\mathbf{MN}}$ Where N and M are at Least p.s.d.

The problem is to find $\mathbf{x}$ such that $\mathbf{x}^*\mathbf{N}\mathbf{x}$ is a minimum subject to $\mathbf{A}^*\mathbf{M}\mathbf{A}\mathbf{x} = \mathbf{A}^*\mathbf{M}\mathbf{y}$. The equations are

$$\mathbf{N}\mathbf{x} + \mathbf{A}^*\mathbf{M}\mathbf{A}\boldsymbol{\lambda} = \mathbf{0}$$

$$\mathbf{A}^*\mathbf{M}\mathbf{A}\mathbf{x} = \mathbf{A}^*\mathbf{M}\mathbf{y}$$

which are obviously consistent. Let

$$\begin{pmatrix} \mathbf{N} & \mathbf{A^*MA} \\ \mathbf{A^*MA} & \mathbf{0} \end{pmatrix}^{-} = \begin{pmatrix} \mathbf{C}_1 & \mathbf{C}_2 \\ \mathbf{C}_3 & \mathbf{C}_4 \end{pmatrix}. \tag{3.3.10}$$

Then $\mathbf{x} = \mathbf{C}_2\mathbf{A^*My}$, in which case one choice of $\mathbf{A}^+_{\mathbf{MN}} = \mathbf{C}_2\mathbf{A^*M}$.

Alternative expressions can be obtained working along the lines of Theorem 3.1.5.

$$\mathbf{A}^+_{\mathbf{MN}} = \mathbf{N}^{-1}\mathbf{A^*MA}(\mathbf{A^*MAN}^{-1}\mathbf{A^*MA})^{-}\mathbf{A^*M}, \quad \text{if } |\mathbf{N}| \neq 0 \tag{3.3.11}$$

$$= \mathbf{N}^{-}\mathbf{A^*MA}(\mathbf{A^*MAN}^{-}\mathbf{A^*MA})^{-}\mathbf{A^*M}, \quad \text{if } \mathscr{M}(\mathbf{A^*MA}) \subset \mathscr{M}(\mathbf{N}) \tag{3.3.12}$$

$$= (\mathbf{N} + \mathbf{A^*MA})^{-}\mathbf{A^*M}[\mathbf{A^*MA}(\mathbf{N} + \mathbf{A^*MA})^{-}\mathbf{A^*M}]^{-}\mathbf{A^*M}$$

$$\text{in the general case.} \tag{3.3.13}$$

3.3.4 Pairs of Norms Leading to Same g-Inverse.

THEOREM 3.3.3 $\mathbf{A}^+ = \mathbf{A}^+_{\mathbf{MN}}$ *iff*

$$\mathbf{M} = \mathbf{A}\mathbf{\Lambda}_1\mathbf{A}^* + \mathbf{A}^{\perp}\mathbf{\Lambda}_2(\mathbf{A}^{\perp})^*, \tag{3.3.14}$$

$$\mathbf{N} = \mathbf{A}^*\mathbf{\Delta}_1\mathbf{A} + (\mathbf{A}^*)^{\perp}\mathbf{\Delta}_2[(\mathbf{A}^*)^{\perp}]^*, \tag{3.3.15}$$

where $\mathbf{\Lambda}_1, \mathbf{\Lambda}_2, \mathbf{\Delta}_1$, *and* $\mathbf{\Delta}_2$ *are arbitrary hermitian matrices subject only to* **M** *and* **N** *being p.d.*

Proof of 'if' part: Let **G** be $\mathbf{A}^+$. Check that

$$\mathbf{MAG} = [\mathbf{A}\mathbf{\Lambda}_1\mathbf{A}^* + \mathbf{A}^{\perp}\mathbf{\Lambda}_2(\mathbf{A}^{\perp})^*]\mathbf{P}_{\mathbf{A}} = \mathbf{A}\mathbf{\Lambda}_1\mathbf{A}^*$$

is hermitian. Similarly $\mathbf{NGA} = \mathbf{A}^*\mathbf{\Delta}_1\mathbf{A}$ is also seen to be hermitian. This shows $\mathbf{G} = \mathbf{A}^+_{\mathbf{MN}}$.

Proof of 'only if' part: Let **G** be both $\mathbf{A}^+$ and $\mathbf{A}^+_{\mathbf{MN}}$. Consider the p.d. matrices **M** and **N** and express these matrices as $\mathbf{M} = \mathbf{EE}^*$ and $\mathbf{N} = \mathbf{FF}^*$, where **E** and **F** are clearly nonsingular. Since $\mathscr{M}(\mathbf{A} \,\vdots\, \mathbf{A}^{\perp}) = \mathscr{E}^m$, we may write $\mathbf{E} = \mathbf{AU}_1 + \mathbf{A}^{\perp}\mathbf{U}_2$, so that

$$\mathbf{M} = \mathbf{A}\mathbf{\Lambda}_1\mathbf{A}^* + \mathbf{A}^{\perp}\mathbf{\Lambda}_2(\mathbf{A}^{\perp})^* + \mathbf{A}\mathbf{\Lambda}_3(\mathbf{A}^{\perp})^* + \mathbf{A}^{\perp}\mathbf{\Lambda}_4\mathbf{A}^*. \tag{3.3.16}$$

where $\mathbf{\Lambda}_1 = \mathbf{U}_1\mathbf{U}_1^*$, $\mathbf{\Lambda}_2 = \mathbf{U}_2\mathbf{U}_2^*$ and $\mathbf{\Lambda}_3 = \mathbf{\Lambda}_4^* = \mathbf{U}_1\mathbf{U}_2^*$. Use now the condition that **MAG** is hermitian, that is, $\mathbf{MAG} = \mathbf{MP}_{\mathbf{A}} = \mathbf{A}\mathbf{\Lambda}_1\mathbf{A}^* + \mathbf{A}^{\perp}\mathbf{\Lambda}_4\mathbf{A}^*$ is hermitian. Hence $\mathbf{A}^{\perp}\mathbf{\Lambda}_4\mathbf{A}^* = \mathbf{A}\mathbf{\Lambda}_3(\mathbf{A}^{\perp})^*$ or $\mathbf{A}\mathbf{\Lambda}_3(\mathbf{A}^{\perp})^* = \mathbf{P}_{\mathbf{A}}\mathbf{A}\mathbf{\Lambda}_3(\mathbf{A}^{\perp})^* = \mathbf{0} \Rightarrow \mathbf{A}^{\perp}\mathbf{\Lambda}_4\mathbf{A}^* = \mathbf{0} \Rightarrow \mathbf{M} = \mathbf{A}\mathbf{\Lambda}_1\mathbf{A}^* + \mathbf{A}^{\perp}\mathbf{\Lambda}_2(\mathbf{A}^{\perp})^*$. The expression for **N** is similarly established. This concludes proof of theorem 3.3.3.

THEOREM 3.3.4 $\mathbf{A}^{+}_{\mathbf{MN}} = \mathbf{A}^{+}_{\mathbf{UV}}$ *iff*

$$\mathbf{U} = \mathbf{MA}\boldsymbol{\Lambda}_1\mathbf{A}^*\mathbf{M} + \mathbf{A}^{\perp}\boldsymbol{\Lambda}_2(\mathbf{A}^{\perp})^* \tag{3.3.17}$$

$$\mathbf{V} = \mathbf{A}^*\boldsymbol{\Delta}_1\mathbf{A} + \mathbf{N}(\mathbf{A}^*)^{\perp}\boldsymbol{\Delta}_2[(\mathbf{A}^*)^{\perp}]^*\mathbf{N}, \tag{3.3.18}$$

where $\boldsymbol{\Lambda}_1$, $\boldsymbol{\Lambda}_2$, $\boldsymbol{\Delta}_1$, *and* $\boldsymbol{\Delta}_2$ *are arbitrary hermitian matrices subject to* $\mathbf{U}$, $\mathbf{V}$ *being p.d.*

Proof: The 'if' part is easy to verify. To prove the 'only if' part we need Lemma 3.3.1.

LEMMA 3.3.1 *If* $\mathbf{M}$ *is p.d.* $\mathcal{M}(\mathbf{MA} : \mathbf{A}^{\perp}) = \mathcal{M}(\mathbf{MA}) + \mathcal{M}(\mathbf{A}^{\perp}) = \mathscr{E}^m$.

Proof of Lemma: Since $\mathcal{M}(\mathbf{MA})$ and $\mathcal{M}(\mathbf{A}^{\perp})$ are subspaces of $\mathscr{E}^m$ of dimension $R(\mathbf{A})$ and $m - R(\mathbf{A})$, respectively, it suffices to show that these two subspaces are virtually disjoint, for which one has to observe that if $\mathbf{MAa} = \mathbf{A}^{\perp}\mathbf{b}$ be a vector common to these two subspaces one must have

$$\mathbf{a}^*\mathbf{A}^*\mathbf{MAa} = \mathbf{b}^*(\mathbf{A}^{\perp})^*\mathbf{Aa} = 0$$

$$\Rightarrow \mathbf{MAa} = \mathbf{A}^{\perp}\mathbf{b} = \mathbf{0}.$$

Thus Lemma 3.3.1 is established.

Consider now the matrix $\mathbf{G} = \mathbf{A}^{+}_{\mathbf{MN}}$ and assume that $\mathbf{G}$ is also $\mathbf{A}^{+}_{\mathbf{UV}}$. This implies $\mathbf{UAG}$ and $\mathbf{VGA}$ are hermitian matrices. Using Lemma 3.3.1 we write

$$\mathbf{U} = \mathbf{MA}\boldsymbol{\Lambda}_1\mathbf{A}^*\mathbf{M} + \mathbf{A}^{\perp}\boldsymbol{\Lambda}_2(\mathbf{A}^{\perp})^* + \mathbf{MA}\boldsymbol{\Lambda}_3(\mathbf{A}^{\perp})^* + \mathbf{A}^{\perp}\boldsymbol{\Lambda}_3^*\mathbf{A}^*\mathbf{M}$$

and, as in the proof of Theorem 3.3.3, show that

$$\mathbf{UAG} = (\mathbf{UAG})^* \Rightarrow \mathbf{U} = \mathbf{MA}\boldsymbol{\Lambda}_1\mathbf{A}^*\mathbf{M} + \mathbf{A}^{\perp}\boldsymbol{\Lambda}_2(\mathbf{A}^{\perp})^*,$$

since $\mathbf{MAG} = \mathbf{G}^*\mathbf{A}^*\mathbf{M}$.

The expression for $\mathbf{V}$ is derived in a similar manner, thus establishing Theorem 3.3.4.

3.4 SOLUTION OF MATRIX EQUATIONS

In Section 2.3, we obtained general solutions to matrix equations of the type $\mathbf{AXB} = \mathbf{C}$ or $\mathbf{AX} = \mathbf{C}$, $\mathbf{XB} = \mathbf{D}$. In this section we solve some special types of equations which have important statistical applications.

3.4.1 The Equation XBX = X

It will be seen that for a matrix $\mathbf{B}$ of order $m \times n$, the most general solution $\mathbf{X}$ of order $n \times m$ of the equation

$$\mathbf{XBX} = \mathbf{X} \tag{3.4.1}$$

will also determine the entire class of matrices which have $\mathbf{B}$ for a generalized inverse.

THEOREM 3.4.1 *A general solution of equation (3.4.1) is*

$$\mathbf{X} = \mathbf{C}(\mathbf{DBC})_r^- \mathbf{D}, \tag{3.4.2}$$

where $\mathbf{C}(n \times p)$ *and* $\mathbf{D}(q \times m)$ *are arbitrary matrices. The solution (3.4.2) has the same rank as the matrix* $\mathbf{DBC}$.

Proof: It is easy to verify that $\mathbf{C}(\mathbf{DBC})_r^- \mathbf{D}$ satisfies (3.4.1). Also, since an idempotent matrix is a reflexive g-inverse of itself, any $\mathbf{X}$ for which (3.4.1) is true can also be expressed as $\mathbf{X}(\mathbf{BX})_r^-$ or $(\mathbf{XB})_r^- \mathbf{X}$. This shows that (3.4.2) with arbitrary choice of $\mathbf{C}$ and $\mathbf{D}$ indeed exhausts all solutions of (3.4.1). Observe that

$$R(\mathbf{X}) \le R(\mathbf{DBC})_r^- = R(\mathbf{DBC}) = R(\mathbf{DBXBC}) \le R(\mathbf{X}).$$

COROLLARY 1: *If* $R(\mathbf{DBC}) = R(\mathbf{D})$ *or* $R(\mathbf{C})$, (3.4.2) *reduces to* $\mathbf{C}(\mathbf{DBC})^- \mathbf{D}$.

Other conditions that $\mathbf{X}$ may be required to satisfy in addition to (3.4.1) may be met by choosing $\mathbf{C}$, $\mathbf{D}$ and the g-inverse of $\mathbf{DBC}$ in a suitable manner as in corollaries 2–6.

COROLLARY 2: *Consider the equation*

$$\mathbf{XBX} = \mathbf{X}, \quad (\mathbf{BX})^* = \mathbf{BX}, \tag{3.4.3}$$

(or $\mathbf{B} = \mathbf{X}_m^-$*). A general solution to (3.4.3) is*

$$\mathbf{X} = \mathbf{C}(\mathbf{DBC})_{lr}^- \mathbf{D}, \tag{3.4.4}$$

where $\mathbf{C}$ *and* $\mathbf{D}$ *are arbitrary except that* $\mathbf{D}^*\mathbf{D} = \mathbf{I}_m$.

Proof: Note that since $\mathbf{F} = \mathbf{DBC}(\mathbf{DBC})_{lr}^-$ is hermitian, $\mathbf{D}^*\mathbf{FD} = \mathbf{BX}$ is also hermitian. This shows (3.4.4) is indeed a solution to (3.4.3). Conversely if $\mathbf{B} = \mathbf{X}_m^-$, $\mathbf{BX}$ is one choice of $(\mathbf{BX})_{lr}^-$. Hence, any solution $\mathbf{X}$ of (3.4.3) can always be expressed as $\mathbf{X}(\mathbf{BX})_{lr}^-$. This shows that (3.4.4) with arbitrary choice of $\mathbf{C}$ and $\mathbf{D}$ subject to $\mathbf{D}^*\mathbf{D} = \mathbf{I}_m$ exhausts all solutions of (3.4.3).

COROLLARY 3: *Consider the equation*

$$\mathbf{XBX} = \mathbf{X}, \qquad (\mathbf{XB})^* = \mathbf{XB}, \tag{3.4.5}$$

(or $\mathbf{B} = \mathbf{X}_l^-$*). A general solution to (3.4.5) is*

$$\mathbf{X} = \mathbf{C}(\mathbf{DBC})_{mr}^- \mathbf{D}, \tag{3.4.6}$$

where $\mathbf{C}$ *and* $\mathbf{D}$ *are arbitrary except that* $\mathbf{CC}^* = \mathbf{I}_n$.

COROLLARY 4: *A general solution to the equation*

$$\mathbf{XBX} = \mathbf{X}, \qquad \mathbf{BXB} = \mathbf{B}, \tag{3.4.7}$$

or $\mathbf{B} = \mathbf{X}_r^-$ *is obtained from (3.4.2) by choosing* $\mathbf{D} = \mathbf{I}_m$ *and* $\mathbf{C} = \mathbf{I}_n$. *This gives* $\mathbf{X} = \mathbf{B}_r^-$.

COROLLARY 5: *A general solution to the equation*

$$\mathbf{XBX} = \mathbf{X}, \mathbf{BXB} = \mathbf{B}, \qquad (\mathbf{BX})^* = \mathbf{BX}, \tag{3.4.8}$$

that is, $\mathbf{B} = \mathbf{X}_{mr}^-$, *is obtained from (3.4.2) either by choosing* $\mathbf{D} = \mathbf{B}^*, \mathbf{C} = \mathbf{I}_n$, *in which case by Corollary 1 one could even replace the reflexive g-inverse of* **DBC** *by any g-inverse. (This gives* $\mathbf{X} = (\mathbf{B}^*\mathbf{B})^-\mathbf{B}^*$), *or, alternatively, by choosing* $\mathbf{D} = \mathbf{I}_m$, $\mathbf{C} = \mathbf{I}_n$ *and* $(\mathbf{DBC})_r^-$ *to be* $(\mathbf{DBC})_{lr}^-$. *(This gives* $\mathbf{X} = \mathbf{B}_{lr}^-$).

COROLLARY 6: *A general solution to the equation*

$$\mathbf{XBX} = \mathbf{X}, \qquad \mathbf{BXB} = \mathbf{B}, \qquad (\mathbf{XB})^* = \mathbf{XB} \tag{3.4.9}$$

that is, $\mathbf{B} = \mathbf{X}_{lr}^-$, *is given by*

$$\mathbf{X} = \mathbf{B}^*(\mathbf{BB}^*)^- = \mathbf{B}_{mr}^-. \tag{3.4.10}$$

COROLLARY 7: *The unique solution to the equation*

$$\mathbf{XBX} = \mathbf{X}, \qquad \mathbf{BXB} = \mathbf{B}, \qquad (\mathbf{XB})^* = \mathbf{XB}, \qquad (\mathbf{BX})^* = \mathbf{BX} \tag{3.4.11}$$

is given by

$$\mathbf{X} = \mathbf{B}^*(\mathbf{B}^*\mathbf{BB}^*)^-\mathbf{B}^* = \mathbf{B}^+, \tag{3.4.12}$$

or equivalently

$$\mathbf{X} = \mathbf{B}^*\mathbf{B}(\mathbf{B}^*\mathbf{BB}^*\mathbf{B})^-\mathbf{B}^* = \mathbf{B}^+. \tag{3.4.13}$$

3.4.2 The Equation $\mathbf{XBX} = \mathbf{0}$

THEOREM 3.4.2 *A general solution to the equation*

$$\mathbf{XBX} = \mathbf{0} \tag{3.4.14}$$

is $\mathbf{X} = \mathbf{YC}$, *where* p *as well as* $\mathbf{C}(p \times m)$ *are arbitrary and* $\mathbf{Y}$ *is an arbitrary solution of the equation* $\mathbf{CBY} = \mathbf{0}$.

Proof: $\mathbf{X} = \mathbf{YC}$ clearly satisfies equation (3.4.14). Conversely, to show that any arbitrary solution $\mathbf{X}$ of (3.4.14) is expressible in this form, let $\mathbf{X} = \mathbf{DC}$ be a rank factorization (1.2.3) of $\mathbf{X}$. Observe that $\mathbf{DCBDC} = \mathbf{0} \Rightarrow \mathbf{CBD} = \mathbf{0}$.

COROLLARY 1: *If, in addition to (3.4.14)* $\mathbf{X}$ *has to satisfy* $\mathbf{WBX} = \mathbf{0}$, *the general solution is modified to the extent that* $\mathbf{Y}$ *is an arbitrary solution of*

$$\begin{pmatrix} \mathbf{C} \\ \mathbf{W} \end{pmatrix} \mathbf{BY} = \mathbf{0}. \tag{3.4.15}$$

3.4.3 The Equation XBXB = XB or BXBX = BX

THEOREM 3.4.3 *A general solution to the equation*

$$\mathbf{XBXB} = \mathbf{XB} \tag{3.4.16}$$

is given by $\mathbf{X} = (\mathbf{DB})^-\mathbf{D} + \mathbf{E}$, *where* $\mathbf{D}(p \times m)$ *and* $\mathbf{E}(n \times m)$ *are arbitrary matrices except that* $\mathbf{EB} = \mathbf{0}$ [*i.e.* $\mathcal{M}(\mathbf{E}^*) \subset \mathcal{O}(\mathbf{B})$].

THEOREM 3.4.4 *A general solution to the equation*

$$\mathbf{BXBX} = \mathbf{BX} \tag{3.4.17}$$

is given by $\mathbf{X} = \mathbf{C}(\mathbf{BC})^- + \mathbf{F}$, *where* $\mathbf{C}(n \times q)$ *and* $\mathbf{F}(n \times m)$ *are arbitrary matrices except that* $\mathbf{BF} = \mathbf{0}$ (*i.e.* $\mathcal{M}(\mathbf{F}) \subset \mathcal{O}(\mathbf{B}^*)$).

Proof of Theorems 3.4.3 and 3.4.4: The proofs are similar to that of Theorem 3.4.1.

Note. If in addition to (3.4.16) it is required that $(\mathbf{XB})^* = \mathbf{XB}$ or in addition to (3.4.17) it is required that $(\mathbf{BX})^* = \mathbf{BX}$, the solutions are modified to $(\mathbf{DB})_m^-\mathbf{D} + \mathbf{E}$ and $\mathbf{C}(\mathbf{BC})_l^- + \mathbf{F}$, respectively.

THEOREM 3.4.5 *If* $\mathbf{B}$ *is hermitian, a general hermitian solution to the equation* (3.4.16) *or* (*here, equivalently, to equation* (3.4.17)) *is given by*

$$\mathbf{X} = \mathbf{C}^*(\mathbf{CBC}^*)^-\mathbf{CBC}^*[(\mathbf{CBC}^*)^-]^*\mathbf{C} + \mathbf{FDF}^*, \tag{3.4.18}$$

where $p, q, \mathbf{C}(p \times m), \mathbf{F}(m \times q)$ *are arbitrary, except that* $\mathbf{BF} = \mathbf{0}$ *and* $\mathbf{D}$ *is an arbitrary diagonal matrix of order* q *with real elements.*

Proof: With $\mathbf{X}$ as determined by (3.4.18), it is clear that $\mathbf{X} = \mathbf{X}^*$ and $\mathbf{XB}$ is idempotent. Conversely, if $\mathbf{XB}$ is idempotent and $\mathbf{X} = \mathbf{X}^*$, $\mathbf{BXB}$ is a hermitian reflexive g-inverse of $\mathbf{XBX}$. Choosing $\mathbf{C} = \mathbf{X}$, $(\mathbf{CBC}^*)^- = \mathbf{BXB}$.

$$\mathbf{C}^*(\mathbf{CBC}^*)^-\mathbf{CBC}^*(\mathbf{CBC}^*)^-\mathbf{C} = \mathbf{XBXBXBXBXBX} = \mathbf{XBX}.$$

The proof is complete once it is noted that

$$\mathbf{B}(\mathbf{X} - \mathbf{XBX}) = \mathbf{0}.$$

3.4.4 The Equation BXBXB = BXB

THEOREM 3.4.6 *A general solution to the equation*

$$\mathbf{BXBXB} = \mathbf{BXB} \tag{3.4.19}$$

is

$$\mathbf{X} = \mathbf{C}(\mathbf{DBC})_r^-\mathbf{D} + \mathbf{E} - \mathbf{B}^-\mathbf{BEBB}^-, \tag{3.4.20}$$

where $\mathbf{C}(n \times p)$, $\mathbf{D}(q \times m)$ *and* $\mathbf{E}(n \times m)$ *are arbitrary matrices.*

Proof: We shall show that a general solution of (3.4.19) is $\mathbf{Y} + \mathbf{W}$, where $\mathbf{Y}$ and $\mathbf{W}$ are general solutions of $\mathbf{Y} = \mathbf{YBY}$, $\mathbf{BWB} = \mathbf{0}$, respectively.

It is easily verified that $\mathbf{Y} + \mathbf{W}$ satisfies (3.4.19). To show that $\mathbf{Y} + \mathbf{W}$ is, in fact, a general solution, it suffices to show that any solution of (3.4.19) can always be expressed in this form. For this note that if $\mathbf{X}$ be a solution of (3.4.19), it also satisfies

$$\mathbf{B}(\mathbf{X} - \mathbf{XBXBX})\mathbf{B} = \mathbf{0}$$

$$(\mathbf{XBXBX})\mathbf{B}(\mathbf{XBXBX}) = \mathbf{XBXBX}.$$

Theorem 3.4.6 follows from Theorem 3.4.1 and the general solution of equation $\mathbf{BWB} = \mathbf{0}$ given by Theorem 2.3.2.

COROLLARY 1: *A general solution of equation (3.4.19) subject to the additional condition*

$$R(\mathbf{X}) = R(\mathbf{BXB}) \tag{3.4.21}$$

is given by $\mathbf{X} = \mathbf{C}(\mathbf{DBC})_r^- \mathbf{D}$ (*that is here*, (3.4.19) $\Rightarrow \mathbf{B} = \mathbf{X}^-$).

3.4.5 The Equation XBXBX = XBX

THEOREM 3.4.7 *A general solution to the equation*

$$\mathbf{XBXBX} = \mathbf{XBX} \tag{3.4.22}$$

is

$$\mathbf{X} = \mathbf{C}(\mathbf{DBC})_r^- \mathbf{D} + \mathbf{YE}, \tag{3.4.23}$$

where $\mathbf{C}(n \times p)$, $\mathbf{D}(c \times m)$ *and* $\mathbf{E}(s \times m)$ *are arbitrary and* $\mathbf{Y}(n \times s)$ *is an arbitrary solution of the equation*

$$\begin{pmatrix} \mathbf{C}(\mathbf{DBC})_r^- \mathbf{D} \\ \mathbf{E} \end{pmatrix} \mathbf{BY} = \mathbf{0}. \tag{3.4.24}$$

Proof: We show $\mathbf{XBXBX} = \mathbf{XBX} \Leftrightarrow \mathbf{X} = \mathbf{W} + \mathbf{Z}$, where $\mathbf{W}$ is a solution of $\mathbf{WBW} = \mathbf{W}$ and $\mathbf{Z}$ satisfies the equations $\mathbf{ZBZ} = \mathbf{0}$, $\mathbf{WBZ} = \mathbf{0}$. The '$\Leftarrow$' part is computational. To prove the '$\Rightarrow$' part, observe that if $\mathbf{X}$ satisfies the equation (3.4.22), then $\mathbf{W} = \mathbf{XBX}$ satisfies the equation $\mathbf{WBW} = \mathbf{W}$ and $\mathbf{Z} = \mathbf{X} - \mathbf{XBX}$ satisfies

$$\mathbf{ZBZ} = \mathbf{XBX} - \mathbf{XBXBX} - \mathbf{XBXBX} + \mathbf{XBXBXBX} = \mathbf{0}$$

$$\mathbf{WBZ} = \mathbf{XBXBX} - \mathbf{XBXBXBX} = \mathbf{0}.$$

Theorem 3.4.7 follows from Theorems 3.4.1 and 3.4.2.

3.4.6 Matrix Decompositions and Solution of Matrix Equations

Jordan decomposition of a square matrix **X** is convenient for solving polynomial equations in **X** with scalar coefficients, since it enables us to transfer the equation to the Jordan matrices occurring in the canonical form of **X**, which involves only one unknown. Thus, for example, it is seen that if $a \neq 0$, and p is a positive integer ≥ 2, the equation

$$a\mathbf{X}^p = \mathbf{X} \tag{3.4.25}$$

admits only semisimple solutions, and a general solution is expressible as $\mathbf{X} = \mathbf{L}^{-1}\mathbf{DL}$, where **D** is a diagonal matrix and the diagonal elements of **D** are either 0 or any one of $(p-1)$th roots of a^{-1} and **L** is an arbitrary non-singular matrix.

Singular decompositions are similarly useful in solving matrix equations of the type

$$a_0(\mathbf{XX}^*)^p\mathbf{X} + a_1(\mathbf{XX}^*)^{p-1}\mathbf{X} + \cdots + a_p\mathbf{X} = \mathbf{M}, \tag{3.4.26}$$

where **M** is a given matrix of order $m \times n$. If $\mathbf{M} = \mathbf{UDV}^*$ be the singular decomposition of **M**, where **U** and **V** are unitary matrices and **D** is diagonal, then a general solution of (3.4.26) is given by

$$\mathbf{X} = \mathbf{UEV}^*, \tag{3.4.27}$$

where **E** is diagonal. If d_i and e_i are, respectively, the ith diagonal elements of **D** and **E**, then e_i could be any root λ of the equation

$$a_0\lambda^{2p+1} + a_1\lambda^{2p-1} + \cdots + a_p\lambda = d_i.$$

There are thus only a finite number of solutions.

3.4.7 The Equation $(\mathbf{XX}^*)^-\mathbf{X} = \mathbf{M}$

It is a simple matter to show that the equation

$$(\mathbf{XX}^*)^+\mathbf{X} = \mathbf{M} \tag{3.3.28}$$

has the unique solution $\mathbf{X} = (\mathbf{M}^*)^+$, and that, if one considers the equation $(\mathbf{XX}^*)^-\mathbf{X} = \mathbf{M}$ where the g-inverse could be arbitrary, a general solution is given by

$$\mathbf{X} = (\mathbf{M}^*)^+_{\mathbf{IP}}, \tag{3.4.29}$$

where **P** is an arbitrary p.d. matrix.

3.4.8 Minimum Norm Least-Squares Solution of AXB = C

As in Theorem 2.3.2, the equation $\mathbf{AXB} = \mathbf{C}$ can be written in the usual form of linear equation

$$(\mathbf{A} \otimes \mathbf{B}')\mathbf{x} = \mathbf{c}, \tag{3.4.30}$$

using the Kronecker product of **A** and **B**′ and expressing **X** and **C** as vectors with columns written one below the other. Using the Moore Penrose inverse, the required solution is

$$\mathbf{x} = (\mathbf{A} \otimes \mathbf{B}')^{+}\mathbf{c} = (\mathbf{A}^{+} \otimes \mathbf{B}^{+\prime})\mathbf{c} \tag{3.4.31}$$

which in matrix form is

$$\mathbf{X} = \mathbf{A}^{+}\mathbf{C}\mathbf{B}^{+}, \tag{3.4.32}$$

the solution given by Penrose (1956).

3.5 MISCELLANEOUS EXPANSIONS OF $\mathbf{A}^{+}$

3.5.1 A Lagrange–Sylvester Interpolation Formula for $\mathbf{A}^{+}$

Let, as in Section 2.10, $\alpha_1^2, \alpha_2^2, \cdots, \alpha_u^2$ and $\alpha_{u+1}^2 = 0$ be the distinct eigenvalues of $\mathbf{A}^*\mathbf{A}$. Since $\mathbf{A}^*\mathbf{A}$ is semisimple, its minimum polynomial is

$$g(\lambda) = \prod_i (\lambda - \alpha_i^2).$$

Write

$$g_i(\lambda) = \prod_{j \neq i} (\lambda - \alpha_j^2).$$

Consider the spectral decomposition of $\mathbf{A}^*\mathbf{A}$

$$\mathbf{A}^*\mathbf{A} = \sum \alpha_i^2 \mathbf{E}_i,$$

where $\mathbf{E}_i = \mathbf{C}_i\mathbf{C}_i^*$, as defined in Section 2.10.

If $P(\lambda)$ be a polynomial in λ, we have

$$P(\mathbf{A}^*\mathbf{A}) = \sum P(\alpha_i^2)\mathbf{E}_i$$

since $\mathbf{E}_i^m = \mathbf{E}_i$ for every positive integer m and $\mathbf{E}_i\mathbf{E}_j = 0 \ \forall\ i \neq j$. Hence

$$g_i(\mathbf{A}^*\mathbf{A}) = g_i(\alpha_i^2)\mathbf{E}_i, \text{ since } g_i(\alpha_j^2) = 0 \ \forall\, j \neq i.$$

Write

$$(\mathbf{A}^*\mathbf{A})^{+} = \sum \alpha_i^{-2}\mathbf{E}_i$$

$$= \sum \alpha_i^{-2} \frac{\prod_{j \neq i} (\mathbf{A}^*\mathbf{A} - \alpha_j^2 I)}{\prod_{j \neq i} (\alpha_i^2 - \alpha_j^2)}.$$

Check

$$\mathbf{A}^+ = (\mathbf{A}^*\mathbf{A})^+\mathbf{A}^*$$

$$= \sum \alpha_i^{-2} \frac{\prod_{j \neq i} (\mathbf{A}^*\mathbf{A} - \alpha_j^2 \mathbf{I})}{\prod_{j \neq i} (\alpha_i^2 - \alpha_j^2)} \mathbf{A}^*. \tag{3.5.1}$$

This may be called the Lagrange–Sylvester interpolation formula for **A**.

3.5.2 Neumann-Type Expansions for $\mathbf{A}^+$

Consider, as in Section 2.10, the representation $\mathbf{A} = \mathbf{BDC}^*$, where **B** and **C** are unitary matrices of order m and n, respectively,

$$\mathbf{D} = \begin{pmatrix} \mathbf{\Lambda} & \mathbf{0} \\ \mathbf{0} & \mathbf{0} \end{pmatrix}.$$

and $\mathbf{\Lambda} = \operatorname{diag}(\alpha_1 \mathbf{I}_{r_1}, \alpha_2 \mathbf{I}_{r_2}, \ldots, \alpha_u \mathbf{I}_{r_u})$, $\Sigma r_i = r$. Check that $\mathbf{A}^+ = \mathbf{CD}^+\mathbf{B}^*$ where,

$$\mathbf{D}^+ = \begin{pmatrix} \mathbf{\Lambda}^{-1} & \mathbf{0} \\ \mathbf{0} & \mathbf{0} \end{pmatrix}$$

and $\mathbf{\Lambda}^{-1} = \operatorname{diag}(\alpha_1^{-1}\mathbf{I}_{r_1}, \quad \alpha_2^{-1}\mathbf{I}_{r_2}, \ldots, \alpha_u^{-1}\mathbf{I}_{r_u})$.

THEOREM 3.5.1 *Let* $0 < \alpha < 2/c$, *where* $c = \max(\alpha_1^2, \alpha_2^2, \ldots, \alpha_n^2)$. *Then* $\alpha\Sigma_{k=0}^{\infty}(\mathbf{I} - \alpha\mathbf{A}^*\mathbf{A})^k\mathbf{A}^*$ *converges and is equal to* $\mathbf{A}^+$.

Proof: Note $\mathbf{A}^*\mathbf{A} = \mathbf{CD}^*\mathbf{DC}^*$, where

$$\mathbf{D}^*\mathbf{D} = \begin{pmatrix} \mathbf{\Lambda}^2 & \mathbf{0} \\ \mathbf{0} & \mathbf{0} \end{pmatrix}$$

$$(\mathbf{I} - \alpha\mathbf{A}^*\mathbf{A})^k = \mathbf{C}(\mathbf{I} - \alpha\mathbf{D}^*\mathbf{D})^k\mathbf{C}^*$$

$$\alpha(\mathbf{I} - \alpha\mathbf{A}^*\mathbf{A})^k\mathbf{A}^* = \mathbf{C}\alpha(\mathbf{I} - \alpha\mathbf{D}^*\mathbf{D})^k\mathbf{D}^*\mathbf{B}^*.$$

Check that $\mathbf{W} = \alpha\Sigma(\mathbf{I} - \alpha\mathbf{D}^*\mathbf{D})^k\mathbf{D}^*$ is a diagonal matrix. If a diagonal element of $\mathbf{D}^*$ is 0, the corresponding diagonal element in **W** is also 0. If a diagonal element in $\mathbf{D}^*$ is α_i, the corresponding diagonal element in **W** is

$$\sum_{k=0}^{\infty} (1 - \alpha\alpha_i^2)^k \alpha_i = \frac{\alpha\alpha_i}{\alpha\alpha_i^2} = \alpha_i^{-1}.$$

This shows $\mathbf{W} = \mathbf{D}^+$. Hence Theorem 3.5.1 is established.

Note. $\alpha\Sigma_{k=0}^{\infty}(\mathbf{I} - \alpha\mathbf{A}^*\mathbf{A})^k\mathbf{A}^*$ is the well-known Neumann expansion for $\mathbf{A}^{-1}$, when **A** is nonsingular.

COROLLARY 1: *With α as in Theorem 3.5.1*

$$\mathbf{A}^+ = \alpha \sum_{k=0}^{\infty} \mathbf{A}^*(\mathbf{I} - \alpha\mathbf{A}\mathbf{A}^*)^k \tag{3.5.2}$$

Proof: From Theorem 3.5.1 we have

$$(\mathbf{A}^*)^+ = \alpha \sum_{k=0}^{\infty} (\mathbf{I} - \alpha\mathbf{A}\mathbf{A}^*)^k\mathbf{A}.$$

$$\mathbf{A}^+ = [(\mathbf{A}^*)^+]^* = \alpha \sum_{k=0}^{\infty} \mathbf{A}^*(\mathbf{I} - \alpha\mathbf{A}\mathbf{A}^*)^k.$$

THEOREM 3.5.8 (*Neumann-Euler expansion*) *With α as in Theorem 3.5.1, let*

$$\mathbf{A}_p^+ = \alpha\left[\{\mathbf{I} + (\mathbf{I} - \alpha\mathbf{A}^*\mathbf{A})\} \prod_{k=1}^{p-1} \{\mathbf{I} + (\mathbf{I} - \alpha\mathbf{A}^*\mathbf{A})^{2^k}\}\right]\mathbf{A}^*$$

then

$$\|\mathbf{A}^+ - \mathbf{A}_p^+\| \le \max_i \frac{(1 - \alpha\alpha_i^2)^{2^p}}{\alpha_i}, \tag{3.5.3}$$

where for a matrix $\mathbf{B}$, $\|\mathbf{B}\|$ *represents the matrix norm defined by*

$$\|\mathbf{B}\| = \max\{\sqrt{\lambda_i} : \lambda_i \in \sigma(\mathbf{B}^*\mathbf{B}), \text{ the spectrum of } \mathbf{B}^*\mathbf{B}\}$$

Proof: Consider the following finite form of an identity due to Euler

$$(1 + x) \prod_{k=1}^{p-1} (1 + x^{2^k}) = \sum_{k=0}^{2^p - 1} x^k$$

and observe that

$$\mathbf{A}^+ - \mathbf{A}_p^+ = \alpha \sum_{k=2^p}^{\infty} (\mathbf{I} - \alpha\mathbf{A}^*\mathbf{A})^k\mathbf{A}^* = \alpha\mathbf{C} \sum_{k=2^p}^{\infty} (\mathbf{I} - \alpha\mathbf{D}^*\mathbf{D})^k\mathbf{D}^*\mathbf{B}^*.$$

Hence

$$\|\mathbf{A}^+ - \mathbf{A}_p^+\| \le \alpha\|\mathbf{C}\| \, \|\sum_{k=2^p}^{\infty} (\mathbf{I} - \alpha\mathbf{D}^*\mathbf{D})^k\mathbf{D}^*\| \, \|\mathbf{B}^*\|$$

$$= \max_i \frac{(1 - \alpha\alpha_i^2)^{2^p}}{\alpha_i}.$$

COROLLARY 1: *With α as in Theorem* 3.5.1, *let*

$$\mathbf{A}_p^+ = \alpha\mathbf{A}^*\left[\{\mathbf{I} + (\mathbf{I} - \alpha\mathbf{A}\mathbf{A}^*)\} \prod_{k=1}^{p} \{\mathbf{I} + (\mathbf{I} - \alpha\mathbf{A}\mathbf{A}^*)^{2^k}\}\right]$$

then

$$\|\mathbf{A}^+ - \mathbf{A}_p^+\| \le \max_i \frac{(1-\alpha\alpha_i^2)^{2^p}}{\alpha_i}. \tag{3.5.4}$$

The following theorem could be proven along the same lines.

THEOREM 3.5.3 *For a complex matrix* $\mathbf{A}$

$$\text{(i) } \mathbf{A}^+ = \sum_{k=1}^{\infty} \mathbf{A}^*(\mathbf{I} + \mathbf{A}\mathbf{A}^*)^{-k} = \sum_{k=1}^{\infty} (\mathbf{I} + \mathbf{A}^*\mathbf{A})^{-k}\mathbf{A}^*, \tag{3.5.5}$$

$$\text{(ii) } \mathbf{A}^+ = \lim_{\lambda \to 0+} \mathbf{A}^*(\lambda\mathbf{I} + \mathbf{A}\mathbf{A}^*)^{-1} = \lim_{\lambda \to 0+} (\lambda\mathbf{I} + \mathbf{A}^*\mathbf{A})^{-1}\mathbf{A}^* \tag{3.5.6}$$

3.6 g-INVERSE OF PARTITIONED MATRICES

In this section we obtain explicit expressions for g-inverses of partitioned matrices, which have applications in many practical problems. We shall first consider the simple partitioned matrix $(\mathbf{A} : \mathbf{a})$ where $\mathbf{A}$ is $m \times n$ matrix and $\mathbf{a}$ is $n \times 1$ vector. Theorem 3.6.1 shows that a g-inverse of $(\mathbf{A} : \mathbf{a})$ can be expressed in the form

$$\mathbf{X} = \begin{pmatrix} \mathbf{G} - \mathbf{d}\mathbf{b}^* \\ \mathbf{b}^* \end{pmatrix} \tag{3.6.1}$$

where $\mathbf{G}$ is a g-inverse of $\mathbf{A}$, $\mathbf{d} = \mathbf{G}\mathbf{a}$, and $\mathbf{b}$ is suitably defined.

THEOREM 3.6.1 *Let* $\mathbf{A}$ *be* $m \times n$ *matrix,* $\mathbf{a}$ *be an n-vector and* $\mathbf{X}$ *be as in (3.6.1).*

Case 1. *Let* $\mathbf{a} \notin \mathscr{M}(\mathbf{A})$ *and* $\mathbf{b} = \mathbf{c}/\mathbf{c}^*\mathbf{a}$, $\mathbf{c} = (\mathbf{I} - \mathbf{A}\mathbf{G})^*(\mathbf{I} - \mathbf{A}\mathbf{G})\mathbf{a}$. *Then:*

$$\begin{aligned} \mathbf{X} &= (\mathbf{A} : \mathbf{a})^- && \text{if} \quad \mathbf{G} = \mathbf{A}^- \\ &= (\mathbf{A} : \mathbf{a})_r^- && \text{if} \quad \mathbf{G} = \mathbf{A}_r^- \\ &= (\mathbf{A} : \mathbf{a})_m^- && \text{if} \quad \mathbf{G} = \mathbf{A}_m^- \\ &= (\mathbf{A} : \mathbf{a})_l^- && \text{if} \quad \mathbf{G} = \mathbf{A}_l^- \\ &= (\mathbf{A} : \mathbf{a})^+ && \text{if} \quad \mathbf{G} = \mathbf{A}^+ \end{aligned} \tag{3.6.2}$$

Case 2. *Let* $\mathbf{a} \in \mathcal{M}(\mathbf{A})$. *Then*

$$\begin{aligned}
\mathbf{X} &= (\mathbf{A} : \mathbf{a})^{-} && \text{if } \mathbf{G} = \mathbf{A}^{-} && \text{and arbitrary } \mathbf{b} \\
&= (\mathbf{A} : \mathbf{a})_r^{-} && \text{if } \mathbf{G} = \mathbf{A}_r^{-} && \text{and } \mathbf{b} = \mathbf{G}^*\alpha \ (\alpha \text{ arbitrary}) \\
&= (\mathbf{A} : \mathbf{a})_l^{-} && \text{if } \mathbf{G} = \mathbf{A}_l^{-} && \text{and arbitrary } \mathbf{b} \\
&= (\mathbf{A} : \mathbf{a})_m^{-} && \text{if } \mathbf{G} = \mathbf{A}_m^{-} && \text{and } \mathbf{b} = \frac{\mathbf{G}^*\mathbf{G}\mathbf{a}}{1 + \mathbf{a}^*\mathbf{G}^*\mathbf{G}\mathbf{a}} \\
&= (\mathbf{A} : \mathbf{a})^{+} && \text{if } \mathbf{G} = \mathbf{A}^{+} && \text{and } \mathbf{b} = \frac{\mathbf{G}^*\mathbf{G}\mathbf{a}}{1 + \mathbf{a}^*\mathbf{G}^*\mathbf{G}\mathbf{A}}.
\end{aligned} \tag{3.6.3}$$

Proof: The results (3.6.2) and (3.6.3) are established by straightforward computation. The results concerning $(\mathbf{A} : \mathbf{a})^{+}$ are due to Greville (1960).

In the next theorem we obtain expressions for a g-inverse of $\mathbf{A}$ when a g-inverse of $(\mathbf{A} : \mathbf{a})$ is known.

THEOREM 3.6.2 *Let* $\mathbf{X} = \begin{pmatrix} \mathbf{G} \\ \mathbf{b}^* \end{pmatrix}$ *be a g-inverse of* $(\mathbf{A} : \mathbf{a})$.

Case 1. $\mathbf{a} \notin \mathcal{M}(\mathbf{A})$. *Then*

$$\begin{aligned}
\mathbf{G} &= \mathbf{A}^{-} && \text{if } \mathbf{X} = (\mathbf{A} : \mathbf{a})^{-} \\
&= \mathbf{A}_r^{-} && \text{if } \mathbf{X} = (\mathbf{A} : \mathbf{a})_r^{-} && \text{and } \mathbf{G}\mathbf{a} = \mathbf{0} \\
&= \mathbf{A}_m^{-} && \text{if } \mathbf{X} = (\mathbf{A} : \mathbf{a})_m^{-} \\
&= \mathbf{A}_l^{-} && \text{if } \mathbf{X} = (\mathbf{A} : \mathbf{a})_l^{-} && \text{and } \mathbf{a} \in \mathcal{O}(\mathbf{A}) \\
&= \mathbf{A}^{+} && \text{if } \mathbf{X} = (\mathbf{A} : \mathbf{a})^{+} && \text{and } \mathbf{a} \in \mathcal{O}(\mathbf{A})
\end{aligned} \tag{3.6.4}$$

Further in Case 1

$$\begin{aligned}
\mathbf{A}^{-} &= \mathbf{G}(\mathbf{I} - \mathbf{a}\mathbf{b}^*) && \text{if } \mathbf{X} = (\mathbf{A} : \mathbf{a})^{-} \\
\mathbf{A}_r^{-} &= \mathbf{G}(\mathbf{I} - \mathbf{a}\mathbf{b}^*) && \text{if } \mathbf{X} = (\mathbf{A} : \mathbf{a})_r^{-} \\
\mathbf{A}_l^{-} &= \mathbf{G}\mathbf{G}^*\left(\mathbf{I} + \frac{\mathbf{A}^*\mathbf{a}\mathbf{b}^*\mathbf{G}^*}{1 - \mathbf{b}^*\mathbf{G}^*\mathbf{A}^*\mathbf{a}}\right)\mathbf{A}^* && \text{if } \mathbf{X} = (\mathbf{A} : \mathbf{a})_l^{-} \\
\mathbf{A}^{+} &= \mathbf{G}\left(\mathbf{I} - \frac{\mathbf{b}\mathbf{b}^*}{\mathbf{b}^*\mathbf{b}}\right) && \text{if } \mathbf{X} = (\mathbf{A} : \mathbf{a})^{+}
\end{aligned} \tag{3.6.5}$$

Case 2. $\mathbf{a} \in \mathcal{M}(\mathbf{A})$ *and* $\mathbf{b}^*\mathbf{a} \neq 1$. *Let* $\mathbf{Y} = \mathbf{G}\left(\mathbf{I} + \dfrac{\mathbf{ab}^*}{1 - \mathbf{b}^*\mathbf{a}}\right)$. *Then*

$$\begin{aligned} \mathbf{Y} &= \mathbf{A}^- && \text{if } \mathbf{X} = (\mathbf{A} : \mathbf{a})^- \\ &= \mathbf{A}_r^- && \text{if } \mathbf{X} = (\mathbf{A} : \mathbf{a})_r^- \\ &= \mathbf{A}_m^- && \text{if } \mathbf{X} = (\mathbf{A} : \mathbf{a})_m^- \\ &= \mathbf{A}_l^- && \text{if } \mathbf{X} = (\mathbf{A} : \mathbf{a})_l^- \\ &= \mathbf{A}^+ && \text{if } \mathbf{X} = (\mathbf{A} : \mathbf{a})^+ \end{aligned} \tag{3.6.6}$$

Proof: The results are obtained by straightforward computation. The results concerning $\mathbf{A}^+$ in (3.6.4) and (3.6.5) are due to Cline (1964).

THEOREM 3.6.3 *Let* $\mathbf{A}$ *be* $m \times n$ *matrix and* $\mathbf{a}$ *be n-vector. Consider the partitioned matrix*

$$\mathbf{B} = \begin{pmatrix} \mathbf{A}^*\mathbf{A} & \mathbf{A}^*\mathbf{a} \\ \mathbf{a}^*\mathbf{A} & \mathbf{a}^*\mathbf{a} \end{pmatrix}. \tag{3.6.7}$$

If $\mathbf{G}$ *is any g-inverse of* $\mathbf{A}^*\mathbf{A}$, *then a g-inverse of* $\mathbf{B}$ *is*

$$\mathbf{B}^- = \begin{pmatrix} \mathbf{G} + \delta\mathbf{HH}^* & -\delta\mathbf{H} \\ -\delta\mathbf{H}^* & \delta \end{pmatrix}, \tag{3.6.8}$$

where $\mathbf{H} = \mathbf{GA}^*\mathbf{a}$, $\delta^{-1} = \mathbf{a}^*(\mathbf{I} - \mathbf{AGA}^*)\mathbf{a}$ *if* $\mathbf{a} \in \mathcal{M}(\mathbf{A})$ *and* $\delta = 0$ *otherwise.*

Proof is omitted.

THEOREM 3.6.4 *Let* $\mathbf{N}$ *be* $n \times n$ *p.s.d. matrix and* $\mathbf{A}$ *be* $m \times n$ *matrix. Define*

$$\mathbf{C}_1 = \mathbf{N}^-[\mathbf{I} - \mathbf{A}^*(\mathbf{AN}^-\mathbf{A}^*)^-\mathbf{AN}^-], \qquad \mathbf{C}_2 = \mathbf{N}^-\mathbf{A}^*(\mathbf{AN}^-\mathbf{A}^*)^-$$
$$\mathbf{C}_3 = \mathbf{C}_2^*, \qquad \mathbf{C}_4 = -(\mathbf{AN}^-\mathbf{A}^*)^-$$

for any choice of the g-inverses. Then

$$\begin{pmatrix} \mathbf{N} & \mathbf{A}^* \\ \mathbf{A} & \mathbf{0} \end{pmatrix}^- = \begin{pmatrix} \mathbf{C}_1 & \mathbf{C}_2 \\ \mathbf{C}_3 & \mathbf{C}_4 \end{pmatrix} \quad \text{if } \mathcal{M}(\mathbf{A}^*) \subset \mathcal{M}(\mathbf{N}) \tag{3.6.9}$$

$$= \begin{pmatrix} \mathbf{E}_1 & \mathbf{E}_2 \\ \mathbf{E}_2^* & \mathbf{E}_2^*\mathbf{A}^* + \mathbf{E}_4 \end{pmatrix} \quad \text{in general,} \tag{3.6.10}$$

where $\mathbf{E}_1$ is obtained from the expression for $\mathbf{C}_1$ by changing $\mathbf{N}$ to $\mathbf{N} + \mathbf{A}^*\mathbf{A}$.

Proof: The results (3.6.9) and (3.6.10) are obtained following the arguments of Theorem 3.1.5.

COMPLEMENTS

1. For a complex matrix $\mathbf{A}(m \times n)$

(i) One choice of $(\mathbf{AA}^*)_m^-$ is $(\mathbf{A}_l^-)^*\mathbf{A}_m^- = (\mathbf{A}^*)_m^-\mathbf{A}_m^-$

(ii) $(\mathbf{AA}^*)_m^-\mathbf{AA}^* = (\mathbf{AA}_m^-)^*$

(iii) One choice of $(\mathbf{UAV})_m^- = \mathbf{V}^*\mathbf{A}_m^-\mathbf{U}^{-1}$, where $\mathbf{U}$ is a nonsingular matrix and $\mathbf{V}$ is unitary.

(iv) If λ is a nonzero scalar $\lambda(\lambda\mathbf{A})_m^-$ is $\mathbf{A}_m^-$.

(v) If $\mathbf{G}_1$ and $\mathbf{G}_2$ are both $\mathbf{A}_m^-$, then $(\mathbf{G}_1 - \mathbf{G}_2)\mathbf{A} = \mathbf{0}$.

(vi) If $R(\mathbf{A}) = n$, every $\mathbf{A}^-$ is $\mathbf{A}_{mr}^-$.

2. For a complex matrix $\mathbf{A}(m \times n)$

(i) If $\mathbf{G}$ is $\mathbf{A}_l^-$, $\mathbf{G}^*$ is $(\mathbf{A}^*)_m^-$ if $\mathbf{G}$ is $\mathbf{A}_m^-$, $\mathbf{G}^*$ is $(\mathbf{A}^*)_l^-$.

(ii) One choice of $(\mathbf{A}^*\mathbf{A})_l^-$ is $\mathbf{A}_l^-(\mathbf{A}_m^-)^* = \mathbf{A}_l^-(\mathbf{A}^*)_l^-$.

(iii) $\mathbf{A}^*\mathbf{A}(\mathbf{A}^*\mathbf{A})_l^- = (\mathbf{A}_l^-\mathbf{A})^*$.

(iv) One choice of $(\mathbf{UAV})_l^-$ is $\mathbf{V}^{-1}\mathbf{A}_l^-\mathbf{U}^*$, where $\mathbf{V}$ is a nonsingular matrix and $\mathbf{U}$ is unitary.

(v) If λ is a nonzero scalar $\lambda(\lambda\mathbf{A})_l^-$ is $\mathbf{A}_l^-$.

(vi) If $\mathbf{G}_1$ and $\mathbf{G}_2$ are both $\mathbf{A}_l^-$, then $\mathbf{A}(\mathbf{G}_1 - \mathbf{G}_2) = \mathbf{0}$.

(vii) If $R(\mathbf{A}) = m$, every $\mathbf{A}_l^-$ is a $\mathbf{A}_{lr}^-$.

3. Each one of the following conditions is n.s. for $(\mathbf{B}_m^-)(\mathbf{A}_m^-)$ to be $(\mathbf{AB})_m^-$:

(i) $\mathbf{A}_m^-\mathbf{ABB}^*\mathbf{A}^* = \mathbf{BB}^*\mathbf{A}^*$

(ii) $\mathscr{M}(\mathbf{BB}^*\mathbf{A}^*) \subset \mathscr{M}(\mathbf{A}^*)$.

(iii) $\mathbf{A}_m^-\mathbf{ABB}^*$ is hermitian.

4. Each one of the following conditions is n.s. for $(\mathbf{B}_l^-)(\mathbf{A}_l^-)$ to be $(\mathbf{AB})_l^-$:

(i) $\mathbf{BB}_l^-\mathbf{A}^*\mathbf{AB} = \mathbf{A}^*\mathbf{AB}$,

(ii) $\mathscr{M}(\mathbf{A}^*\mathbf{AB}) \subset \mathscr{M}(\mathbf{B})$,

(iii) $\mathbf{A}^*\mathbf{ABB}_l^-$ is hermitian.

5. The following statements are true:

(i) $(\mathbf{A}^+)^+ = \mathbf{A}$,

(ii) $(\mathbf{A}^*)^+ = (\mathbf{A}^+)^*$,

(iii) $(\mathbf{AA}^*)^+ = (\mathbf{A}^+)^*\mathbf{A}^+$,

(iv) $\mathbf{A}^+\mathbf{A} = \mathbf{AA}^+$ if $\mathbf{A}$ is normal,

(v) $(\mathbf{A}^n)^+ = (\mathbf{A}^+)^n$ if $\mathbf{A}$ is normal,

(vi) $(\mathbf{AA}^*)^+\mathbf{AA}^* = \mathbf{AA}^+$,

(vii) $(\mathbf{UAV})^+ = \mathbf{V}^*\mathbf{A}^+\mathbf{U}^*$ when $\mathbf{U}$ and $\mathbf{V}$ are unitary matrices.

(viii) $(\lambda\mathbf{A})^+ = \lambda^{-1}\mathbf{A}^+$ if λ is a scalar $\neq 0$,

(ix) $\mathbf{A}^+ = (\mathbf{A}^*\mathbf{A})^+\mathbf{A}^*$,

(x) $\mathbf{B}$ is $m \times r$ of rank r and $\mathbf{C}$ is $r \times n$ of rank r. Then $(\mathbf{BC})^+ = \mathbf{C}^+\mathbf{B}^+$,

(xi) If $\mathbf{A} = \Sigma\mathbf{A}_i$, where $\mathbf{A}_i\mathbf{A}_j^* = \mathbf{0}$ and $\mathbf{A}_i^*\mathbf{A}_j = \mathbf{0}$, then $\mathbf{A}^+ = \Sigma\mathbf{A}_i^+$,

(xii) $\mathbf{A}^+ = (\mathbf{A}^*\mathbf{A})^+\mathbf{A}^* = \mathbf{A}^*(\mathbf{AA}^*)^+$.

6. Let $\mathbf{A}$ and $\mathbf{B}$ be matrices with the product $\mathbf{AB}$ defined. Furthermore, let $\mathbf{B}_1 = \mathbf{A}^+\mathbf{AB}$ and $\mathbf{A}_1 = \mathbf{AB}_1\mathbf{B}_1^+$. Then $\mathbf{AB} = \mathbf{A}_1\mathbf{B}_1$ and $(\mathbf{AB})^+ = (\mathbf{A}_1\mathbf{B}_1)^+ = \mathbf{B}_1^+\mathbf{A}_1^+$. (Cline, 1964.)

7. Each one of the following conditions are n.s. for $\mathbf{B}^+\mathbf{A}^+$ to be $(\mathbf{AB})^+$

(i) $\mathbf{A}^+\mathbf{ABB}^*\mathbf{A}^* = \mathbf{BB}^*\mathbf{A}^*$, $\mathbf{BB}^+\mathbf{A}^*\mathbf{AB} = \mathbf{A}^*\mathbf{AB}$.

(ii) $\mathcal{M}(\mathbf{BB}^*\mathbf{A}^*) \subset \mathcal{M}(\mathbf{A}^*)$, $\mathcal{M}(\mathbf{A}^*\mathbf{AB}) \subset \mathcal{M}(\mathbf{B})$.

(iii) $\mathbf{A}^+\mathbf{ABB}^*$ and $\mathbf{A}^*\mathbf{ABB}^+$ are hermitian.

(iv) $\mathbf{A}^+\mathbf{ABB}^*\mathbf{A}^*\mathbf{ABB}^+ = \mathbf{BB}^*\mathbf{A}^*\mathbf{A}$.

(v) $\mathbf{A}^+\mathbf{AB} = \mathbf{B}(\mathbf{AB})^+\mathbf{AB}$, $\mathbf{BB}^+\mathbf{A}^* = \mathbf{A}^*\mathbf{AB}(\mathbf{AB})^+$. (Greville, 1966.)

8. (i) $\mathbf{AB} = \mathbf{0} \Rightarrow \mathbf{B}^+\mathbf{A}^+ = \mathbf{0}$.

(ii) If $\mathbf{P} = \mathbf{P}^* = \mathbf{P}^2$, $(\mathbf{PQ})^+ = \mathbf{Q}^+\mathbf{P}$ whenever either $\mathbf{PQ} = \mathbf{Q}$ or $\mathbf{P}$ commutes with $\mathbf{Q}$, $\mathbf{Q}^+\mathbf{Q}$ and $\mathbf{QQ}^+$.

9. Let $\mathbf{G}$ be a reflexive g-inverse of $\mathbf{A}$. By Lemma 2.7.1 there exist p.d. matrices $\mathbf{M}$ and $\mathbf{N}$ such that

$$(\mathbf{GA})^* = \mathbf{N}(\mathbf{GA})\mathbf{N}^{-1}$$

$$(\mathbf{AG})^* = \mathbf{M}(\mathbf{AG})\mathbf{M}^{-1}.$$

Hence each reflexive g-inverse of $\mathbf{A}$ is of the type $\mathbf{A}^+_{\mathbf{MN}}$, for a suitable choice of $\mathbf{M}$ and $\mathbf{N}$.

10. Let the norm of a matrix $\mathbf{X}$ be defined by

$$\|\mathbf{X}\| = \{tr\, \mathbf{X}^*\mathbf{X}\}^{1/2}.$$

The matrix $\mathbf{X}_0$ is a best approximate solution of the matrix equation $f(\mathbf{X}) = \mathbf{G}$ if for all $\mathbf{X}$ either

(i) $\|f(\mathbf{X}) - \mathbf{G}\| \geq \|f(\mathbf{X}_0) - \mathbf{G}\|$, or

(ii) $\|f(\mathbf{X}) - \mathbf{G}\| = \|f(\mathbf{X}_0) - \mathbf{G}\|$ and $\|\mathbf{X}\| \geq \|\mathbf{X}_0\|$

(a) $\mathbf{A}^+\mathbf{C}$ is the unique best approximate solution of the matrix equation $\mathbf{AX} = \mathbf{C}$.

(b) $\mathbf{A}^+\mathbf{CB}^+$ is the unique best approximate solution of the matrix equation $\mathbf{AXB} = \mathbf{C}$.‡ (Penrose, 1956).

(c) $\|\mathbf{AX} - \mathbf{I}\|$ is minimum if $\mathbf{X} = \mathbf{A}^-_\ell$

(d) $\|\mathbf{XA} - \mathbf{I}\|$ is minimum if $\mathbf{X} = \mathbf{A}^-_m$

(e) $\|\mathbf{AX} - \mathbf{I}\|$ and $\|\mathbf{XA} - \mathbf{I}\|$ are minimum if $\mathbf{X} = \mathbf{A}^-_{\ell m}$, i.e.

$$\mathbf{X} = \mathbf{A}^+ + [\mathbf{I} - \mathbf{A}^+\mathbf{A}]\mathbf{U}[\mathbf{I} - \mathbf{AA}^+],$$

where $\mathbf{U}$ is arbitrary. Observe that for a singular matrix $\mathbf{A}$, $\mathbf{A}^-_{\ell m}$ in this sense comes closest to satisfying the conditions of the regular inverse of a nonsingular matrix.

11. Let $\mathbf{M}$ and $\mathbf{N}$ be both p.d. Then $\mathbf{A}^+_{\mathbf{MN}}\mathbf{A} = \mathbf{AA}^+_{\mathbf{MN}}$ if

(a) $\mathcal{M}(\mathbf{A}) = \mathcal{M}(\mathbf{N}^{-1}\mathbf{A}^*)$

(b) $\mathbf{M} = [\mathbf{A}\Lambda_1\mathbf{A}^* + \mathbf{N}^{-1}\mathbf{A}^\perp\Lambda_2(\mathbf{A}^\perp)^*\mathbf{N}^{-1}]^{-1}$,

where Λ_1 and Λ_2 are arbitrary hermitian matrices subject to $\mathbf{M}$ being p.d. For (a) it is sufficient that $\mathbf{A}$ be $\mathbf{N}$-normal, that is, $\mathbf{NAA}^* = \mathbf{A}^*\mathbf{AN}$

If (a) and (b) are true, then $(\mathbf{A}^n)^+_{\mathbf{MN}} = (\mathbf{A}^+_{\mathbf{MN}})^n$.

12. The following statements are true for p.d. $\mathbf{M}$ and $\mathbf{N}$.

(i) $(\mathbf{UMAVN})^+_{\mathbf{MN}} = \mathbf{V}^*\mathbf{NA}^+_{\mathbf{MN}}\mathbf{U}^*\mathbf{M}$, if

$\mathbf{U}^*\mathbf{MUM} = \mathbf{UMU}^*\mathbf{M} = \mathbf{I}_m$

$\mathbf{V}^*\mathbf{NVN} = \mathbf{VNV}^*\mathbf{N} = \mathbf{I}_n$.

(ii) $\mathbf{A}^+_{\mathbf{MN}} = (\mathbf{A}^*\mathbf{MA})^+_{\mathbf{IN}}\mathbf{A}^*\mathbf{M}$.

13. Let $\mathbf{M}$ and $\mathbf{N}$ be p.d. matrices. If $\mathbf{G}$ is a matrix such that it is the minimum $\mathbf{N}$-norm g-inverse of $\mathbf{A}$ while $\mathbf{A}$ is the minimum $\mathbf{M}$-norm g-inverse of $\mathbf{G}$, then $\mathbf{G} = \mathbf{A}^+_{\mathbf{MN}}$. If $\mathbf{G}$ is a matrix such that it is the $\mathbf{M}$-least-squares g-inverse of $\mathbf{A}$ while $\mathbf{A}$ is the $\mathbf{N}$-least-squares g-inverse of $\mathbf{G}$, then $\mathbf{G} = \mathbf{A}^+_{\mathbf{MN}}$.

14. (Expressions for $\mathbf{A}^+_{\mathbf{MN}}$ and $\mathbf{A}^+$)

(a) Let $\mathbf{M}$ and $\mathbf{N}$ be p.d. matrices with $\mathbf{M}^{1/2}$, $\mathbf{N}^{1/2}$ as Hermitian square roots. Then

$$\mathbf{A}^+_{\mathbf{MN}} = \mathbf{N}^{-1/2}(\mathbf{M}^{1/2}\mathbf{A}\mathbf{N}^{-1/2})^+\mathbf{M}^{1/2}.$$

(b) Let $\mathbf{A} = \mathbf{DE}$ be a rank factorization (1.2.3) of $\mathbf{A}$. Then

$$\mathbf{A}^+ = \mathbf{E}^*(\mathbf{EE}^*)^{-1}(\mathbf{D}^*\mathbf{D})^{-1}\mathbf{D}^* \text{ (Greville, 1960)}$$

and $\mathbf{A}^+_{\mathbf{MN}} = \mathbf{N}^{-1}\mathbf{E}^*(\mathbf{EN}^{-1}\mathbf{E}^*)^{-1}(\mathbf{D}^*\mathbf{MD})^{-1}\mathbf{D}^*\mathbf{M}$, if $\mathbf{M}$ and $\mathbf{N}$ are p.d. matrices.

(c) Let $\mathbf{A} = \lambda_1\boldsymbol{\delta}_1\boldsymbol{\varepsilon}_1^* + \cdots + \lambda_r\boldsymbol{\delta}_r\boldsymbol{\varepsilon}_r^*$ be a singular value decomposition (1.2.6) of $\mathbf{A}$. Then

$$\mathbf{A}^+ = \lambda_1^{-1}\boldsymbol{\varepsilon}_1\boldsymbol{\delta}_1^* + \cdots + \lambda_r^{-1}\boldsymbol{\varepsilon}_r\boldsymbol{\delta}_r^*.$$

(d) $\mathbf{A}^+ = \mathbf{A}^*(\mathbf{A}^*\mathbf{A}\mathbf{A}^*)^-\mathbf{A}^* = \mathbf{A}^*\mathbf{A}(\mathbf{A}^*\mathbf{A}\mathbf{A}^*\mathbf{A})^-\mathbf{A}^*$
for any choice of the g-inverses of $(\mathbf{A}^*\mathbf{A}\mathbf{A}^*)$ and $(\mathbf{A}^*\mathbf{A}\mathbf{A}^*\mathbf{A})$.

(e) Let $\mathbf{W}$ be any solution of $\mathbf{WAA}^* = \mathbf{A}^*$ and $\mathbf{Y}$ any solution of $\mathbf{A}^*\mathbf{AY} = \mathbf{A}^*$. Then
(i) $\mathbf{AWA} = \mathbf{A}$, (ii) $\mathbf{AYA} = \mathbf{A}$, (iii) $(\mathbf{WA})^* = \mathbf{WA}$, (iv) $(\mathbf{AY})^* = \mathbf{AY}$, and $\mathbf{A}^+ = \mathbf{WAY}$. (Decell, 1965a)

15. Let $\mathbf{A}, \mathbf{U}, \mathbf{V}$ be matrices of order $n \times n$, $n \times r$ and $n \times r$, respectively, such that $R(\mathbf{A}) = n - r$, $R(\mathbf{U}) = R(\mathbf{V}) = r$, $\mathscr{M}(\mathbf{A}) \oplus \mathscr{M}(\mathbf{U}) = \mathscr{M}(\mathbf{A}^*) \oplus \mathscr{M}(\mathbf{V}) = \mathscr{E}^n$, then

(a) $\begin{pmatrix} \mathbf{A} & \mathbf{U} \\ \mathbf{V}^* & \mathbf{0} \end{pmatrix}$ is nonsingular.

(b) If $\begin{pmatrix} \mathbf{A} & \mathbf{U} \\ \mathbf{V}^* & \mathbf{0} \end{pmatrix}^{-1} = \begin{pmatrix} \mathbf{B} & \mathbf{S} \\ \mathbf{T}^* & \mathbf{R} \end{pmatrix}$

(i) $\mathbf{ABA} = \mathbf{A}$, (ii) $\mathbf{BAB} = \mathbf{B}$, (iii) $\mathbf{BU} = \mathbf{0}$, (iv) $\mathbf{V}^*\mathbf{B} = \mathbf{0}$. (Blattner, 1962)

(c) If $\mathbf{M}$ and $\mathbf{N}$ are p.d. matrices such that $\mathbf{V}^*\mathbf{N}^{-1}\mathbf{A}^* = \mathbf{0}$, $\mathbf{A}^*\mathbf{MU} = \mathbf{0}$, then $\mathbf{B} = \mathbf{A}^+_{\mathbf{MN}}$.

16. Let $\mathbf{A}$ and $\mathbf{B}$ be any two matrices with the product $\mathbf{AB}$ defined.
Let $\mathbf{B}_1 = \mathbf{A}^-_{\mathbf{MI}}\mathbf{AB}$ $\quad \mathbf{A}_1 = \mathbf{A}(\mathbf{B}_1)^+_{\mathbf{IN}}\mathbf{B}_1$. Then

$$\mathbf{AB} = \mathbf{A}_1\mathbf{B}_1 \quad \text{and } (\mathbf{AB})^+_{\mathbf{MN}} = (\mathbf{B}_1)^+_{\mathbf{IN}}(\mathbf{A}_1)^+_{\mathbf{MI}}.$$

17. (i) Let $\mathbf{M}$ and $\mathbf{N}$ be nonsingular matrices, not necessarily hermitian. Further let $\mathbf{G}$ be a matrix such that

$$\mathbf{AGA} = \mathbf{A}, \qquad \mathbf{GAG} = \mathbf{G} \quad (\mathbf{GA})^*\mathbf{N} = \mathbf{NGA} \quad (\mathbf{AG})^*\mathbf{M} = \mathbf{MAG}$$

which are analogous to the conditions satisfied by $\mathbf{A}^+_{\mathbf{MN}}$ (defined with respect to p.d. matrices $\mathbf{M}$ and $\mathbf{N}$). A matrix $\mathbf{G}$ satisfying the above conditions exists and is unique if $R(\mathbf{AN}^{-1}\mathbf{A}^*) = R(\mathbf{A})$ and $R(\mathbf{A}^*\mathbf{MA}) = R(\mathbf{A})$. We may denote such an inverse by the same symbol, $\mathbf{A}_{\mathbf{MN}}^{+}$.

(ii) Suppose, in the example, the reflexive condition $\mathbf{GAG} = \mathbf{G}$ is omitted. Then a general solution for a matrix $\mathbf{G}$ satisfying the other conditions can be written in the form

$$\mathbf{A}_{\mathbf{MN}}^{+} + (\mathbf{I} - \mathbf{A}_{\mathbf{MN}}^{+}\mathbf{A})\mathbf{T}(\mathbf{I} - \mathbf{AA}_{\mathbf{MN}}^{+}),$$

where $\mathbf{T}$ is arbitrary. (Khatri, 1970).

18. Let $\mathbf{M}$ and $\mathbf{N}$ be nonsingular matrices not necessarily hermitian. Then there exists a matrix $\mathbf{G}$ satisfying the following conditions

$$\mathbf{AGA} = \mathbf{A} \quad \mathbf{GAG} = \mathbf{G} \quad (\mathbf{NGA})^* = \mathbf{NGA} \quad (\mathbf{MAG})^* = \mathbf{MAG}$$

provided $R(\mathbf{AN}^{-1}\mathbf{A}^*) = R(\mathbf{A})$ and $R(\mathbf{A}^*\mathbf{MA}) = R(\mathbf{A})$.

19. Let $\mathbf{N}$ be a nonsingular matrix not necessarily hermitian. Then there exists a matrix $\mathbf{G}$ satisfying the conditions

$$\mathbf{AGA} = \mathbf{A} \text{ and } \mathbf{A}^*\mathbf{G}^*\mathbf{NGA} = \mathbf{NGA},$$

provided $R(\mathbf{AN}^{-1}\mathbf{A}^*) = R(\mathbf{A})$.

Further, if $\mathbf{M}$ be a nonsingular matrix such that $R(\mathbf{A}^*\mathbf{MA}) = R(\mathbf{A})$, then there exists a matrix $\mathbf{G}$ satisfying the conditions

$$\mathbf{AGA} = \mathbf{A}, \qquad \mathbf{GAG} = \mathbf{G}, \qquad \mathbf{A}^*\mathbf{G}^*\mathbf{NGA} = \mathbf{NGA}, \qquad \mathbf{G}^*\mathbf{A}^*\mathbf{MAG} = \mathbf{MAG}.$$

20. Minimization of hermitian forms: Let $\mathbf{x}^*\mathbf{Ax}$ be a p.d. hermitian form and $\mathbf{Bx} = \mathbf{u}$, a set of consistent equations. Then min $\mathbf{x}^*\mathbf{Ax}$ subject to $\mathbf{Bx} = \mathbf{u}$ is attained at $\mathbf{x} = \mathbf{B}_{m(\mathbf{A})}^{-}\mathbf{u}$, where $\mathbf{B}_{m(\mathbf{A})}^{-}$ is the minimum $\mathbf{A}$-norm g-inverse of $\mathbf{B}$. Let $\mathbf{x}_1^*\mathbf{Cx}_1$ be a p.d. hermitian form on a subset $\mathbf{x}_1$ of the variables $\mathbf{x}$. Then min $\mathbf{x}_1^*\mathbf{Cx}_1$ subject to $\mathbf{Bx} = \mathbf{u}$ is attained at $\mathbf{x} = \mathbf{B}_{m(\mathbf{A})}^{-}\mathbf{u}$, where $\mathbf{A}$ is the p.s.d. matrix defined by

$$\mathbf{A} = \begin{pmatrix} \mathbf{C} & \mathbf{0} \\ \mathbf{0} & \mathbf{0} \end{pmatrix},$$

so that $\mathbf{B}_{m(\mathbf{A})}^{-}$ is the minimum $\mathbf{A}$-seminorm g-inverse of $\mathbf{B}$.

21. A $m \times n$ incidence matrix $\mathbf{T}$ is a matrix which has exactly two nonzero entries that are 1 and -1 in each of its n columns, and none of its m rows is a nullrow. Rows i and j of $\mathbf{T}$ are 'directly connected' if there is a column which has nonzero entries in rows i and j. Rows i and j of $\mathbf{T}$ are 'indirectly connected' if there is a sequence of rows $i, k_1, k_2, \ldots, k_e, j$ in which any two adjacent rows are directly connected. $\mathbf{T}$ is a 'connected incidence matrix' if each pair of its rows are connected either directly or indirectly. If $\mathbf{T}$ is a connected incidence matrix, then

$$\mathbf{I} - \mathbf{TT}^{+} = \frac{1}{m}\mathbf{E},$$

where $\mathbf{E}$ is a $m \times n$ matrix whose elements are all equal to 1. (Ijiri, 1965).

22. Let $\mathbf{A}$ be $m \times n$, $\mathbf{B}$ be $m \times r$, $\mathbf{C}$ be $n \times r$ and $\mathbf{D}$ be $r \times r$ matrices. Further let $\mathbf{D}$ be nonsingular, $\mathbf{G}$ be a g-inverse of $\mathbf{A}$, and $\mathbf{W} = \mathbf{C}^*\mathbf{GB} + \mathbf{D}^{-1}$ be nonsingular. Define

$X = G - GBW^{-1}C^*G$. Then

(i) $X = (A + BDC^*)^-$ if $\mathscr{M}(B) \subset \mathscr{M}(A)$ or $\mathscr{M}(C) \subset \mathscr{M}(A^*)$.

(ii) $X = (A + BDC^*)_r^-$ if $\mathscr{M}(B) \subset \mathscr{M}(A)$ or $\mathscr{M}(C) \subset \mathscr{M}(A^*)$ and $G = A_r^-$.

(iii) $X = (A + BDC^*)_m^-$ if $\mathscr{M}(C) \subset \mathscr{M}(A^*)$ and $G = A_m^-$.

(iv) $X = (A + BDC^*)_l^-$ if $\mathscr{M}(B) \subset \mathscr{M}(A)$ and $G = A_l^-$

(v) If in (iii) and (iv) $G = A_r^-$, then the corresponding g-inverse of $A + BDC^*$ is also reflexive.

(vi) $X = (A + BDC^*)^+$ if $\mathscr{M}(B) \subset \mathscr{M}(A)$, $\mathscr{M}(C) \subset \mathscr{M}(A^*)$ and $G = A^+$.

23.
$$[U \vdots V]^+ = \begin{bmatrix} U^+ - U^+VC^+ - U^+V(I - C^+C)KV^*U^{+*}U^+(I - VC^+) \\ C^+ + (I - C^+C)KV^*U^{+*}U^+(I - VC^+) \end{bmatrix}$$
$$= \begin{bmatrix} U^+ - U^+VC^+ - U^+V(I - C^+C)KV^*U^{+*}U^+(I - VC^+) \\ V^+ - V^+U\tilde{C}^+ - V^+U(I - \tilde{C}^+\tilde{C})\tilde{K}U^*V^{+*}V^+(I - U\tilde{C}^+) \end{bmatrix}$$
$$= \begin{bmatrix} \tilde{C}^+ + (I - \tilde{C}^+\tilde{C})\tilde{K}U^*V^{+*}V^+(I - U\tilde{C}^+) \\ C^+ + (I - C^+C)KV^*U^{+*}U^+(I - VC^+) \end{bmatrix}$$

where

$$C = (I - UU^+)V$$
$$\tilde{C} = (I - VV^+)U$$
$$K = [I + (I - C^+C)V^*U^{+*}U^+V(I - C^+C)]^{-1}$$
$$\tilde{K} = [I + (I - \tilde{C}^+\tilde{C})U^*V^{+*}V^+U(I - \tilde{C}^+\tilde{C})]^{-1}.$$ (Cline, 1964).

24. Let $[U \vdots V]^+ = \begin{pmatrix} G \\ \cdots \\ H \end{pmatrix}$ then

(a) $U^+ = G[I + V(I - HV)^+H][I - L^+L]$, where $L = H - (I - HV)(I - HV)^+H$

(b) $U^+ = G[I + V(I - HV)^+H]$ if $VC^+V = C$

(c) $U^+ = G[I - H^+H]$ iff $VC^+V = V$, or iff $(HV)^2 = HV$

(d) $U^+ = G$ iff $C = V$ or iff $(HV)^2 = HV$ and $(VH)^* = VH$

25. Let A be $m \times m$ hermitian matrix of rank r and B be $(m - r) \times m$ matrix of rank $m - r$, such that $BA = 0$. Then

(a) $A + B^*B$ is nonsingular,

(b) $A + B^+B$ is nonsingular,

(c) $AA^+ + B^+B$ is idempotent,

(d) $A^+ + B^+B = (A + B^+B)^{-1}$.

26. If A is a singular complex matrix of order $n \times n$, the following statements are equivalent

(a) $R(A) = R(A^2)$.

(b) $I - A^+A + A$ is nonsingular.

(c) $\mathscr{M}(I - A^+A)$ and $\mathscr{M}(A)$ are virtually disjoint.

(For other equivalent conditions, see example 11 in complements to Chapter 1).

CHAPTER 4

Other Special Types of g-Inverse

4.1 g-INVERSE IN SPECIFIED LINEAR MANIFOLDS

We raise the question: Given matrices $\mathbf{A}, \mathbf{P}, \mathbf{Q}$ of order $m \times n, n \times p, q \times m$, does there exist a g-inverse $\mathbf{G}$ of $\mathbf{A}$ such that $\mathcal{M}(\mathbf{G}) \subset \mathcal{M}(\mathbf{P})$ and $\mathcal{M}(\mathbf{G}^*) \subset \mathcal{M}(\mathbf{Q}^*)$, that is, a g-inverse of the form, $\mathbf{G} = \mathbf{PCQ}$. The answer to this question is contained in Lemma 4.1.1.

LEMMA 4.1.1 *Given matrices $\mathbf{A}, \mathbf{P}, \mathbf{Q}$ of order $m \times n$, $n \times p$, $q \times m$, respectively, a necessary and sufficient condition for $\mathbf{A}$ to have a g-inverse of the form $\mathbf{G} = \mathbf{PCQ}$ is that*

$$R(\mathbf{QAP}) = R(\mathbf{A}), \tag{4.1.1}$$

in which case the only choice for $\mathbf{C}$ is $(\mathbf{QAP})^-$. An inverse with the required property is unique, if further $R(\mathbf{P}) = R(\mathbf{Q}) = R(\mathbf{A})$.

Proof: Necessity follows from the definition of a g-inverse. Let $\mathbf{G} = \mathbf{PCQ}$ be a g-inverse of $\mathbf{A}$. Then

$$\mathbf{A} = \mathbf{AGA} = \mathbf{AGAGA} = \mathbf{APCQAPCQA}.$$

Therefore, $R(\mathbf{A}) \geq R(\mathbf{QAP}) \geq R(\mathbf{A}) \Rightarrow R(\mathbf{QAP}) = R(\mathbf{A})$.

If $R(\mathbf{QAP}) = R(\mathbf{A})$, from Lemma 2.2.5(d) it follows that $\mathbf{P}(\mathbf{QAP})^-\mathbf{Q}$ is a g-inverse of $\mathbf{A}$, which proves the sufficiency part.

Now let $\mathbf{PCQ}$ be a g-inverse of $\mathbf{A}$, that is,

$$\mathbf{A} = \mathbf{APCQA} \Rightarrow \mathbf{QAP} = \mathbf{QAPCQAP} \Rightarrow \mathbf{C} = (\mathbf{QAP})^-.$$

Observe now that $R(\mathbf{Q}) = R(\mathbf{P}) = R(\mathbf{A}) = R(\mathbf{QAP}) \Rightarrow \mathcal{M}(\mathbf{Q}) = \mathcal{M}(\mathbf{QAP})$, $\mathcal{M}(\mathbf{P}^*) = \mathcal{M}(\mathbf{P}^*\mathbf{A}^*\mathbf{Q}^*) \Rightarrow \mathbf{Q} = \mathbf{QAPD}$, $\mathbf{P} = \mathbf{EQAP}$ for some $\mathbf{D}$ and $\mathbf{E}$. Hence

$$\mathbf{G} = \mathbf{PCQ} = \mathbf{P}(\mathbf{QAP})^-\mathbf{Q} = \mathbf{EQAP}(\mathbf{QAP})^-\mathbf{QAPD} = \mathbf{EQAPD} = \mathbf{PD} = \mathbf{EQ}.$$

Thus the uniqueness of **G** is established, and the proof of Lemma 4.1.1 is concluded.

4.2 g-INVERSE OF A WITH $\mathcal{M}(\mathbf{G}) \subset \mathcal{M}(\mathbf{A})$

From Lemma 4.1.1 it follows on taking $\mathbf{P} = \mathbf{A}$ and $\mathbf{Q} = \mathbf{I}$ that a g-inverse with a column space identical with the column space of **A** will exist, iff **A** is square and $R(\mathbf{A}) = R(\mathbf{A}^2)$, and that such a g-inverse is always expressible in the form $\mathbf{G} = \mathbf{A}(\mathbf{A}^2)^-$. A g-inverse with this property will be denoted by $\mathbf{A}_\chi^-$. Some properties of $\mathbf{A}_\chi^-$ are described in Lemmas 4.2.1 and 4.2.2.

LEMMA 4.2.1

(i) **A** *is one choice for* $(\mathbf{A}_\chi^-)_\chi^-$.
(ii) $\mathbf{A}(\mathbf{A}_\chi^-)^2 = \mathbf{A}_\chi^-$, $\mathbf{A}_\chi^-\mathbf{A}^2 = \mathbf{A}$.
(iii) *For every positive integer m,* $(\mathbf{A}_\chi^-)^m$ *is a reflexive g-inverse of* $\mathbf{A}^m$.
(iv) (*A partial law of indices*). *Let* $r_1, r_2, \ldots, r_n$ *be a sequence of positive integers,* $s_1 = r_1 + r_3 + \cdots$, $s_2 = r_2 + r_4 + \cdots$, *the summation in* s_1 *is taken over all* r_i *with an odd subscript, and the summation in* s_2 *is taken over all* r_i *in the sequence with an even subscript. If n is odd and* $s_1 > s_2$, *then*

$$\mathbf{A}^{r_1}(\mathbf{A}_\chi^-)^{r_2}\mathbf{A}^{r_3}\cdots(\mathbf{A}_\chi^-)^{r_{n-1}}\mathbf{A}^{r_n} = \mathbf{A}^{s_1-s_2}$$

$$(\mathbf{A}_\chi^-)^{r_1}\mathbf{A}^{r_2}(\mathbf{A}_\chi^-)^{r_3}\cdots\mathbf{A}^{r_{n-1}}(\mathbf{A}_\chi^-)^{r_n} = (\mathbf{A}_\chi^-)^{s_1-s_2}.$$

If n is even and $s_1 < s_2$, *then*

$$\mathbf{A}^{r_1}(\mathbf{A}_\chi^-)^{r_2}\mathbf{A}^{r_3}\cdots\mathbf{A}^{r_{n-1}}(\mathbf{A}_\chi^-)^{r_n} = (\mathbf{A}_\chi^-)^{s_2-s_1}$$

$$(\mathbf{A}_\chi^-)^{r_1}\mathbf{A}^{r_2}(\mathbf{A}_\chi^-)^{r_3}\cdots(\mathbf{A}_\chi^-)^{r_{n-1}}\mathbf{A}^{r_n} = \mathbf{A}^{s_2-s_1}.$$

Proof: Clearly $R(\mathbf{A}_\chi^-) = R(\mathbf{A})$. Hence

$$\mathbf{A}_\chi^- = \mathbf{A}(\mathbf{A}^2)^- \Rightarrow \mathcal{M}(\mathbf{A}_\chi^-) = \mathcal{M}(\mathbf{A}) \Rightarrow \mathbf{A} = \mathbf{A}_\chi^-\mathbf{D}$$

for some **D**. Also from Lemma 2.5.1 it is seen that **A** is a g-inverse of $\mathbf{A}_\chi^-$. Hence (i) is established. To prove (ii) observe that

$$\mathbf{A}_\chi^-\mathbf{A}^2 = \mathbf{A}_\chi^-\mathbf{A}\mathbf{A}_\chi^-\mathbf{D} = \mathbf{A}_\chi^-\mathbf{D} = \mathbf{A},$$

$$\mathbf{A}(\mathbf{A}_\chi^-)^2 = \mathbf{A}\mathbf{A}_\chi^-\mathbf{A}(\mathbf{A}^2)^- = \mathbf{A}(\mathbf{A}^2)^- = \mathbf{A}_\chi^-.$$

While proving (i) we have noted that

$$\mathbf{A}\mathbf{A}_\chi^-\mathbf{A} = \mathbf{A}, \qquad \mathbf{A}_\chi^-\mathbf{A}\mathbf{A}_\chi^- = \mathbf{A}_\chi^-.$$

Check that for any positive integer m, repeated application of (ii) gives

$$\mathbf{A}^m(\mathbf{A}_\chi^-)^m\mathbf{A}^m = \mathbf{A}^m\mathbf{A}_\chi^-\mathbf{A} = \mathbf{A}^m,$$

$$(\mathbf{A}_\chi^-)^m\mathbf{A}^m(\mathbf{A}_\chi^-)^m = (\mathbf{A}_\chi^-)^m\mathbf{A}\mathbf{A}_\chi^- = (\mathbf{A}_\chi^-)^m,$$

thus establishing (iii). Proof of (iv) is similar.

LEMMA 4.2.2

(i) *If* $\mathbf{x}$ *is a (right) eigenvector of* $\mathbf{A}$ *corresponding to a nonnull eigenvalue* λ, *then* $\mathbf{x}$ *is also a (right) eigenvector of* $\mathbf{A}_\chi^-$ *corresponding to its eigenvalue* $1/\lambda$. *The converse of this result is also true.*

(ii) *Let* $\mathbf{x}_1, \mathbf{x}_2 = (\mathbf{A} - \lambda\mathbf{I})\mathbf{x}_1, \ldots, \mathbf{x}_k = (\mathbf{A} - \lambda\mathbf{I})\mathbf{x}_{k-1}$ *be a chain of generalized eigenvectors of* $\mathbf{A}$ *corresponding to its eigenvalue* λ, *where* k *is the least integer such that* $(\mathbf{A} - \lambda\mathbf{I})^k\mathbf{x}_1 = \mathbf{0}$ *and let* $\mathscr{M}(\mathbf{x}_1, \mathbf{x}_2, \ldots, \mathbf{x}_k)$ *be the linear space spanned by* $\mathbf{x}_1, \mathbf{x}_2, \ldots, \mathbf{x}_k$. *Then* $\mathscr{M}(\mathbf{x}_1, \mathbf{x}_2, \ldots, \mathbf{x}_k)$ *is invariant under* $\mathbf{A}_\chi^-$.

Proof: Use Lemma 4.2.1 (ii) to note that, since $\mathbf{A}_\chi^-\mathbf{A}^2 = \mathbf{A}$,

$$\mathbf{A}\mathbf{x} = \lambda\mathbf{x} \Rightarrow \mathbf{A}_\chi^-(\lambda^2\mathbf{x}) = \lambda\mathbf{x} \Rightarrow \mathbf{A}_\chi^-\mathbf{x} = \lambda^{-1}\mathbf{x}.$$

The converse similarly follows from $\mathbf{A}(\mathbf{A}_\chi^-)^2 = \mathbf{A}_\chi^-$. This proves (i).

(ii) can be proven along the same lines, noting that for $i < k$, $\mathbf{A}\mathbf{x}_i = \mathbf{x}_{i+1} + \lambda\mathbf{x}_i$, $\mathbf{A}\mathbf{x}_k = \lambda\mathbf{x}_k$. The Lemma 4.2.2 is proved.

Consider now the Jordan canonical representation of matrix $\mathbf{A}$

$$\mathbf{A} = \mathbf{L}\begin{pmatrix} \mathbf{J}_1 & \mathbf{0} & \cdots & \mathbf{0} \\ \mathbf{0} & \mathbf{J}_2 & \cdots & \mathbf{0} \\ \vdots & \vdots & & \vdots \\ \mathbf{0} & \mathbf{0} & \cdots & \mathbf{J}_k \end{pmatrix}\mathbf{L}^{-1},$$

where $\mathbf{J}_i$ is a lower Jordan matrix of order r_i, corresponding to some eigenvalue λ_i of $\mathbf{A}$. It is known that $\mathbf{J}_i$ is nonsingular, if $\lambda_i \neq 0$ and, if $\lambda_i = 0$, $R(\mathbf{J}_i^m) = \max(0, r_i - m)$. Also

$$R(\mathbf{A}) = \Sigma R(\mathbf{J}_i), \qquad R(\mathbf{A}^2) = \Sigma R(\mathbf{J}_i^2).$$

Hence $R(\mathbf{A}) = R(\mathbf{A}^2) \Rightarrow R(\mathbf{J}_i) = R(\mathbf{J}_i^2)$ which is trivially true, if $\mathbf{J}_i$ is nonsingular (i.e., $\lambda_i \neq 0$) and for a singular $\mathbf{J}_i$, if $r_i = 1$, that is, $\mathbf{J}_i$ is a null matrix of order 1. Let us therefore write

$$\mathbf{A} = \mathbf{L}\begin{pmatrix} \mathbf{J}_1 & \mathbf{0} & \cdots & \mathbf{0} & \mathbf{0} \\ \mathbf{0} & \mathbf{J}_2 & \cdots & \mathbf{0} & \mathbf{0} \\ \vdots & \vdots & & \vdots & \vdots \\ \mathbf{0} & \mathbf{0} & \cdots & \mathbf{J}_m & \mathbf{0} \\ \mathbf{0} & \mathbf{0} & \cdots & \mathbf{0} & \mathbf{0} \end{pmatrix}\mathbf{L}^{-1},$$

where each $\mathbf{J}_i$ can now be taken to be nonsingular. Consider the corresponding partitions $(\mathbf{L}_1 \vdots \mathbf{L}_2 \vdots \cdots \vdots \mathbf{L}_m \vdots \mathbf{L}_{m+1})$ of $\mathbf{L}$ and

$$(\mathbf{M}_1 \vdots \mathbf{M}_2 \vdots \cdots \vdots \mathbf{M}_m \vdots \mathbf{M}_{m+1})$$

of $(\mathbf{L}')^{-1}$. For $i \leq m$, $\mathbf{AL}_i = \mathbf{L}_i\mathbf{J}_i \Rightarrow (\mathbf{A}_\chi^-)\mathbf{L}_i\mathbf{J}_i^2 = \mathbf{L}_i\mathbf{J}_i \Rightarrow \mathbf{A}_\chi^-\mathbf{L}_i = \mathbf{L}_i\mathbf{J}_i^{-1}$. Also, $\mathbf{M}'_{m+1}\mathbf{AL} = \mathbf{0} \Rightarrow \mathbf{M}'_{m+1}\mathbf{A} = \mathbf{0} \Rightarrow \mathbf{M}'_{m+1}\mathbf{A}_\chi^- = \mathbf{0}$. This shows $\mathbf{A}_\chi^-$ is of the form

$$\mathbf{A}_\chi^- = \mathbf{L}\begin{pmatrix} \mathbf{J}_1^{-1} & \mathbf{0} & \cdots & \mathbf{0} & \mathbf{K}_1 \\ \mathbf{0} & \mathbf{J}_2^{-1} & \cdots & \mathbf{0} & \mathbf{K}_2 \\ \vdots & \vdots & & \vdots & \vdots \\ \mathbf{0} & \mathbf{0} & \cdots & \mathbf{J}_m^{-1} & \mathbf{K}_m \\ \mathbf{0} & \mathbf{0} & \cdots & \mathbf{0} & \mathbf{0} \end{pmatrix}\mathbf{L}^{-1},$$

where $\mathbf{K}_i$ are certain matrices possibly non-null.

4.3 g-INVERSE OF A WITH $\mathcal{M}(G) \subset \mathcal{M}(A)$ AND $\mathcal{M}(G^*) \subset \mathcal{M}(A^*)$

A g-inverse with coincident row space is denoted by $\mathbf{A}_\rho^-$. $\mathbf{A}_\rho^-$ exists whenever $R(\mathbf{A}) = R(\mathbf{A}^2)$ and has properties similar to those of $\mathbf{A}_\chi^-$ as enunciated in Lemmas 4.2.1 and 4.2.2. For example $\mathbf{A}$ is one choice for $(\mathbf{A}_\rho^-)_\rho^-$, $\mathbf{A}_\rho^- = (\mathbf{A}^2)^-\mathbf{A}$, $(\mathbf{A}_\rho^-)^2\mathbf{A} = \mathbf{A}_\rho^-$, $\mathbf{A}^2\mathbf{A}_\rho^- = \mathbf{A}$ and Lemma 4.2.2 remains true with right eigenvectors replaced by left.

Using Frobenius inequality it is seen that $R(\mathbf{A}) = R(\mathbf{A}^2) \Rightarrow R(\mathbf{A}) = R(\mathbf{A}^m)$ for any positive integer $m \geq 3$. Hence from Lemma 4.1.1 we have

$$\mathbf{A}_{\rho\chi}^- = \mathbf{A}(\mathbf{A}^3)^-\mathbf{A} \tag{4.3.1}$$

as the unique g-inverse with row and column spaces coincident with those of $\mathbf{A}$.

Repeating the argument of the previous section we observe that

$$\mathbf{A}_{\rho\chi}^{-} = \mathbf{L}\begin{pmatrix} \mathbf{J}_1^{-1} & \mathbf{0} & \cdots & \mathbf{0} & \mathbf{0} \\ \mathbf{0} & \mathbf{J}_2^{-1} & \cdots & \mathbf{0} & \mathbf{0} \\ \vdots & \vdots & & \vdots & \vdots \\ \mathbf{0} & \mathbf{0} & \cdots & \mathbf{J}_m^{-1} & \mathbf{0} \\ \mathbf{0} & \mathbf{0} & \cdots & \mathbf{0} & \mathbf{0} \end{pmatrix}\mathbf{L}^{-1}.$$

This shows that $\mathbf{A}_{\rho\chi}^{-}$ exists whenever $R(\mathbf{A}) = R(\mathbf{A}^2)$ and is identical with the Scroggs-Odell (1966) pseudo-inverse (see Section 4.7.2).

$\mathbf{A}_{\rho\chi}^{-}$ *and the principal idempotents of* $\mathbf{A}$. The principal idempotents of a square matrix were studied in Section 2.9. Let $\lambda_0 = 0, \lambda_1, \ldots, \lambda_u$ be the distinct eigenvalues of $\mathbf{A}$. We write $\mathbf{E}_i$ for the idempotent $\mathbf{K}_{\lambda_i}$, $\mathbf{N}_i$ for the nilpotent $(\mathbf{A} - \lambda_i\mathbf{I})\mathbf{K}_{\lambda_i}$, $\mathbf{A}_i$ for $\mathbf{A}\mathbf{K}_{\lambda_i}$, so that $\mathbf{A} = \Sigma\mathbf{A}_i$ and $\mathbf{A}_i = \lambda_i\mathbf{E}_i + \mathbf{N}_i = \lambda_i(\mathbf{E}_i + \mathbf{M}_i)$, where $\mathbf{M}_i = \lambda_i^{-1}\mathbf{N}_i$.

LEMMA 4.3.1 *For every positive integer p,*

$$R(\mathbf{A}^p) = \sum R(\mathbf{A}_i^p).$$

Proof: Since $\mathbf{E}_i$ commutes with $\mathbf{A}$ and $\mathbf{E}_i\mathbf{E}_j = \mathbf{0}\ \forall\, i \neq j$, it is easily seen that

$$\mathbf{A}^p = (\sum \mathbf{A}_i)^p = \sum \mathbf{A}_i^p \tag{4.3.2}$$

$(4.3.2) \Rightarrow R(\mathbf{A}^p) \leq \Sigma\, R(\mathbf{A}_i^p)$

$$(\mathbf{E}_o' \vdots \cdots \vdots \mathbf{E}_u')'\mathbf{A}^p(\mathbf{E}_0 \vdots \mathbf{E}_1 \vdots \cdots \vdots \mathbf{E}_u) = \operatorname{diag}(\mathbf{A}_0^p, \mathbf{A}_1^p, \ldots, \mathbf{A}_u^p)$$

$$\Rightarrow R(\mathbf{A}^p) \geq R[\operatorname{diag}(\mathbf{A}_0^p, \mathbf{A}_1^p, \ldots, \mathbf{A}_u^p)] = \sum R(\mathbf{A}_i^p).$$

Hence $R(\mathbf{A}^p) = \Sigma\, R(\mathbf{A}_i^p)$.

LEMMA 4.3.2 $R(\mathbf{A}) = R(\mathbf{A}^2) \Rightarrow \mathbf{A}_0 = \mathbf{N}_0 = \mathbf{0}$

Proof: We note that in general $R(\mathbf{A}_i^2) \leq R(\mathbf{A}_i)$. Hence $R(\mathbf{A}) = R(\mathbf{A}^2) \Rightarrow R(\mathbf{A}_i) = R(\mathbf{A}_i^2)\ \forall\, i$. However, $\mathbf{A}_0 = \mathbf{N}_0$ is nilpotent. Hence $R(\mathbf{A}_0) = R(\mathbf{A}_0^2) \Rightarrow \mathbf{A}_0 = \mathbf{0}$.

LEMMA 4.3.3 Let $\mathbf{P} = (\mathbf{E}_1 \vdots \cdots \vdots \mathbf{E}_u), \mathbf{Q}^* = (\mathbf{E}_1^* \vdots \cdots \vdots \mathbf{E}_u^*)$. $R(\mathbf{A}) = R(\mathbf{A}^2) \Rightarrow \mathcal{M}(\mathbf{A}) = \mathcal{M}(\mathbf{P}), \mathcal{M}(\mathbf{A}^*) = \mathcal{M}(\mathbf{Q}^*)$.

Proof: $\mathbf{A}_i = \mathbf{E}_i\mathbf{A} = \mathbf{A}\mathbf{E}_i$, $\mathbf{A} = \Sigma\,\mathbf{A}_i \Rightarrow \mathcal{M}(\mathbf{A}) \subset \mathcal{M}(\mathbf{P})$, $\mathcal{M}(\mathbf{A}^*) \subset \mathcal{M}(\mathbf{Q}^*)$. Lemma 4.3.3 therefore follows from $R(\mathbf{A}) = R(\mathbf{P}) = R(\mathbf{Q})$, which is a consequence of Lemmas 4.3.1, 4.3.2 and

$$\mathbf{QAP} = \operatorname{diag}(\mathbf{A}_1, \ldots, \mathbf{A}_u). \tag{4.3.3}$$

LEMMA 4.3.4 $\mathbf{A}_{\rho\chi}^{-} = \Sigma\,(\mathbf{A}_i)_{\rho\chi}^{-}$.

Proof: Lemma 4.3.3 and (4.3.3) ⇒

$$\mathbf{A}_{\rho\chi}^{-} = \mathbf{P}(\mathbf{QAP})^{-}\mathbf{Q} = \sum \mathbf{E}_i\mathbf{A}_i^{-}\mathbf{E}_i.$$

The identification of $\mathbf{G}_i = \mathbf{E}_i\mathbf{A}_i^{-}\mathbf{E}_i$ as $(\mathbf{A}_i)_{\rho\chi}^{-}$ requires straightforward verification of the following:

$$\mathbf{A}_i\mathbf{G}_i\mathbf{A}_i = \mathbf{A}_i,\ \mathscr{M}(\mathbf{E}_i) = \mathscr{M}(\mathbf{A}_i), \qquad \mathscr{M}(\mathbf{E}_i^*) = \mathscr{M}(\mathbf{A}_i^*).$$

LEMMA 4.3.5 $\mathbf{A}_{\rho\chi}^{-} = \Sigma\,\lambda_i^{-1}\{\mathbf{E}_i + \Sigma\,(-\mathbf{M}_i)^k\}$.

Proof: Since $\mathbf{M}_i$ is nilpotent, $\lambda_i\{\mathbf{E}_i + \Sigma\,(-\mathbf{M}_i)^k\}$ involves only a finite number of terms. Direct multiplication shows

$$\lambda_i^{-1}\{\mathbf{E}_i + \sum(-\mathbf{M}_i)^k\}\mathbf{A}_i = \{\mathbf{E}_i + \sum(-\mathbf{M}_i)^k\}\{\mathbf{E}_i + \mathbf{M}_i\} = \mathbf{E}_i$$
$$\Rightarrow \lambda_i^{-1}\{\mathbf{E}_i + \sum(-\mathbf{M}_i)^k\}$$

is a g-inverse of $\mathbf{A}_i$. Further, $\mathscr{M}(\mathbf{E}_i) = \mathscr{M}(\mathbf{A}_i)$, $\mathscr{M}(\mathbf{E}_i^*) = \mathscr{M}(\mathbf{A}_i^*) \Rightarrow \lambda_i^{-1}\{\mathbf{E}_i + \Sigma\,(-\mathbf{M}_i)^k\} = (\mathbf{A}_i)_{\rho\chi}^{-}$.

Note. It is interesting to observe that (4.3.2) could be rewritten as

$$\mathbf{A}^p = \sum \lambda_i^p\left\{\mathbf{E}_i + \sum\binom{p}{k}\mathbf{M}_i^k\right\} \tag{4.3.4}$$

and that a formal substitution of $p = -1$ in (4.3.4) gives $\mathbf{A}_{\rho\chi}^{-}$.

Lemma 4.3.5 and the interesting observation recorded in the note are due to Englefield (1966).

4.4 g-INVERSE WITH THE POWER PROPERTY

We have seen in Lemma 4.2.1 (iii) that, when $\mathbf{G}$ is $\mathbf{A}_\chi^-$

$$\mathbf{A}^m\mathbf{G}^m\mathbf{A}^m = \mathbf{A}^m, \qquad \mathbf{G}^m\mathbf{A}^m\mathbf{G}^m = \mathbf{G}^m \tag{4.4.1}$$

for every positive integer m. The same is also true if $\mathbf{G}$ is $\mathbf{A}_\rho^-$. We show below that the converse of this proposition is not necessarily true, even when $\mathbf{A}_\chi^-$ and $\mathbf{A}_\rho^-$ would otherwise exist unless $R(\mathbf{A}) = n - 1$, where n is the order of the square matrix $\mathbf{A}$. Henceforth we shall find it convenient to refer to a g-inverse satisfying (4.4.1) as an inverse with the power property.

LEMMA 4.4.1 *Let* $\mathbf{A}$ *be a square matrix of order* $n \times n$*, such that* $R(\mathbf{A}) = R(\mathbf{A}^2)$*. Then* $\mathbf{G}$ *is a g-inverse of* $\mathbf{A}$ *with the power property iff*

$$\mathbf{G} = \mathbf{L}\begin{pmatrix}\mathbf{C}^{-1} & \mathbf{J}\\ \mathbf{F} & \mathbf{FCJ}\end{pmatrix}\mathbf{L}^{-1}, \tag{4.4.2}$$

where **C** *and* **L** *are nonsingular matrices connected with* **A** *by the representation*

$$\mathbf{A} = \mathbf{L}\begin{pmatrix} \mathbf{C} & \mathbf{0} \\ \mathbf{0} & \mathbf{0} \end{pmatrix}\mathbf{L}^{-1} \tag{4.4.3}$$

F *and* **J** *are otherwise arbitrary, except that* $\mathbf{JF} = \mathbf{0}$.

Proof: The 'if' part is trivial. To prove the 'only if' part we observe as in the proof of Lemma 4.2.2 that the representation (4.4.3) is valid whenever $R(\mathbf{A}) = R(\mathbf{A}^2)$. The class of g-inverses of **A** has therefore the representation

$$\mathbf{G} = \mathbf{L}\begin{pmatrix} \mathbf{C}^{-1} & \mathbf{J} \\ \mathbf{F} & \mathbf{H} \end{pmatrix}\mathbf{L}^{-1},$$

where **J**, **F**, and **H** are arbitrary matrices. It is a simple matter to check that $\mathbf{G} = \mathbf{GAG} \Rightarrow \mathbf{H} = \mathbf{FCJ}$. Further, $\mathbf{A}^2 = \mathbf{A}^2\mathbf{G}^2\mathbf{A}^2 \Rightarrow \mathbf{JF} = \mathbf{0}$.

COROLLARY: *Let* **A** *be a square matrix of order* $n \times n$, *such that* $R(\mathbf{A}) = R(\mathbf{A}^2) = n - 1$. *Then* **G** *is a g-inverse of* **A** *with the power property iff* **G** *is either* $\mathbf{A}_\chi^-$ *or* $\mathbf{A}_\rho^-$ *or both.*

Proof: The corollary follows from Lemma 4.4.1 since, when $R(\mathbf{A}) = n - 1$, $\mathbf{JF} = \mathbf{0} \Rightarrow \mathbf{J} = \mathbf{0}$ or $\mathbf{F} = \mathbf{0}$ or both $\Rightarrow \mathbf{H} = \mathbf{FCJ} = \mathbf{0}$.

Even though a g-inverse of **A** with the power property is not necessarily of the type $\mathbf{A}_\chi^-$ or $\mathbf{A}_\rho^-$, it enjoys to a certain extent the eigenvalue property true for such inverses. Compare Lemma 4.4.2 with Lemma 4.2.2 (i) in this connection.

LEMMA 4.4.2 *If* **G** *is a g-inverse of* **A** *with the power property, then every nonnull eigenvalue of* **G** *is the reciprocal of an eigenvalue of* **A** *and vice versa.*

Proof: Let $\lambda^k + a_1\lambda^{k-1} + \cdots + a_k$ be the minimum polynomial for **G**. Then

$$\mathbf{A}^k(\mathbf{G}^k + a_1\mathbf{G}^{k-1} + \cdots + a_k\mathbf{I})\mathbf{A}^k = \mathbf{0}.$$

Hence using $\mathbf{A}^m\mathbf{G}^m\mathbf{A}^m = \mathbf{A}^m$ we have

$$\mathbf{A}^k + a_1\mathbf{A}^{k+1} + \cdots + a_k\mathbf{A}^{2k} = \mathbf{0}. \tag{4.4.4}$$

Let **x** be an eigenvector of **A** corresponding to its nonnull eigenvalue λ, then

$$\lambda^k(1 + a_1\lambda + \cdots + a_k\lambda^k)\mathbf{x} = \mathbf{0} \tag{4.4.5}$$

which implies $1 + a_1\lambda + \cdots + a_k\lambda^k = 0$, since $\lambda^k \neq 0$ and $\mathbf{x} \neq \mathbf{0}$. Hence $\mu = 1/\lambda$ satisfies the equation

$$\mu^k + a_1\mu^{k-1} + \cdots + a_k = 0. \tag{4.4.6}$$

This suggests $1/\lambda$ is an eigenvalue of $\mathbf{G}$. If (4.4.1) holds good for all positive integers m, every nonnull eigenvalue of $\mathbf{A}$ is the reciprocal of an eigenvalue of $\mathbf{G}$, and vice versa.

4.5 A g-INVERSE THAT COMMUTES WITH THE MATRIX

Observe now that, since $\mathbf{A}_{\rho\chi}^{-}$ is both $\mathbf{A}_{\rho}^{-}$ and $\mathbf{A}_{\chi}^{-}$

$$\mathbf{A}(\mathbf{A}_{\rho\chi}^{-})^2\mathbf{A} = \mathbf{A}_{\rho\chi}^{-}\mathbf{A} = \mathbf{A}\mathbf{A}_{\rho\chi}^{-}, \tag{4.5.1}$$

using Lemma 4.2.1 (ii) and the corresponding result for $\mathbf{A}_{\rho}^{-}$. This shows $\mathbf{A}_{\rho\chi}^{-}$ commutes with $\mathbf{A}$. Also, if $\mathbf{G}$ is a g-inverse that commutes with $\mathbf{A}$, we have $\mathbf{A} = \mathbf{AGA} = \mathbf{A}^2\mathbf{G} \Rightarrow R(\mathbf{A}) = R(\mathbf{A}^2)$. Thus we have:

LEMMA 4.5.1 *A necessary and sufficient condition for the existence of a commuting g-inverse of* $\mathbf{A}$ *is that*

$$R(\mathbf{A}) = R(\mathbf{A}^2). \tag{4.5.2}$$

LEMMA 4.5.2 *The most general form of a commuting g-inverse of* $\mathbf{A}$ *is given by*

$$\mathbf{G} = \mathbf{A}_{\rho\chi}^{-} + (\mathbf{I} - \mathbf{A}_{\rho\chi}^{-}\mathbf{A})\mathbf{D}(\mathbf{I} - \mathbf{A}\mathbf{A}_{\rho\chi}^{-}), \tag{4.5.3}$$

where $\mathbf{D}$ *is an arbitrary* $n \times m$ *matrix.*

Proof: Check that with $\mathbf{G}$, as defined in the Lemma 4.5.2, we have $\mathbf{AG} = \mathbf{A}\mathbf{A}_{\rho\chi}^{-} = \mathbf{A}_{\rho\chi}^{-}\mathbf{A} = \mathbf{GA}$. This shows $\mathbf{G}$ is a commuting g-inverse.

Conversely let $\mathbf{A}_{\rho\chi}^{-} + \mathbf{E}$ be a commuting g-inverse. This implies

$$\mathbf{A}^2(\mathbf{A}_{\rho\chi}^{-} + \mathbf{E}) = \mathbf{A} = \mathbf{A}^2\mathbf{A}_{\rho\chi}^{-} \Rightarrow \mathbf{A}^2\mathbf{E} = \mathbf{0} \Rightarrow \mathbf{AE} = \mathbf{0},$$

since $R(\mathbf{A}) = R(\mathbf{A}^2)$. Similarly, we have $\mathbf{EA} = \mathbf{0}$, which shows that

$$\mathbf{E} = (\mathbf{I} - \mathbf{A}_{\rho\chi}^{-}\mathbf{A})\mathbf{D}(\mathbf{I} - \mathbf{A}\mathbf{A}_{\rho\chi}^{-})$$

for some $\mathbf{D}$. Note that in this expression for $\mathbf{E}$ one could have used any other g-inverse of $\mathbf{A}$ in place of $\mathbf{A}_{\rho\chi}^{-}$. Further, if $\mathbf{E}$ is nonnull, following an argument similar to that of Theorem 2.7.1, we note that $R(\mathbf{A}_{\rho\chi}^{-} + \mathbf{E}) > R(\mathbf{A}_{\rho\chi}^{-}) = R(\mathbf{A})$. This shows that $\mathbf{A}_{\rho\chi}^{-}$ is indeed the unique commuting reflexive g-inverse of $\mathbf{A}$.

4.6 A g-INVERSE IN THE SUBALGEBRA GENERATED BY THE MATRIX

LEMMA 4.6.1 *$\mathbf{A}_{\rho\chi}^{-}$ is a polynomial of finite degree in $\mathbf{A}$.*

Proof: Let the polynomial equation of minimum degree (p) satisfied by $\mathbf{A}$ be written in the form

$$b_0\mathbf{I} + b_1\mathbf{A} + b_2\mathbf{A}^2 + \cdots + b_p\mathbf{A}^p = \mathbf{0}, \tag{4.6.1}$$

where $b_p \neq 0$. If $b_0 \neq 0$, this equation implies

$$-\frac{1}{b_0}(b_1\mathbf{I} + b_2\mathbf{A} + \cdots + b_p\mathbf{A}^{p-1}) = \mathbf{A}^{-1}.$$

Hence if $\mathbf{A}$ is singular, one must have $b_0 = 0$. Let r be the smallest integer for which $b_r \neq 0$. Putting $a_i = -b_i/b_r (i \geq r + 1)$, we can rewrite the above equation in the form

$$\mathbf{A}^r = a_{r+1}\mathbf{A}^{r+1} + a_{r+2}\mathbf{A}^{r+2} + \cdots + a_p\mathbf{A}^p, \tag{4.6.2}$$

where clearly $r \geq 1$. Multiplying both sides by $(\mathbf{A}_{\rho\chi}^{-})^{r+1}$, we have

$$\mathbf{A}_{\rho\chi}^{-} = a_{r+1}\mathbf{A}\mathbf{A}_{\rho\chi}^{-} + a_{r+2}\mathbf{A} + \cdots + a_p\mathbf{A}^{p-r-1}$$

Hence

$$\mathbf{A}\mathbf{A}_{\rho\chi}^{-} = a_{r+1}\mathbf{A} + a_{r+2}\mathbf{A}^2 + \cdots + a_p\mathbf{A}^{p-r}$$

and

$$\mathbf{A}_{\rho\chi}^{-} = (a_{r+1}^2 + a_{r+2})\mathbf{A} + (a_{r+1}a_{r+2} + a_{r+3})\mathbf{A}^2$$
$$+(a_{r+1}a_{p-1} + a_p)\mathbf{A}^{p-r-1} + a_p\mathbf{A}^{p-r}. \tag{4.6.3}$$

This completes proof of Lemma 4.6.1.

Incidentally, $\mathbf{A} = \mathbf{A}\mathbf{A}_{\rho\chi}^{-}\mathbf{A}$ is a polynomial equation of degree $p - r + 1$ satisfied by $\mathbf{A}$, which contradicts our assumption regarding the degree of the minimum equation for $\mathbf{A}$ unless $r = 1$. Hence, $r = 1$ and one may rewrite the minimum equation in the form $\mathbf{A} = \mathbf{A}^2P(\mathbf{A})$, where

$$P(\mathbf{A}) = a_2\mathbf{I} + a_3\mathbf{A} + \cdots + a_p\mathbf{A}^{p-2}. \tag{4.6.4}$$

We have

$$\mathbf{A}_{\rho\chi}^{-} = P(\mathbf{A}) - a_2[\mathbf{I} - \mathbf{A}P(\mathbf{A})].$$

LEMMA 4.6.2 *Let $\mathscr{S}(\mathbf{A})$ denote the subalgebra generated by $\mathbf{A}$. A necessary and sufficient condition for $\mathscr{S}(\mathbf{A})$ to contain a g-inverse of $\mathbf{A}$ is that $R(\mathbf{A}) = R(\mathbf{A}^2)$.*

Proof: Since each member of $\mathscr{S}(\mathbf{A})$ commutes with $\mathbf{A}$, the necessity part is seen to follow from Lemma 4.5.1. Lemma 4.6.1 shows the condition is sufficient.

LEMMA 4.6.3 *If $R(\mathbf{A}) = R(\mathbf{A}^2)$ each g-inverse in $\mathscr{S}(\mathbf{A})$ can be expressed in the form*

$$P(\mathbf{A}) + c[\mathbf{I} - \mathbf{A}P(\mathbf{A})],$$

where c is a scalar and $P(\mathbf{A})$ is as defined in (4.6.4).

It is easy to establish the result of lemma 4.6.3.

Computation of the characteristic polynomial of a matrix (Leverrier-Faddeev method): The computation of $\mathbf{A}_{\rho\chi}^- = \mathbf{A}(\mathbf{A}^3)^-\mathbf{A}$, which gives just one g-inverse in $\mathscr{S}(\mathbf{A})$ is fairly straightforward and does not require the knowledge of either the minimum polynomial or the characteristic polynomial of $\mathbf{A}$. If, however, the explicit polynomial representation of $\mathbf{A}_{\rho\chi}^-$ is required, one can use Lemma 4.6.1, which incidentally presupposes a knowledge of the minimum polynomial. Unless one is interested in obtaining a representation of the least possible degree, the use of the minimum polynomial could be avoided. We show below how one could use the characteristic polynomial for this purpose.

Let us write the characteristic polynomial of $\mathbf{A}$ in the form

$$\lambda^n - p_1\lambda^{n-1} - \cdots - p_n, \tag{4.6.5}$$

where $p_1 = tr_1\mathbf{A}$, $p_2 = -tr_2\mathbf{A}, \ldots, p_n = (-1)^{n-1}tr_n\mathbf{A} = (-1)^{n-1}|\mathbf{A}|$. If $\mathbf{A}$ is singular, $p_n = 0$. If, however, $p_{n-1} \neq 0$, in view of Cayley-Hamilton theorem, one can write

$$\mathbf{A} = p_{n-1}^{-1}(\mathbf{A}^n - p_1\mathbf{A}^{n-1} - \cdots - p_{n-2}\mathbf{A}^2), \tag{4.6.6}$$

which shows that

$$p_{n-1}^{-1}(\mathbf{A}^{n-2} - p_1\mathbf{A}^{n-3} - \cdots - p_{n-2}\mathbf{I}) \tag{4.6.7}$$

is a g-inverse of $\mathbf{A}$, though not equal to $\mathbf{A}_{\rho\chi}^-$, unless $p_{n-2} = 0$. If $p_{n-2} \neq 0$, one could use (4.6.6) itself to express the right-hand side as a polynomial not involving the $\mathbf{A}^2$ term. We have thus

$$\mathbf{A} = \frac{1}{p_{n-1}^2}[-p_{n-2}\mathbf{A}^{n+1} + (p_{n-1} + p_1p_{n-2})\mathbf{A}^n - (p_1p_{n-1} - p_2p_{n-2})\mathbf{A}^{n-1}$$
$$- \cdots - (p_{n-1}p_{n-3} - p_{n-2}^2)\mathbf{A}^3].$$

Observe that

$$\mathbf{A}_{\rho\chi}^- = \frac{1}{p_{n-1}^2}[-p_{n-2}\mathbf{A}^{n-1} + (p_{n-1} + p_1p_{n-2})\mathbf{A}^{n-2} - (p_1p_{n-1} - p_2p_{n-2})\mathbf{A}^{n-3}$$
$$- \cdots - (p_{n-1}p_{n-3} - p_{n-2}^2)\mathbf{A}].$$

If $p_{n-1} = 0$, let k be the smallest positive integer for which $p_{n-k} \neq 0$. We write

$$\mathbf{A}^k = \frac{1}{p_{n-k}}(\mathbf{A}^n - p_1\mathbf{A}^{n-1} - \cdots - p_{n-k-1}\mathbf{A}^{k+1}) \tag{4.6.8}$$

and proceeding as above, obtain a g-inverse of $\mathbf{A}^k$, which can be represented as a polynomial $p(\mathbf{A})$ of finite degree in $\mathbf{A}$. If conditions of Lemma 4.6.2 are satisfied, we have $\mathbf{A}_{\rho\chi}^- = \mathbf{A}^{k-1}p(\mathbf{A})$.

One can use any method convenient to compute the characteristic polynomial. We describe here an elegant method due to Leverrier and Faddeev. Define $\mathbf{A}_i$ recursively by the relations

$$\mathbf{A}_1 = \mathbf{A}$$

$$\mathbf{A}_i = \mathbf{A}(\mathbf{A}_{i-1} - q_{i-1}\mathbf{I}) \quad \text{for } i \geq 2,$$

where $q_i = tr\mathbf{A}_i/i$.

Using Newton's identities connecting the coefficients p_i of a polynomial $\lambda^n - p_1\lambda^{n-1} - \cdots - p_n$ with the sum s_k of the k-th power of the roots of the same polynomial, namely

$$kp_k = s_k - s_{k-1}p_1 - \cdots - s_1p_{k-1}$$

it is a simple matter to check that $q_i = p_i$ $(i = 1, 2, \ldots, n)$. Also, if $\mathbf{B}_i = \mathbf{A}_i - q_i\mathbf{I}$, then $\mathbf{B}_i = \mathbf{0}$ for $i \geq n$. Hence, $q_n = p_n \neq 0$, $\mathbf{AB}_{n-1} = q_n\mathbf{I} \Rightarrow \mathbf{B}_{n-1}/q_n = \mathbf{A}^{-1}$. If $q_n = p_n = 0$, but $q_{n-1} = p_{n-1} \neq 0$, $\mathbf{B}_{n-2}/q_{n-1}$ is a g-inverse of $\mathbf{A}$ and is in fact identical with (4.6.7).

As already observed, whenever $\mathbf{A}$ is hermitian $\mathbf{A}_{\rho\chi}^-$ exists and coincides with $\mathbf{A}^+$. Further if $\mathbf{A}$ is n.n.d., $R(\mathbf{A}) = n - k$ if and only if $p_{n-k} \neq 0$ and $p_{n-i} = 0$ for $i = 0, 1, 2, \ldots, k - 1$.

This gives us a neat method of computing $\mathbf{B}^+$, where $\mathbf{B}$ is any arbitrary matrix. The method is as follows. Compute $\mathbf{A} = \mathbf{B}^*\mathbf{B}$ and observe that it is n.n.d. Obtain $\mathbf{A}^+$ by the above method. Postmultiply $\mathbf{A}^+$ by $\mathbf{B}^*$ to get $\mathbf{B}^+$.

4.7 g-INVERSE WITH THE EIGENVALUE PROPERTY

4.7.1 Cyclic Decomposition of a Linear Space

Let $\mathbf{A}$ be a square matrix of order n and $\mathbf{x}_1$ an arbitrary n-vector. The vectors $\mathbf{x}_1, \mathbf{Ax}_1, \ldots, \mathbf{A}^n\mathbf{x}_1$ obviously cannot be linearly independent. Let k be the largest integer for which $\mathbf{x}_1, \mathbf{x}_2 = \mathbf{Ax}_1, \ldots, \mathbf{x}_k = \mathbf{A}^{k-1}\mathbf{x}_1$ are linearly independent. By implication,

$$\mathbf{A}\mathbf{x}_k = a_k\mathbf{x}_k + a_{k-1}\mathbf{x}_{k-1} + \cdots + a_1\mathbf{x}_1. \tag{4.7.1}$$

We shall consider two mutually exclusive situations: (I) $\mathbf{A}\mathbf{x}_k = \mathbf{0}$, (II) $\mathbf{A}\mathbf{x}_k \neq \mathbf{0}$. In the latter case we shall assume without any loss of generality that $a_1 \neq 0$.† The linear space generated by $\mathbf{x}_1, \mathbf{x}_2, \ldots, \mathbf{x}_k$ is called the cyclic subspace with generator $\mathbf{x}_1$. Corresponding to (I) and (II) above, we have therefore two types of cyclic subspaces, cyclic subspaces of type (I) where for the generator $\mathbf{x}_1$ we have $\mathbf{A}^k\mathbf{x}_1 = \mathbf{0}$ and those of type (II), where $\mathbf{A}^k\mathbf{x}_1 \neq \mathbf{0}\ \forall\, k$. Observe that if the subspace $\mathscr{S}$ is cyclically generated by $\mathbf{x}_1$, then

$$\mathbf{A}\mathbf{X}_1 = \mathbf{X}_1\mathbf{D}_1, \tag{4.7.2}$$

where the matrix $\mathbf{X}_1(n \times k) = (\mathbf{x}_1 \vdots \mathbf{x}_2 \vdots \cdots \vdots \mathbf{x}_k)$ and

$$\mathbf{D}_1 = \begin{pmatrix} 0 & 0 & 0 & \cdots & 0 & 0 \\ 1 & 0 & 0 & \cdots & 0 & 0 \\ 0 & 1 & 0 & \cdots & 0 & 0 \\ \cdot & \cdot & \cdot & & \cdot & \cdot \\ \cdot & \cdot & \cdot & & \cdot & \cdot \\ \cdot & \cdot & \cdot & & \cdot & \cdot \\ 0 & 0 & 0 & & 1 & 0 \end{pmatrix}$$

if $\mathscr{S}$ is of type (I), and

$$\mathbf{D}_1 = \begin{pmatrix} 0 & 0 & 0 & \cdots & 0 & a_1 \\ 1 & 0 & 0 & \cdots & 0 & a_2 \\ 0 & 1 & 0 & \cdots & 0 & a_3 \\ \cdot & \cdot & \cdot & & \cdot & \cdot \\ \cdot & \cdot & \cdot & & \cdot & \cdot \\ \cdot & \cdot & \cdot & & \cdot & \cdot \\ 0 & 0 & 0 & \cdots & 1 & a_k \end{pmatrix}$$

if $\mathscr{S}$ is of type (II). Since here $|\mathbf{D}_1| = (-1)^{k+1}a_1 \neq 0$, we have the following result.

LEMMA 4.7.1 *If the cyclic subspace $\mathscr{S}$ is of type (II), the operator* $\mathbf{A}$ *restricted to $\mathscr{S}$ is a nonsingular operator.*

† For, if $a_1 = 0$, let i be smallest integer for which $a_{i+1} \neq 0$. Then choosing $\mathbf{x}_{i+1}$ as the initial vector $\mathbf{y}_1$ and proceeding as before we may cyclically generate $k - i$ vectors: $\mathbf{y}_1, \mathbf{y}_2 = \mathbf{A}\mathbf{y}_1, \ldots, \mathbf{y}_{k-i} = \mathbf{A}\mathbf{y}_{k-i-1}$ such that

$$\mathbf{A}\mathbf{y}_{k-i} = b_{k-i}\mathbf{y}_{k-i} + \cdots + b_1\mathbf{y}_1$$

and $b_i = a_{i+1} \neq 0$.

The matrix $\mathbf{D}_1$ in either case is called the companion matrix of the corresponding subspace, relative to the chosen basis.

LEMMA 4.7.2 *If a subspace $\mathscr{S}$ is invariant under* $\mathbf{A}$, *$\mathscr{S}$ is a cyclic subspace of type (II).*

Proof: For a proof of this lemma see Pease (1965, Theorem 2, p. 198) which in fact establishes the existence of a cyclic generator of $\mathscr{S}$ of type II.

LEMMA 4.7.3 *If $\mathscr{S}$ and $\mathscr{T}$ are cyclic subspaces, of types (I) and (II) respectively, then $\mathscr{S}$ and $\mathscr{T}$ are virtually disjoint, that is, have only the null vector in common.*

Proof: Let $\mathbf{y}$ be a nonnull vector common to $\mathscr{S}$ and $\mathscr{T}$. By Lemma 4.7.1 $\mathbf{y} \in \mathscr{T} \Rightarrow \mathbf{A}^i\mathbf{y} \neq \mathbf{0}$ for every positive integer i. On the other hand $\mathbf{y} \in \mathscr{S} \Rightarrow \mathbf{A}^i\mathbf{y} = \mathbf{0}$ for some positive integer i. This is a contradiction.

LEMMA 4.7.4 *If $\mathscr{S}$ and $\mathscr{T}$ are cyclic subspaces both of type (II), then so is $\mathscr{S} + \mathscr{T}$.*

Proof: We first show that $\mathscr{S} \cap \mathscr{T}$ is a cyclic subspace of type (II), for which observe that $\mathbf{x} \in \mathscr{S} \cap \mathscr{T} \Rightarrow \mathbf{Ax} \in \mathscr{S}$. Also $\mathbf{x} \in \mathscr{S} \cap \mathscr{T} \Rightarrow \mathbf{x} \in \mathscr{T} \Rightarrow \mathbf{Ax} \in \mathscr{T}$. Hence $\mathbf{x} \in \mathscr{S} \cap \mathscr{T} \Rightarrow \mathbf{Ax} \in \mathscr{S} \cap \mathscr{T}$. Since $\mathscr{S} \cap \mathscr{T} \subset \mathscr{S}$, $\mathbf{A}$ is a nonsingular operator in $\mathscr{S} \cap \mathscr{T}$. This shows $\mathscr{S} \cap \mathscr{T}$ is invariant under $\mathbf{A}$, and hence a cyclic subspace of type (II) by lemma 4.7.2.

Let $\mathbf{x} \in \mathscr{S} + \mathscr{T}$. We can express $\mathbf{x} = \mathbf{y} + \mathbf{z} \ni \mathbf{y} \in \mathscr{S}, \mathbf{z} \in \mathscr{T}$. This implies $\mathbf{Ax} = \mathbf{Ay} + \mathbf{Az}$ where $\mathbf{Ay} \in \mathscr{S}$, $\mathbf{Az} \in \mathscr{T}$. Hence $\mathbf{Ax} \in \mathscr{S} + \mathscr{T}$. $\mathbf{Ax} = \mathbf{0} \Rightarrow \mathbf{Ay} = -\mathbf{Az} \in \mathscr{S} \cap \mathscr{T} \Rightarrow \mathbf{y} = -\mathbf{z} \in \mathscr{S} \cap \mathscr{T} \Rightarrow \mathbf{x} = \mathbf{y} + \mathbf{z} = \mathbf{0}$. This shows $\mathscr{S} + \mathscr{T}$ is invariant under $\mathbf{A}$ and hence a cyclic subspace of type (II) by lemma 4.7.2.

LEMMA 4.7.5 *Let $\mathscr{S}_1, \mathscr{S}_2, \ldots, \mathscr{S}_m$ be virtually disjoint cyclic subspaces each of type (I) and $\mathscr{T}$ be a cyclic subspace of type (I) not wholly contained in $\mathscr{S}_1 \oplus \mathscr{S}_2 \oplus \cdots \oplus \mathscr{S}_m$. Then, one of the following is true,*

(a) *for some i, $1 \leq i \leq m$, $\mathscr{S}_i \subset \mathscr{T}$ in which case*

$$\mathscr{S}_1 \oplus \mathscr{S}_2 \oplus \cdots \oplus \mathscr{S}_m + \mathscr{T} = \mathscr{S}_1 \oplus \cdots \oplus \mathscr{S}_{i-1} \oplus \mathscr{S}_{i+1} \oplus \cdots \oplus \mathscr{S}_m + \mathscr{T}$$

(b) *$\mathscr{S}_1 \oplus \mathscr{S}_2 \oplus \cdots \oplus \mathscr{S}_m + \mathscr{T}$ can be expressed as the direct sum of $m + 1$ cyclic subspaces each of type (I).*

Proof: We indicate a proof of the lemma for the case $m = 2$, the general case being similar except for details. Let $\mathscr{S}_1, \mathscr{S}_2$ and $\mathscr{T}$ be of dimensions s_1, s_2 and t with $\mathbf{x}_1, \mathbf{x}_2$ and $\mathbf{y}$ as their respective cyclic generators. The (a) part of lemma 4.7.5 is easy. Also the (b) part is trivially true if $\mathscr{T}$ is virtually disjoint with $\mathscr{S}_1 \oplus \mathscr{S}_2$. Let us therefore consider the case where $\{\mathscr{S}_1 \oplus \mathscr{S}_2\} \cap \mathscr{T}$ is of a positive dimension d. The intersection is clearly a cyclic subspace of type

(I) cyclically generated by $\mathbf{A}^{t-d}\mathbf{y}$, so that we can write

$$\mathbf{A}^{t-d}y = c_1\mathbf{A}^{\alpha_1}\boldsymbol{\xi}_1 + c_2\mathbf{A}^{\alpha_2}\boldsymbol{\xi}_2$$

where ξ_i is a cyclic generator of $\mathscr{S}_i$ not necessarily identical with x_i and α_i is nonnegative integer $\leq s_i$. Further, for each i, $\alpha_i \geq s_i - d$ and the equality holds for at least one i. We assume that the subspaces $\mathscr{S}_i$ are so numbered that $\alpha_1 \leq \alpha_2$, and consider the following four mutually exclusive and exhaustive possibilities.

Case 1: $t - d \leq \alpha_1$. Here, write $\mathbf{u} = \mathbf{y} - c_1\mathbf{A}^{\alpha_1 - t + d}\boldsymbol{\xi}_1 - c_2\mathbf{A}^{\alpha_2 - t + d}\boldsymbol{\xi}_2$, define $\mathscr{U}$ as the cyclic subspace of dimension $t - d$ generated by $\mathbf{u}$ and check that

$$\mathscr{S}_1 \oplus \mathscr{S}_2 + \mathscr{T} = \mathscr{S}_1 \oplus \mathscr{S}_2 \oplus \mathscr{U}$$

Case 2: $t - d > \alpha_1 = s_1 - d$. Hence write $\boldsymbol{\eta}_1 = \mathbf{A}^{t-d-\alpha_1}\mathbf{y} - c_1\boldsymbol{\xi}_1 - c_2\mathbf{A}^{\alpha_2 - \alpha_1}\boldsymbol{\xi}_2$ define $\mathscr{S}'_1$ as the cyclic subspace generated by $\boldsymbol{\eta}_1$ of dimension α_1 (here $= s_1 - d$) and check that

$$\mathscr{S}_1 \oplus \mathscr{S}_2 + \mathscr{T} = \mathscr{S}'_1 \oplus \mathscr{S}_2 \oplus \mathscr{T}$$

Case 3: $t - d > \alpha_1, s_1 > \alpha_1 > s_1 - d, t - d, \leq \alpha_2$. Here put $\alpha_1 - s_1 + d = e$ and observe that

$$\mathbf{A}^{t-e}\mathbf{y} = c_2\mathbf{A}^{\alpha_2 + d - e}\boldsymbol{\xi}_2$$

Put $\mathbf{v} = \mathbf{y} - c_2\mathbf{A}^{\alpha_2 - t + d}\boldsymbol{\xi}_2$ define $\mathscr{V}$ as the cyclic subspace of dimension $t - e$, generated by $\mathbf{v}$ and check that

$$\mathscr{S}_1 \oplus \mathscr{S}_2 + \mathscr{T} = \mathscr{S}'_1 \oplus \mathscr{S}_2 \oplus \mathscr{V}$$

Case 4: $t - d > \alpha_2,\ s_1 > \alpha_1 > s_1 - d,\ \alpha_2 = s_2 - d$. We put $\boldsymbol{\eta}_2 = \mathbf{A}^{t-d-\alpha_2}\mathbf{y} - c_2\boldsymbol{\xi}_2$, define $\mathscr{S}'_2$ as the cyclic subspace generated by $\boldsymbol{\eta}_2$ of dimension $\alpha_2 + d - e$ (here $s_2 - e$) and check that

$$\mathscr{S}_1 \oplus \mathscr{S}_2 + \mathscr{T} = \mathscr{S}'_1 \oplus \mathscr{S}'_2 \oplus \mathscr{T}.$$

This completes proof of lemma 4.7.5.

LEMMA 4.7.6 *If* $\mathbf{x} = \mathbf{X}_1\mathbf{c} \in \mathscr{S}$ *is an eigenvector of* $\mathbf{A}$ *corresponding to its eigenvalue* λ, *then* $\mathbf{c}$ *is an eigenvector of* $\mathbf{D}_1$ *corresponding to the same eigenvalue* λ *of* $\mathbf{D}_1$.

Proof: $\mathbf{Ax} = \mathbf{AX}_1\mathbf{c} = \mathbf{X}_1\mathbf{D}_1\mathbf{c}$

$$= \lambda\mathbf{x} = \mathbf{X}_1(\lambda\mathbf{c}).$$

Since $R(\mathbf{X}_1) = k$, we have $\mathbf{D}_1\mathbf{c} = \lambda\mathbf{c}$. This proves Lemma 4.7.6.

Observe that the characteristic polynomial of $\mathbf{D}_1$ is λ^k if $\mathscr{S}$ is of type (I) and $\lambda^k - a_k\lambda^{k-1} - \cdots - a_1$ if $\mathscr{S}$ is of type (II). λ corresponds to a null eigenvalue, if $\mathscr{S}$ is type (I), and to a nonnull eigenvalue otherwise.

LEMMA 4.7.7 *A cyclic subspace $\mathscr{S}$ always contains an eigenvector of* $\mathbf{A}$.

Proof: If $\mathscr{S}$ is of type (*I*), $\mathbf{x}_k$ is an eigenvector of $\mathbf{A}$, since $\mathbf{A}\mathbf{x}_k = \mathbf{0}$. If $\mathscr{S}$ is of type (II), consider a factorization

$$\lambda^k - a_k\lambda^{k-1} - \cdots - a_1 = (\lambda - \lambda_1)(\lambda - \lambda_2)\ldots(\lambda - \lambda_k) \qquad (4.7.4)$$

Such a factorization into linear factors is always possible in the complex field. Equation (4.7.4) can then be rewritten as

$$(\mathbf{A} - \lambda_1\mathbf{I})(\mathbf{A} - \lambda_2\mathbf{I})\ldots(\mathbf{A} - \lambda_k\mathbf{I})\mathbf{x}_1 = \mathbf{0}. \qquad (4.7.5)$$

Observe that $(\mathbf{A} - \lambda_2\mathbf{I})\ldots(\mathbf{A} - \lambda_k\mathbf{I})\mathbf{x}_1 = \mathbf{y} \neq \mathbf{0}$, for $\mathbf{y} = \mathbf{0}$ would imply that $\mathbf{x}_1, \mathbf{x}_2, \ldots, \mathbf{x}_k$ are linearly dependent which is a contradiction. $\mathbf{y}$ belongs to $\mathscr{S}$ and is an eigenvector of $\mathbf{A}$, corresponding to the eigenvalue λ_1.

It is clear that using lemmas 4.7.4 and 4.7.5, wherever necessary, the whole space can be decomposed into virtually disjoint subspaces each either of type (I) or (II). Thus let $\mathscr{E}^n = \mathscr{S}_1 \oplus \mathscr{S}_2 + \cdots \oplus \mathscr{S}_u$ and $\mathbf{X}_i$ be a matrix of full rank such that $\mathscr{M}(\mathbf{X}_i) = \mathscr{S}_i, i = 1, 2, \ldots, u$. If $\mathscr{S}_i$ is of type (**I**) we assume that columns of $\mathbf{X}_i$ form a basis of $\mathscr{S}_i$ cyclically obtained as in (4.7.2). For subspaces of type (II) this restriction is not essential though we use the same notation $\mathbf{D}_i$ to express the relation $\mathbf{A}\mathbf{X}_i = \mathbf{X}_i\mathbf{D}_i$. Here $\mathbf{D}_i$ need not be a companion matrix in the sense understood earlier. Let us assume that these subspaces have been so renumbered that the first p of them $\mathscr{M}(\mathbf{X}_i), \ldots, \mathscr{M}(\mathbf{X}_p)$ are the type (II), while the rest are of type (I). Check that since

$$\mathscr{E}^n = \mathscr{M}(\mathbf{X}_1) \oplus \mathscr{M}(\mathbf{X}_2) \oplus \cdots \oplus \mathscr{M}(\mathbf{X}_u) = \mathscr{M}(\mathbf{X}),$$

where $\mathbf{X} = (\mathbf{X}_1 \vdots \mathbf{X}_2 \vdots \cdots \vdots \mathbf{X}_u)$, the matrix $\mathbf{X}$ is nonsingular and $\mathbf{A}\mathbf{X} = \mathbf{X}\mathbf{D}$, where

$$\mathbf{D} = \begin{pmatrix} \mathbf{D}_1 & \mathbf{0} & \cdots & \mathbf{0} \\ \mathbf{0} & \mathbf{D}_2 & \cdots & \mathbf{0} \\ \cdot & \cdot & \cdots & \cdot \\ \cdot & \cdot & \cdots & \cdot \\ \mathbf{0} & \mathbf{0} & \cdots & \mathbf{D}_u \end{pmatrix}. \qquad (4.7.6)$$

Let us now consider an arbitrary eigenvector $\mathbf{y}$ of $\mathbf{A}$ corresponding to an eigenvalue λ. Since

$$\mathscr{E}^n = \mathscr{M}(\mathbf{X}_1) \oplus \mathscr{M}(\mathbf{X}_2) \oplus \cdots \oplus \mathscr{M}(\mathbf{X}_u)$$

one can express

$$\mathbf{y} = \mathbf{y}_1 + \mathbf{y}_2 + \cdots + \mathbf{y}_u, \tag{4.7.7}$$

where $\mathbf{y}_i \in \mathcal{M}(\mathbf{X}_i)$ and such a representation is necessarily unique. We shall now prove

LEMMA 4.7.8 *Let $\mathbf{y}_i$ be defined as in (4.7.7). If $\mathbf{y}$ is an eigenvector of $\mathbf{A}$ corresponding to the eigenvalue λ, and $\mathbf{y}_i \neq \mathbf{0}$, then $\mathbf{y}_i$ is also an eigenvector of $\mathbf{A}$ corresponding to the same eigenvalue λ.*

Proof: If $\lambda \neq 0$, observe

$$\mathbf{y} = \frac{1}{\lambda}\mathbf{A}\mathbf{y} = \frac{1}{\lambda}(\mathbf{A}\mathbf{y}_1 + \cdots + \mathbf{A}\mathbf{y}_u).$$

Uniqueness of representation (4.7.7) implies $\mathbf{y}_i = \lambda^{-1}\mathbf{A}\mathbf{y}_i$. If $\lambda = 0$, $\mathbf{A}\mathbf{y} = \mathbf{0} \Rightarrow \mathbf{A}\mathbf{y}_1 + \mathbf{A}\mathbf{y}_2 + \cdots + \mathbf{A}\mathbf{y}_u = \mathbf{0} \Rightarrow \mathbf{A}\mathbf{y}_i = \mathbf{0}\ \forall\, i$, for otherwise $\mathbf{X}$ cannot be a nonsingular matrix. Proof of Lemma 4.7.8 is thus concluded.

From the discussion following the proof of Lemma 4.7.6 it is clear that $\lambda = 0 \Rightarrow \mathbf{y}_i = \mathbf{0}\ \forall\, i = 1, 2, \ldots, p$ and $\lambda \neq 0 \Rightarrow \mathbf{y}_i = \mathbf{0}\ \forall\, i = p+1, p+2, \ldots, u$.

4.7.2 g-Inverse with the Eigenvalue Property

For the matrix $\mathbf{D}$, as defined by (4.7.6), consider the Moore-Penrose inverse $\mathbf{D}^+$ which is easily seen to be as follows:

$$\mathbf{D}^+ = \begin{pmatrix} \mathbf{D}_1^{-1} & \mathbf{0} & \mathbf{0} & \mathbf{0} & \mathbf{0} \\ \mathbf{0} & \mathbf{D}_2^{-1} & \mathbf{0} & \mathbf{0} & \mathbf{0} \\ \cdot & \cdot & \cdot & \cdot & \cdot \\ \mathbf{0} & \mathbf{0} & \mathbf{D}_p^{-1} & \mathbf{0} & \mathbf{0} \\ \mathbf{0} & \mathbf{0} & \mathbf{0} & \mathbf{D}_{p+1}^* & \mathbf{0} \\ \cdot & \cdot & \cdot & \cdot & \cdot \\ \mathbf{0} & \mathbf{0} & \mathbf{0} & \mathbf{0} & \mathbf{D}_u^* \end{pmatrix}.$$

$\mathbf{X}\mathbf{D}^+\mathbf{X}^{-1}$ is a g-inverse of $\mathbf{A} = \mathbf{X}\mathbf{D}\mathbf{X}^{-1}$, and from Lemma 4.7.6 and Lemma 4.7.8 it is clear that the g-inverse $\mathbf{X}\mathbf{D}^+\mathbf{X}^{-1}$ has the eigenvalue property in the sense that, if $\mathbf{y}$ is an eigenvector of $\mathbf{A}$ corresponding to a nonnull eigenvalue λ, it is also an eigenvector of $\mathbf{A}_{\mathbf{X}}^{\oplus} = \mathbf{X}\mathbf{D}^+\mathbf{X}^{-1}$ corresponding to its eigenvalue $1/\lambda$. This is because, if $\lambda \neq 0$, $\mathbf{c}$, as defined in Lemma 4.7.6, is also an eigenvector of $\mathbf{D}_1^{-1}$ corresponding to its eigenvalue $1/\lambda$.

Also, if $\mathbf{Y}^* = \mathbf{X}^{-1}$, one could take $\mathbf{A}^*\mathbf{Y} = \mathbf{Y}\mathbf{D}^*$, where $\mathbf{D}^* = \operatorname{diag}(\mathbf{D}_1^*, \mathbf{D}_2^*, \ldots, \mathbf{D}_u^*)$. This shows that what has been claimed about right eigenvectors of $\mathbf{A}$ and $\mathbf{X}\mathbf{D}^+\mathbf{X}^{-1}$ would also be true about their left eigenvectors.

$\mathbf{D}^+$ involves a special choice $\mathbf{D}_i^+ = \mathbf{D}_i^*$ for the g-inverse of the singular companion matrix $\mathbf{D}_i$, $i = p+1, p+2, \ldots, u$. It should also be clear that no matter how $\mathbf{D}_i^-$ is chosen in such cases, an eigenvector of $\mathbf{D}_i$, which necessarily has to correspond to a zero eigenvalue, can never be an eigenvector of $\mathbf{D}_i^-$.

Put $\mathbf{E}_1 = \text{diag}(\mathbf{D}_1, \mathbf{D}_2, \ldots, \mathbf{D}_p)$ and $\mathbf{E}_2 = \text{diag}(\mathbf{D}_{p+1}, \mathbf{D}_{p+2}, \ldots, \mathbf{D}_u)$. From the discussion in the preceding paragraphs it is seen that the following result is true.

LEMMA 4.7.9 *A g-inverse* $\mathbf{A}^-$ *of* $\mathbf{A}$ *has the eigenvalue property in respect of right and left eigenvectors of* $\mathbf{A}$ *if it can be expressed as*

$$\mathbf{A}^- = \mathbf{X}\mathbf{D}^-\mathbf{X}^{-1} \tag{4.7.8}$$

where $\mathbf{D}^- = \text{diag}(\mathbf{E}_1^{-1}, \mathbf{E}_2^-)$, *where* $\mathbf{E}_2^-$ *is any arbitrary g-inverse of* $\mathbf{E}_2$.

A g-inverse of $\mathbf{A}$ with the eigenvalue property is denoted by $\mathbf{A}_e^-$.

Numerical Illustration. Let us consider

$$\mathbf{A} = \begin{pmatrix} 4 & 2 & -3 \\ -2 & 0 & 1 \\ 2 & 2 & 2 \end{pmatrix}.$$

Take $\mathbf{x}_1' = (0, 1, 0)$, $(\mathbf{A}\mathbf{x}_1)' = \mathbf{x}_2' = (2, 0, 2)$, $\mathbf{x}_3' = (\mathbf{A}\mathbf{x}_2)' = (2, -2, 0)$ so that $\mathbf{x}_1, \mathbf{x}_2$ and $\mathbf{x}_3$ are clearly linearly independent. Observe that $\mathbf{x}_3$ is an eigenvector of $\mathbf{A}$, corresponding to eigenvalue 2. Consider now $\mathbf{y}_1' = \mathbf{x}_2' - 2\mathbf{x}_1' = (2, -2, 2)$ and observe $\mathbf{y}_2' = \mathbf{x}_3' - 2\mathbf{x}_2' = (-2, 2, 4)$. Obviously, $\mathbf{A}\mathbf{y}_2 = \mathbf{0}$. Choosing

$$\mathbf{X}_1 = \begin{pmatrix} 1 \\ -1 \\ 0 \end{pmatrix}, \qquad \mathbf{X}_2 = \begin{pmatrix} 1 & -1 \\ -1 & -1 \\ 1 & -2 \end{pmatrix}$$

we have

$$\mathbf{A}_{\mathbf{X}} = \mathbf{X}\mathbf{D}^+\mathbf{X}^{-1} = \begin{pmatrix} 1 & 1 & -1 \\ -1 & -1 & -1 \\ 0 & 1 & -2 \end{pmatrix} \begin{pmatrix} 1/2 & 0 & 0 \\ 0 & 0 & 1 \\ 0 & 0 & 0 \end{pmatrix} \begin{pmatrix} 1 & 1 & -1 \\ -1 & -1 & -1 \\ 0 & 1 & -2 \end{pmatrix}^{-1}$$

$$= \begin{pmatrix} 1/2 & 0 & 1 \\ -1/2 & 0 & -1 \\ 0 & 0 & 1 \end{pmatrix} \begin{pmatrix} 1 & 1 & -1 \\ -1 & -1 & -1 \\ 0 & 1 & -2 \end{pmatrix}^{-1}$$

$$= \begin{pmatrix} 1/2 & 0 & 1 \\ -1/2 & 0 & -1 \\ 0 & 0 & 1 \end{pmatrix} \begin{pmatrix} 3/2 & 1/2 & -1 \\ -1 & -1 & 1 \\ -1/2 & -1/2 & 0 \end{pmatrix}$$

$$= \begin{pmatrix} 1/4 & -1/4 & -1/2 \\ -1/4 & 1/4 & 1/2 \\ -1/2 & -1/2 & 0 \end{pmatrix}.$$

The approach in this section is parallel to that of Scroggs and Odell (1966) except for two major differences. First, while in the canonical form of the matrix $\mathbf{A}$, considered by these authors the diagonal cells are Jordan matrices, the canonical form obtained as $\mathbf{AX} = \mathbf{XD}$ involves companion matrices in diagonal positions as seen in (4.7.6). Second, to obtain unique inverse with the eigenvalue property, Scroggs and Odell impose further conditions on the matrices $\mathbf{X}$ of the transformation. A generalized eigen vector† corresponding to a zero eigenvalue of $\mathbf{A}$ will be called briefly a generalized null vector of $\mathbf{A}$. $\mathbf{x}$ is a generalized null vector of height k, if $\mathbf{A}^{k-1}\mathbf{x} \neq \mathbf{0}$ and $\mathbf{A}^k\mathbf{x} = \mathbf{0}$. $\mathbf{x}$ is a generalized null vector of maximal height k, if $\mathbf{A}^{k-1}\mathbf{x} \neq \mathbf{0}$, $\mathbf{A}^k\mathbf{x} = \mathbf{0}$ and $\mathbf{x}$ is outside $\mathcal{M}(\mathbf{A})$. The Scroggs-Odell condition on matrix $\mathbf{X}$ is as follows:

Each generalized null vector of maximal height included as a column of $\mathbf{X}$, is orthogonal to every other generalized null vector of the same maximal height, if any are included also as columns of $\mathbf{X}$ and to every generalized null vectors of lesser height, regardless of whether they are included as columns of $\mathbf{X}$ or not.

For the matrix $\mathbf{A}$ of the numerical illustration, the inverse of Scroggs and Odell is

$$\frac{1}{36} \times \begin{pmatrix} 23 & 5 & -26 \\ -19 & -1 & 34 \\ -2 & -2 & -4 \end{pmatrix}.$$

Let $\mathcal{M}(\mathbf{X}_1) \oplus \mathcal{M}(\mathbf{X}_2) \oplus \cdots \oplus \mathcal{M}(\mathbf{X}_u)$ and $\mathcal{M}(\mathbf{V}_1) \oplus \mathcal{M}(\mathbf{V}_2) \oplus \cdots \oplus \mathcal{M}(\mathbf{V}_t)$ be alternative cyclic decompositions of $\mathcal{E}^n$ that is

$$\mathbf{A} = \mathbf{XDX}^{-1} = \mathbf{VCV}^{-1}.$$

We shall now prove

LEMMA 4.7.10 *If the columns of both* $\mathbf{X}$ *and* $\mathbf{V}$ *satisfy the Scroggs-Odell condition, then* $\mathbf{A}_{\mathbf{X}}^{\oplus} = \mathbf{A}_{\mathbf{V}}^{\oplus}$, *that is*

$$\mathbf{XD}^{+}\mathbf{X}^{-1} = \mathbf{VC}^{+}\mathbf{V}^{-1}. \tag{4.7.9}$$

† See Lemma 4.2.2 (ii) in this connection.

Proof: Let $(\mathbf{Y}_1^* \vdots \mathbf{Y}_2^* \vdots \cdots \vdots \mathbf{Y}_u^*)'$ and $(\mathbf{W}_1^* \vdots \mathbf{W}_2^* \vdots \cdots \vdots \mathbf{W}_t^*)'$ represent conformable partitions of $\mathbf{X}^{-1}$ and $\mathbf{V}^{-1}$, respectively.

It is well known that the number as well as dimensions of cyclic subspaces of type (I) are identical in the two representations, though the same need not be true in respect of cyclic subspaces of type (II).

Let there be precisely v subspaces of type (I) in both representations of dimensions k each, namely, $\mathscr{M}(\mathbf{X}_{i_1}), \ldots, \mathscr{M}(\mathbf{X}_{i_v})$ for one, $\mathscr{M}(\mathbf{V}_{j_1}), \ldots, \mathscr{M}(\mathbf{V}_{j_v})$ for the other. The corresponding contributions to $\mathbf{A}$ are

$$\mathbf{A}_{k,\mathbf{X}} = \mathbf{X}_{i_1}\mathbf{E}\mathbf{Y}_{i_1}^* + \cdots + \mathbf{X}_{i_v}\mathbf{E}\mathbf{Y}_{i_v}^*$$

and

$$\mathbf{A}_{k,\mathbf{V}} = \mathbf{V}_{j_1}\mathbf{E}\mathbf{W}_{j_1}^* + \cdots + \mathbf{V}_{j_v}\mathbf{E}\mathbf{W}_{j_v}^*,$$

respectively, and to $\mathbf{A}^{\oplus}$ are

$$\mathbf{A}_{k,\mathbf{X}}^{\oplus} = \mathbf{X}_{i_1}\mathbf{E}^*\mathbf{Y}_{i_1}^* + \cdots + \mathbf{X}_{i_v}\mathbf{E}^*\mathbf{Y}_{i_v}^*$$

and

$$\mathbf{A}_{k,\mathbf{V}}^{\oplus} = \mathbf{V}_{j_1}\mathbf{E}^*\mathbf{W}_{j_1}^* + \cdots + \mathbf{V}_{j_v}\mathbf{E}^*\mathbf{W}_{j_v}^*,$$

where $\mathbf{E}$ is the companion matrix of type (I) and order k. We show that if both $\mathbf{X}$ and $\mathbf{V}$ satisfy the Scroggs-Odell condition, then

$$\mathbf{A}_{k,\mathbf{X}} = \mathbf{A}_{k,\mathbf{V}} \quad \text{and} \quad \mathbf{A}_{k,\mathbf{X}}^{\oplus} = \mathbf{A}_{k,\mathbf{V}}^{\oplus}. \tag{4.7.10}$$

Let $\mathbf{P}_i(n \times v)$ be a matrix whose α-th column is identical with the i-th column of $\mathbf{X}_{i_\alpha}$ and $\mathbf{R}_i(n \times v)$ be a matrix whose α-th column is identical with the i-th column of $\mathbf{V}_{j_\alpha}$. Let $\mathbf{Q}_i^*(v \times n)$, $\mathbf{S}_i^*(v \times n)$ be similarly obtained from the rows of $\mathbf{Y}_{i_\alpha}^*$'s and $\mathbf{W}_{j_\alpha}^*$'s, respectively. Observe

$$\begin{aligned}
\mathbf{A}_{k,\mathbf{X}} &= \mathbf{P}_2\mathbf{Q}_1^* + \mathbf{P}_3\mathbf{Q}_2^* + \cdots + \mathbf{P}_k\mathbf{Q}_{k-1}^* \\
\mathbf{A}_{k,\mathbf{V}} &= \mathbf{R}_2\mathbf{S}_1^* + \mathbf{R}_3\mathbf{S}_2^* + \cdots + \mathbf{R}_k\mathbf{S}_{k-1}^* \\
\mathbf{A}_{k,\mathbf{X}}^{\oplus} &= \mathbf{P}_1\mathbf{Q}_2^* + \mathbf{P}_2\mathbf{Q}_3^* + \cdots + \mathbf{P}_{k-1}\mathbf{Q}_k^* \\
\mathbf{A}_{k,\mathbf{V}}^{\oplus} &= \mathbf{R}_1\mathbf{S}_2^* + \mathbf{R}_2\mathbf{S}_3^* + \cdots + \mathbf{R}_{k-1}\mathbf{S}_k^*.
\end{aligned}$$

The columns of $\mathbf{P}_1$ are generalized null vectors of maximal height k, while for $i \geq 2$ the columns of $\mathbf{P}_i$ are generalized null vectors of lesser height. Similarly the columns of $\mathbf{R}_1$ are generalized null vectors of maximal height k, while for $i \geq 2$ the columns of $\mathbf{R}_i$ are generalized null vectors of lesser height. It is easily seen that if both $\mathbf{X}$ and $\mathbf{V}$ satisfy the Scroggs-Odell condition, then $\mathbf{R}_1 = \mathbf{P}_1\mathbf{U}$, where $\mathbf{U}$ is nonsingular. Hence $\mathbf{R}_i = \mathbf{A}^{i-1}\mathbf{R}_1 = \mathbf{A}^{i-1}\mathbf{P}_1\mathbf{U} = \mathbf{P}_i\mathbf{U}$. Also, since the rows of the $\mathbf{Q}_i^*$'s and $\mathbf{S}_i^*$'s are obtained from the rows of the inverse matrix $\mathbf{X}^{-1}$ and $\mathbf{V}^{-1}$, respectively, we have $\mathbf{S}_i^* = \mathbf{U}^{-1}\mathbf{Q}_i^*$. Thus

(4.7.10) is established.

$$(4.7.10) \Rightarrow \sum_{k\geq 1} \mathbf{A}_{k,\mathbf{X}} = \sum_{k\geq 1} \mathbf{A}_{k,\mathbf{V}} = \mathbf{A}_0$$

$$\Rightarrow \mathbf{A} - \sum_{k\geq 1} \mathbf{A}_{k,\mathbf{X}} = \mathbf{A} - \sum_{k\geq 1} \mathbf{A}_{k,\mathbf{V}} = \mathbf{A}_1.$$

Further

$$(4.7.10) \Rightarrow (\mathbf{A}_0)_{\mathbf{X}}^{\oplus} = \sum_{k\geq 1} \mathbf{A}_{k,\mathbf{X}}^{\oplus} = \sum_{k\geq 1} \mathbf{A}_{k,\mathbf{V}}^{\oplus} = (\mathbf{A}_0)_{\mathbf{V}}^{\oplus}$$

$$\Rightarrow \mathbf{A}_{\mathbf{X}}^{\oplus} - (\mathbf{A}_0)_{\mathbf{X}}^{\oplus} = (\mathbf{A}_1)_{\mathbf{X}}^{\oplus}$$

$$\mathbf{A}_{\mathbf{V}}^{\oplus} - (\mathbf{A}_0)_{\mathbf{V}}^{\oplus} = (\mathbf{A}_1)_{\mathbf{V}}^{\oplus}.$$

Let $\mathscr{I}$ denote the set of indices $i\,(1 \leq i \leq v)$ for which $\mathscr{M}(\mathbf{X}_i)$ is of type (II). Since

$$\mathbf{A}_1 = \sum_{i\in\mathscr{I}} \mathbf{X}_i \mathbf{D}_i \mathbf{Y}_i^*,$$

and for $i \in \mathscr{I}$, the corresponding companion matrices $\mathbf{D}_i$ are nonsingular, it is seen that

$$R(\mathbf{A}_1) = R(\mathbf{A}_1^2).$$

Hence following the argument of Section 4.3 we observe that both $(\mathbf{A}_1)_{\mathbf{X}}^{\oplus}$ and $(\mathbf{A}_1)_{\mathbf{V}}^{\oplus}$ are identical with the unique inverse $(\mathbf{A}_1)_{\rho\chi}^{-}$. We have therefore

$$\mathbf{A}_{\mathbf{X}}^{\oplus} = (\mathbf{A}_1)_{\mathbf{X}}^{\oplus} + (\mathbf{A}_0)_{\mathbf{X}}^{\oplus} = (\mathbf{A}_1)_{\mathbf{V}}^{\oplus} + (\mathbf{A}_0)_{\mathbf{V}}^{\oplus} = \mathbf{A}_{\mathbf{V}}^{\oplus}.$$

This completes the proof of Lemma 4.7.10.

4.8 INVERSES COMMUTING WITH A POWER OF THE MATRIX

In this section we consider a g-inverse $\mathbf{G}$ of a square matrix $\mathbf{A}$, satisfying the following conditions:

$$\mathbf{A} = \mathbf{AGA} \tag{4.8.1}$$

$$\mathbf{GA}^k = \mathbf{A}^k\mathbf{G} \text{ for some positive integer } k. \tag{4.8.2}$$

LEMMA 4.8.1 *Let d denote the least positive integer for which*

$$R(\mathbf{A}^d) = R(\mathbf{A}^{d+1}). \tag{4.8.3}$$

If $\mathbf{G}$ satisfies (4.8.1) *and* (4.8.2), *then $k \geq d$.*

Proof. (4.8.1) and (4.8.2) $\Rightarrow \mathbf{GA}^{k+1} = \mathbf{A}^k \Rightarrow R(\mathbf{A}^k) = R(\mathbf{A}^{k+1})$. Hence, $k \geq d$.

LEMMA 4.8.2 *If a g-inverse $\mathbf{G}$ of $\mathbf{A}$ commutes with $\mathbf{A}^k$ it also commutes with $\mathbf{A}^{k+1}$.*

Proof: (4.8.1) and (4.8.2) $\Rightarrow \mathbf{GA}^{k+1} = \mathbf{A}^k = \mathbf{A}^{k+1}\mathbf{G}$.

LEMMA 4.8.3 *If a g-inverse* $\mathbf{G}$ *of* $\mathbf{A}$ *commutes with* $\mathbf{A}^k$ *for some* $k > d$, *as defined in (4.8.3), it also commutes with* $\mathbf{A}^{k-1}$.

Proof: Consider $\mathbf{B} = \mathbf{A}^d$ and observe $R(\mathbf{B}) = R(\mathbf{B}^2)$. Hence $\mathbf{B}_{\rho\chi}^-$ exists. Postmultiplying (4.8.2) on both sides by $\mathbf{A}^{2d-1}(\mathbf{B}^4)^-\mathbf{B}^2$ we have

$$\mathbf{GA}^{k-d-1}\mathbf{B}^3(\mathbf{B}^4)^-\mathbf{B}^2 = \mathbf{GA}^{k-d-1}\mathbf{BB}_{\rho\chi}^-\mathbf{B} = \mathbf{GA}^{k-d-1}\mathbf{B}$$

$$= \mathbf{GA}^{k-1} = \mathbf{A}^k\mathbf{GA}^{2d-1}(\mathbf{B}^4)^-\mathbf{B}^2 = \mathbf{A}^{k+2d-2}(\mathbf{B}^4)^-\mathbf{B}^2 = \mathbf{A}^{k-2}\mathbf{B}_{\rho\chi}^-.$$

Similarly premultiplying both sides of (4.8.2) by $\mathbf{B}^2(\mathbf{B}^4)^-\mathbf{A}^{2d-1}$, we have

$$\mathbf{A}^{k-1}\mathbf{G} = \mathbf{B}_{\rho\chi}^-\mathbf{A}^{k-2}.$$

By Lemma 4.6.1, $\mathbf{B}_{\rho\chi}^-$ is a polynomial in $\mathbf{B} = \mathbf{A}^d$. We have therefore

$$\mathbf{B}_{\rho\chi}^-\mathbf{A}^{k-2} = \mathbf{A}^{k-2}\mathbf{B}_{\rho\chi}^- \Rightarrow \mathbf{A}^{k-1}\mathbf{G} = \mathbf{GA}^{k-1}.$$

Lemmas 4.8.1 and 4.8.2 show that it suffices to consider solutions $\mathbf{G}$ of (4.8.1) and (4.8.2) for $k = d$.

LEMMA 4.8.4 *If* $\mathbf{x}$ *is an eigenvector, corresponding to a non-null eigenvalue* λ *of* $\mathbf{A}$, *and* $\mathbf{G}$ *satisfies (4.8.1) and (4.8.2), then* $\mathbf{x}$ *is also an eigenvector of* $\mathbf{G}$, *corresponding to its eigenvalue* $1/\lambda$.

Proof: We have $\mathbf{Ax} = \lambda\mathbf{x}$. Also

$$\mathbf{GA}^k\mathbf{x} = \lambda^k\mathbf{Gx} = \mathbf{A}^k\mathbf{Gx} = \frac{1}{\lambda}\mathbf{A}^k\mathbf{GAx} = \frac{1}{\lambda}\mathbf{A}^k\mathbf{x} = \lambda^{k-1}\mathbf{x}.$$

Since $\lambda \neq 0$, we have $\mathbf{Gx} = \lambda^{-1}\mathbf{x}$. If $\mathbf{G}$ satisfies (4.8.1) and (4.8.2), $\mathbf{G}^*$ also satisfies (4.8.1) and (4.8.2) with $\mathbf{A}$ replaced by $\mathbf{A}^*$. Hence

$$\mathbf{x}^*\mathbf{A} = \lambda\mathbf{x}^* \Rightarrow \mathbf{x}^*\mathbf{G} = \lambda^{-1}\mathbf{x}^*.$$

LEMMA 4.8.5 *If* $\mathbf{G}$ *satisfies (4.8.1) and (4.8.2),* $p \geq k$ *and* e *are arbitrary positive integers then*

$$\mathbf{A}^p = \mathbf{A}^{p+e}\mathbf{G}^e = \mathbf{G}^e\mathbf{A}^{p+e}.$$

Proof: The result is easily proved. In fact from what we have seen earlier, the lemma is true for any positive integer $p \geq d$.

Generalized inverses satisfying the two conditions

$$\text{(i) } \mathbf{GAG} = \mathbf{G} \qquad \text{and} \qquad \text{(ii) } \mathbf{AG}^k = \mathbf{G}^k\mathbf{A} \tag{4.8.4}$$

in addition to (4.8.1) and (4.8.2) were considered by Erdelyi (1966, 1967). It is easy to check that $\mathbf{A}_{\mathbf{X}}^{\oplus}$ as defined in Section 4.7.2 satisfies all the four conditions. This shows that the solution is not necessarily unique, as it could very much depend on the choice of $\mathbf{X}$, unless of course $d = 1$. We had noted in Section 4.3 that in this case $\mathbf{A}_{\rho\chi}^{-}$ is the unique reflexive commuting g-inverse of $\mathbf{A}$.

A g-inverse satisfying all the four conditions obviously satisfies the quasi-commutative relations

$$(\mathbf{AG} - \mathbf{GA})\mathbf{A}^{k-1} = \mathbf{A}^{k-1}(\mathbf{AG} - \mathbf{GA})$$

$$(\mathbf{AG} - \mathbf{GA})\mathbf{G}^{k-1} = \mathbf{G}^{k-1}(\mathbf{AG} - \mathbf{GA}).$$

In this sense such inverses could be called quasi-commutative.

4.9 SOME USEFUL MATRIX DECOMPOSITION THEOREMS†

THEOREM 4.9.1 *A square matrix* $\mathbf{A}$ *has a decomposition*

$$\mathbf{A} = \mathbf{A}_1 + \mathbf{A}_2 \tag{4.9.1}$$

with the properties

(a) $R(\mathbf{A}_1) = R(\mathbf{A}_1^2)$, (4.9.2)

(b) $\mathbf{A}_2$ *is nilpotent, and* (4.9.3)

(c) $\mathbf{A}_1\mathbf{A}_2 = \mathbf{A}_2\mathbf{A}_1 = \mathbf{0}$. (4.9.4)

Proof: Let p be a positive integer, such that $R(\mathbf{A}^p) = R(\mathbf{A}^{p+1})$. By Frobenius inequality, for such an integer p, we have

$$R(\mathbf{A}^p) = R(\mathbf{A}^{p+\nu}), \qquad \nu = 0, 1, 2, \ldots \tag{4.9.5}$$

To prove the theorem we first establish some lemmas.

LEMMA 4.9.1 *Let* $\mathbf{K} = \mathbf{A}^p(\mathbf{A}^{2p+1})^-\mathbf{A}^p$. *Then* $R(\mathbf{K}) = R(\mathbf{K}^2) = R(\mathbf{A}^p)$.

Proof: Lemma 4.9.1 follows from the fact that

$$\begin{aligned}\mathbf{K}^2\mathbf{A} &= \mathbf{A}^p(\mathbf{A}^{2p+1})^-\mathbf{A}^p\mathbf{A}^p(\mathbf{A}^{2p+1})^-\mathbf{A}^{p+1} \\ &= \mathbf{A}^p(\mathbf{A}^{2p+1})^-\mathbf{A}^p\mathbf{G}\mathbf{A}^{p+1} = \mathbf{A}^p(\mathbf{A}^{2p+1})^-\mathbf{A}^p = \mathbf{K},\end{aligned}$$

since $\mathbf{G} = \mathbf{A}^p(\mathbf{A}^{2p+1})^-$ is a g-inverse of $\mathbf{A}^{p+1}$ and $\mathbf{KA}^{p+1} = \mathbf{A}^p(\mathbf{A}^{2p+1})^-\mathbf{A}^{2p+1} = \mathbf{A}^p$. Lemma 4.9.1 shows that $\mathbf{K}_{\rho\chi}^-$ exists. We write

$$\mathbf{A}_1 = \mathbf{K}_{\rho\chi}^-, \qquad \mathbf{A}_2 = \mathbf{A} - \mathbf{A}_1.$$

† Theorem 4.9.1 and Lemma 4.9.3 are known results. The results of this section are easy to prove using Jordan canonical form. In this section we give an exposition of these results free of Jordan canonical form.

$\mathbf{A}_1$ so defined clearly satisfies (4.9.2) and in fact

$$\mathbf{K}^d = (\mathbf{A}_1^d)^-_{\rho\chi}, \qquad \mathbf{A}_1^d = (\mathbf{K}^d)^-_{\rho\chi} \tag{4.9.6}$$

for every positive integer d (see Example 2 in the complements to Chapter 4).

LEMMA 4.9.2 *For every positive integer* $d \geq p$, $\mathbf{A}^d = (\mathbf{K}^d)^-_{\rho\chi}$.

Proof: Since for $d \geq p$, we have trivially

$$\mathscr{M}(\mathbf{A}^d) = \mathscr{M}(\mathbf{K}^d), \qquad \mathscr{M}[(\mathbf{A}^d)^*] = \mathscr{M}[(\mathbf{K}^d)^*].$$

To prove Lemma 4.9.2 it suffices to show that $\mathbf{A}^d$ is a g-inverse of $\mathbf{K}^d$ for every positive integer d for which observe that

$$\begin{aligned}\mathbf{K}^d\mathbf{A}^d\mathbf{K}^d &= \mathbf{K}^{d-1}\mathbf{A}^p(\mathbf{A}^{2p+1})^-\mathbf{A}^{2p+d}(\mathbf{A}^{2p+1})^-\mathbf{A}^p\mathbf{K}^{d-1}\\ &= \mathbf{K}^{d-1}\mathbf{A}^{p+d-1}(\mathbf{A}^{2p+1})^-\mathbf{A}^p\mathbf{K}^{d-1}\\ &= \mathbf{K}^{d-1}\mathbf{A}^{d-1}\mathbf{K}^d.\end{aligned}$$

Repetition of this argument gives

$$\begin{aligned}\mathbf{K}^d\mathbf{A}^d\mathbf{K}^d &= \mathbf{K}^{d-2}\mathbf{A}^{d-2}\mathbf{K}^d = \cdots = \mathbf{K}\mathbf{A}\mathbf{K}^d\\ &= \mathbf{A}^p(\mathbf{A}^{2p+1})^-\mathbf{A}^{2p+1}(\mathbf{A}^{2p+1})^-\mathbf{A}^p\mathbf{K}^{d-1}\\ &= \mathbf{A}^p(\mathbf{A}^{2p+1})^-\mathbf{A}^p\mathbf{K}^{d-1} = \mathbf{K}^d.\end{aligned}$$

This concludes the proof of Lemma 4.9.2.

Equations (4.9.5) and (4.9.6) together with the uniqueness of a $\rho\chi$ g-inverse imply $\mathbf{A}^d = \mathbf{A}_1^d$ for each positive integer $d \geq p$. Observe that

$$\mathbf{A}^p = \mathbf{A}_1^p$$

and

$$\mathbf{A}^{p+1} = (\mathbf{A}_1 + \mathbf{A}_2)\mathbf{A}^p = (\mathbf{A}_1 + \mathbf{A}_2)\mathbf{A}_1^p = \mathbf{A}_1^{p+1}.$$

Hence $\mathbf{A}_2\mathbf{A}_1^p = \mathbf{0} \Rightarrow \mathbf{A}_2\mathbf{A}_1 = \mathbf{0}$, since $\mathscr{M}(\mathbf{A}_1^p) = \mathscr{M}(\mathbf{A}_1)$. Similarly it is seen that $\mathbf{A}_1\mathbf{A}_2 = \mathbf{0}$. Further,

$$\mathbf{A}_1\mathbf{A}_2 = \mathbf{A}_2\mathbf{A}_1 = \mathbf{0} \Rightarrow \mathbf{A}_1^p = \mathbf{A}^p = (\mathbf{A}_1 + \mathbf{A}_2)^p = \mathbf{A}_1^p + \mathbf{A}_2^p.$$

Hence $\mathbf{A}_2^p = \mathbf{0}$. With (4.9.3) and (4.9.4) thus verified the proof of Theorem 4.9.1 is concluded.

LEMMA 4.9.3 *The decomposition* (4.9.1) *is unique.*

Proof: Let $\mathbf{A} = \mathbf{B}_1 + \mathbf{B}_2$ be an alternative decomposition also satisfying (4.9.2), (4.9.3), and (4.9.4). These conditions imply that for a positive integer p sufficiently large

$$\mathbf{A}^p = \mathbf{A}_1^p = \mathbf{B}_1^p.$$

Hence $\mathscr{M}(\mathbf{A}_1) = \mathscr{M}(\mathbf{A}_1^p) = \mathscr{M}(\mathbf{B}_1^p) = \mathscr{M}(\mathbf{B}_1)$. Similarly, $\mathscr{M}(\mathbf{A}_1^*) = \mathscr{M}(\mathbf{B}_1^*)$. Thus in addition to

$$\mathbf{A}_1\mathbf{A}_2 = \mathbf{A}_2\mathbf{A}_1 = \mathbf{B}_1\mathbf{B}_2 = \mathbf{B}_2\mathbf{B}_1 = \mathbf{0}$$

we have

$$\mathbf{A}_1\mathbf{B}_2 = \mathbf{B}_2\mathbf{A}_1 = \mathbf{B}_1\mathbf{A}_2 = \mathbf{B}_2\mathbf{A}_1 = \mathbf{0}.$$

Since $\mathscr{M}(\mathbf{A}_1 - \mathbf{B}_1) \subset \mathscr{M}(\mathbf{A}_1)$

$$\mathbf{A}_1 - \mathbf{B}_1 = \mathbf{A}_1(\mathbf{A}_1^2)^-\mathbf{A}_1(\mathbf{A}_1 - \mathbf{B}_1) = \mathbf{0}.$$

Hence $\mathbf{A}_1 = \mathbf{B}_1$ and $\mathbf{A}_2 = \mathbf{B}_2$.

THEOREM 4.9.2 *With $\mathbf{A}_1$ and $\mathbf{A}_2$ as defined in Theorem 4.9.1, every g-inverse $\mathbf{A}^-$ of $\mathbf{A}$ has the representation*

$$\mathbf{A}^- = (\mathbf{A}_1)_{\rho\chi}^- + \mathbf{A}_2^-. \tag{4.9.7}$$

Proof: Let us write $\mathbf{M} = \mathbf{A}^- - (\mathbf{A}_1)_{\rho\chi}^-$. Since $\mathbf{A}_1$ and $(\mathbf{A}_1)_{\rho\chi}^-$ have identical row and column spaces, we have

$$(\mathbf{A}_1 + \mathbf{A}_2)[(\mathbf{A}_1)_{\rho\chi}^- + \mathbf{M}](\mathbf{A}_1 + \mathbf{A}_2)$$
$$= \mathbf{A}_1(\mathbf{A}_1)_{\rho\chi}^-\mathbf{A}_1 + \mathbf{A}_1\mathbf{M}\mathbf{A}_1 + \mathbf{A}_1\mathbf{M}\mathbf{A}_2 + \mathbf{A}_2\mathbf{M}\mathbf{A}_1 + \mathbf{A}_2\mathbf{M}\mathbf{A}_2 = \mathbf{A}_1 + \mathbf{A}_2$$
$$\Rightarrow \mathbf{A}_1\mathbf{M}\mathbf{A}_1 + \mathbf{A}_1\mathbf{M}\mathbf{A}_2 + \mathbf{A}_2\mathbf{M}\mathbf{A}_1 + \mathbf{A}_2\mathbf{M}\mathbf{A}_2 = \mathbf{A}_2.$$

Multiplying to the left on both sides by $\mathbf{A}_1$ we have

$$\mathbf{A}_1^2\mathbf{M}\mathbf{A}_1 + \mathbf{A}_1^2\mathbf{M}\mathbf{A}_2 = \mathbf{0} \Rightarrow \mathbf{A}_1\mathbf{M}\mathbf{A}_1 + \mathbf{A}_1\mathbf{M}\mathbf{A}_2 = \mathbf{0}$$
$$\Rightarrow \mathbf{A}_2\mathbf{M}\mathbf{A}_1 + \mathbf{A}_2\mathbf{M}\mathbf{A}_2 = \mathbf{A}_2.$$

Multiplying to the right on both sides by $\mathbf{A}_1$ we have

$$\mathbf{A}_2\mathbf{M}\mathbf{A}_1^2 = \mathbf{0} \Rightarrow \mathbf{A}_2\mathbf{M}\mathbf{A}_1 = \mathbf{0}.$$

Therefore $\mathbf{A}_2\mathbf{M}\mathbf{A}_2 = \mathbf{A}_2$ or $\mathbf{M} = \mathbf{A}_2^-$. This completes the proof of Theorem 4.9.2.

4.10 A COMMUTING PSEUDOINVERSE AND RELATED RESULTS

4.10.1 Definition and Properties

For a square matrix $\mathbf{A}$ of order n, Drazin (1958) defined a pseudoinverse $\mathbf{G}$ by the following conditions:

(i) $\mathbf{AG} = \mathbf{GA}$,
(ii) $\mathbf{A}^k = \mathbf{A}^{k+1}\mathbf{G}$ for some positive integer k,
(iii) $\mathbf{G} = \mathbf{G}^2\mathbf{A}$.

The smallest value of k for which (ii) is true is called the index of **A**. A pseudo-inverse of **A** satisfying the above conditions will be denoted by $\mathbf{A}_{\text{com}}$. We notice from (i) and (ii) that **A** is a g-inverse of $\mathbf{A}_{\text{com}}$, but $\mathbf{A}_{\text{com}}$ is not a g-inverse of **A** as defined in this book. We prove

THEOREM 4.10.1 $\mathbf{A}_{\text{com}}$ *exists and is unique.*

Proof: In order that condition (ii) of the definition is true it is necessary that

$$R(\mathbf{A}^k) = R(\mathbf{A}^{k+1}). \tag{4.10.1}$$

Let d be the smallest integer for which (4.10.1) is true. Consider an integer $p \geq d$. We show that the matrix $\mathbf{K} = \mathbf{A}^p(\mathbf{A}^{2p+1})^-\mathbf{A}^p$, as studied in Lemma 4.9.1, is one choice of $\mathbf{A}_{\text{com}}$.

Proof of (i): Since $\mathscr{M}(\mathbf{A}^p) = \mathscr{M}(\mathbf{A}^{2p+1})$, by (4.9.5), the matrix **K** does not depend on the choice of $(\mathbf{A}^{2p+1})^-$. Hence

$$\mathbf{K} = \mathbf{A}^p(\mathbf{A}^{2p+1})^-_{\rho\chi}\mathbf{A}^p.$$

We observe that the existence of $(\mathbf{A}^{2p+1})^-_{\rho\chi}$ is again guaranteed by (4.9.5). Since $(\mathbf{A}^{2p+1})^-_{\rho\chi}$ is a polynomial of finite degree in $\mathbf{A}^{2p+1}$, **K** is clearly a polynomial of finite degree in **A**. Hence **K** commutes with **A**. Thus condition (i) is verified.

Proof of (ii): Observe that by (4.9.5) and Lemma 2.2.4(c), $\mathbf{A}^{p+1}(\mathbf{A}^{2p+1})^-\mathbf{A}^{p-d}$ is one choice for $(\mathbf{A}^d)^-$. Hence $\mathbf{A}^{d+1}\mathbf{K} = \mathbf{A}^d$. Thus (ii) is satisfied for $k = d$.

Proof of (iii): $\mathbf{KAK} = \mathbf{A}^p(\mathbf{A}^{2p+1})^-\mathbf{A}^{2p+1}(\mathbf{A}^{2p+1})^-\mathbf{A}^p = \mathbf{A}^p(\mathbf{A}^{2p+1})^-\mathbf{A}^p = \mathbf{K}$, since $\mathscr{M}(\mathbf{A}^p) = \mathscr{M}(\mathbf{A}^{2p+1})$. Hence condition (iii) of definition is also satisfied by the matrix **K**. The existence part of Theorem 4.10.1 is thus established.

To prove uniqueness, let **K** and **G** be distinct choices for $\mathbf{A}_{\text{com}}$. Consider the expression $\mathbf{K}^{d+1}\mathbf{A}^{2d+1}\mathbf{G}^{d+1}$. Repeated application of (ii) and (iii) gives

$$\mathbf{K}^{d+1}\mathbf{A}^{2d+1}\mathbf{G}^{d+1} = \mathbf{K}^{d+1}\mathbf{A}^d = \mathbf{K}.$$

Consider now alternative equivalent versions of (ii) and (iii) which are true in view of (i), namely

$$\text{(ii)}' \ \mathbf{A}^k = \mathbf{G}\mathbf{A}^{k+1}, \qquad \text{(iii)}' \ \mathbf{G} = \mathbf{A}\mathbf{G}^2.$$

Making repeated application of (ii)′ and (iii)′, we have

$$\mathbf{K}^{d+1}\mathbf{A}^{2d+1}\mathbf{G}^{d+1} = \mathbf{A}^d\mathbf{G}^{d+1} = \mathbf{G}.$$

Hence $\mathbf{K} = \mathbf{G}$ and the proof of Theorem 4.10.1 is concluded. We have already established the following interesting results as a byproduct.

LEMMA 4.10.1 $\mathbf{A}_{\text{com}}$ *is a polynomial of finite degree in* $\mathbf{A}$.

We also recall from Lemma 4.9.2 that for any integer $m \geq d$ (the index of $\mathbf{A}$) $\mathbf{A}_{\text{com}}^m$ is a reflexive inverse of $\mathbf{A}^m$. The following result is easy to prove.

LEMMA 4.10.2

(a) $(\mathbf{A}_{\text{com}})^m = (\mathbf{A}^m)_{\text{com}}$
(b) $(\mathbf{A}^m)_{\text{com}}\mathbf{A}^m = \mathbf{A}_{\text{com}}\mathbf{A}$
(c) $\mathbf{A}^m(\mathbf{A}^m)_{\text{com}} = \mathbf{A}\mathbf{A}_{\text{com}} \quad \forall\, m = 1, 2, \ldots.$

4.10.2 Two Other Unique Inverses

For a square matrix $\mathbf{A}$ of index 1 [that is, $R(\mathbf{A}) = R(\mathbf{A}^2)$] we have so far specifically listed two unique g-inverses, the Moore-Penrose inverse $\mathbf{A}^+$ and the inverse $\mathbf{A}_{\rho\chi}^-$. Actually, two other unique hybrid inverses could be defined by adopting some conditions from $\mathbf{A}^+$ and some from $\mathbf{A}_{\rho\chi}^-$. We have seen in Lemma 2.5.5 that a reflexive g-inverse is uniquely determined by its left and right idempotents. This suggests the following definition of the two hybrid inverses.

Notation	Left Idempotent	Right Idempotent	Algebraic Expression
$\mathbf{A}_{\rho^*\chi}^-$	$\mathbf{A}_{\rho\chi}^-\mathbf{A}$	$\mathbf{A}\mathbf{A}^+ = \mathbf{P}_{\mathbf{A}}$	$\mathbf{A}_{\rho\chi}^-\mathbf{A}\mathbf{A}^+$
$\mathbf{A}_{\rho\chi^*}^-$	$\mathbf{A}^+\mathbf{A} = \mathbf{P}_{\mathbf{A}^*}$	$\mathbf{A}\mathbf{A}_{\rho\chi}^-$	$\mathbf{A}^+\mathbf{A}\mathbf{A}_{\rho\chi}^-$

LEMMA 4.10.3 $\mathbf{A}_{\rho^*\chi}^- = \mathbf{A}(\mathbf{A}^*\mathbf{A}^2)^-\mathbf{A}^*$, $\mathbf{A}_{\rho\chi^*}^- = \mathbf{A}^*(\mathbf{A}^2\mathbf{A}^*)^-\mathbf{A}$.

Proof: It is clear from the definition that $\mathbf{A}_{\rho^*\chi}^-$ and $\mathbf{A}_{\rho\chi^*}^-$ are matrices of the type $\mathbf{AWA}^*$ and $\mathbf{A}^*\mathbf{VA}$, respectively. Lemma 4.10.3 follows from Lemma 2.2.5(d) and the uniqueness implicit in Lemma 2.2.6(g).

LEMMA 4.10.4

(a) *If the matrix* $\mathbf{A}$ *is EPr then* $\mathbf{A}_{\rho^*\chi}^- = \mathbf{A}_{\rho\chi^*}^- = \mathbf{A}^+$; *conversely, if any two of three g-inverses* $\mathbf{A}^+$, $\mathbf{A}_{\rho^*\chi}^-$, *and* $\mathbf{A}_{\rho\chi^*}^-$ *are equal then* $\mathbf{A}$ *is EPr.*
(b) $\mathbf{A}_{\rho\chi^*}^- = (\mathbf{A}^{n+1})_m^-\mathbf{A}^n, \quad \mathbf{A}_{\rho^*\chi}^- = \mathbf{A}^n(\mathbf{A}^{n+1})_l^- \quad \forall\, n = 1, 2, \ldots$
(c) $(\mathbf{A}_{\rho\chi^*}^-)^n = (\mathbf{A}^n)_{\rho\chi^*}^-, \quad (\mathbf{A}_{\rho^*\chi}^-)^n = (\mathbf{A}^n)_{\rho^*\chi}^- \quad \forall\, n = 1, 2, \ldots$
(d) $(\mathbf{A}_{\rho\chi^*}^-)_{\rho\chi^*}^- = (\mathbf{A}_{\rho\chi^*}^-)_{\rho^*\chi}^- = (\mathbf{A}_{\rho\chi^*}^-)^+$
$(\mathbf{A}_{\rho^*\chi}^-)_{\rho^*\chi}^- = (\mathbf{A}_{\rho^*\chi}^-)_{\rho\chi^*}^- = (\mathbf{A}_{\rho^*\chi}^-)^+$
(e) $(\mathbf{A}_{\rho\chi^*}^-)(\mathbf{A}_{\rho\chi^*}^-)^+ = (\mathbf{A}_{\rho\chi^*}^-)^+(\mathbf{A}_{\rho\chi^*}^-) = \mathbf{A}^+\mathbf{A}$
$(\mathbf{A}_{\rho^*\chi}^-)(\mathbf{A}_{\rho^*\chi}^-)^+ = (\mathbf{A}_{\rho^*\chi}^-)^+(\mathbf{A}_{\rho^*\chi}^-) = \mathbf{A}\mathbf{A}^+$
(f) $(\mathbf{A}_{\rho\chi^*}^-)(\mathbf{A}_{\rho^*\chi}^-) = (\mathbf{A}^2)^+$.

Proof: (a) follows from the definition of an EPr matrix. $\mathbf{A}$ is EPr iff $\mathcal{M}(\mathbf{A}) = \mathcal{M}(\mathbf{A}^*)$.

Proof of the rest of Lemma 4.10.4 is computational.

4.10.3 Left and Right Power Inverses

For a matrix $\mathbf{A}$ of index d, and an integer $m \geq d$, $(\mathbf{A}^m)^-_{\rho\chi^*}$ is called by Cline (1968) the unique left power inverse of $\mathbf{A}^m$, $(\mathbf{A}^m)^-_{\rho^*\chi}$ the unique right power inverse of $\mathbf{A}^m$ and denoted, respectively, by $(\mathbf{A}^m)_L$ and $(\mathbf{A}^m)_R$. It is easy to establish the properties of such inverses as given in Lemmas 4.10.5 and 4.10.6.

LEMMA 4.10.5

$$(\mathbf{A}^k)^+ = (\mathbf{A}^k)_L\mathbf{A}^k(\mathbf{A}^k)_R$$

and

$$\mathbf{A}_{\text{com}} = (\mathbf{A}^k)_R\mathbf{A}^{2k-1}(\mathbf{A}^k)_L,$$

where k is an arbitrary integer $\geq d$.

LEMMA 4.10.6

(a) $(\mathbf{A}^k)_L = (\mathbf{A}^{d+k})^+\mathbf{A}^d, \quad (\mathbf{A}^k)_R = \mathbf{A}^d(\mathbf{A}^{d+k})^+$
(b) $(\mathbf{A}^{k+m})_L = (\mathbf{A}^k)_L(\mathbf{A}^m)_L, \quad (\mathbf{A}^{k+m})_R = (\mathbf{A}^k)_R(\mathbf{A}^m)_R$
(c) $[(\mathbf{A}^k)_L]^n = (\mathbf{A}^{nk})_L, \quad [(\mathbf{A}^k)_R]^n = (\mathbf{A}^{nk})_R$
for every positive integer $n \geq 1$
(d) $(\mathbf{A}^k)_L(\mathbf{A}^m)_R = (\mathbf{A}^{k+m})^+$
(e) $[(\mathbf{A}^k)_L]^+ = (\mathbf{A}^d)^+\mathbf{A}^{d+k}, \quad [(\mathbf{A}^k)_R]^+ = \mathbf{A}^{d+k}(\mathbf{A}^d)^+$
(f) $(\mathbf{A}^k)_L[(\mathbf{A}^k)_L]^+ = [(\mathbf{A}^k)_L]^+(\mathbf{A}^k)_L = (\mathbf{A}^d)^+\mathbf{A}^d$
$(\mathbf{A}^k)_R[(\mathbf{A}^k)_R]^+ = [(\mathbf{A}^k)_R]^+(\mathbf{A}^k)_R = \mathbf{A}^d(\mathbf{A}^d)^+.$

In (a)–(f) above k and m are arbitrary integers $\geq d$.

4.11 CONSTRAINED INVERSES

In many electrical and mechanical problems the state of a system is determined by the stability of a characteristic such as energy or power which is a function of certain variables. These variables are often subject to a given number of homogeneous constraints.

Let $E = \mathbf{v}^*\mathbf{A}\mathbf{v} - 2\mathbf{b}^*\mathbf{v}$ be a function of n variables $v_1, \ldots, v_n$, represented by a vector $\mathbf{v}$, where $\mathbf{A}$ and $\mathbf{b}$ are given $n \times n$ matrix and n-vector, respectively. Further, let $\mathbf{C}\mathbf{v} = \mathbf{0}$ represent homogeneous constraints on $\mathbf{v}$.

If there are no constraints, a value of $\mathbf{v}$ at which E is stable is determined by equating the derivative of E with respect to $\mathbf{v}$ to zero.

$$\mathbf{A}\mathbf{v} - \mathbf{b} = \mathbf{0}. \tag{4.11.1}$$

If Equation (4.11.1) is consistent, then $\mathbf{v} = \mathbf{A}^{-}\mathbf{b}$, where $\mathbf{A}^{-}$ is a g-inverse of $\mathbf{A}$.

If there are constraints on $\mathbf{v}$, it may be possible to replace $\mathbf{v}$ by $\mathbf{q}$ with a smaller number of unconstrained coordinates and obtain E in the form $\mathbf{q}^*\mathbf{B}\mathbf{q} - 2\mathbf{f}^*\mathbf{q}$, in which case the method described above is applicable. However, neither the matrix $\mathbf{B}$ nor the coordinates of $\mathbf{q}$ have physical significance and it is desirable to work directly with $\mathbf{A}$ and $\mathbf{v}$. With this motivation Bott and Duffin (1953) introduced what is called a constrained inverse of a square matrix $\mathbf{A}$, denoted by $\mathbf{T}$, with the property that a stable point can be obtained as $\mathbf{v} = \mathbf{T}\mathbf{b}$ when there are constraints (as in the unconstrained case, $\mathbf{v} = \mathbf{A}^{-}\mathbf{b}$).

Such a matrix $\mathbf{T}$ defined by Bott and Duffin is not a g-inverse of $\mathbf{A}$ in our sense. In this section we develop the theory of constrained inverses, extend the concepts to a general matrix, not necessarily square as in the discussion by Bott and Duffin, introduce new types of constraints and consider some applications.

Let $\mathbf{A}$ be a matrix of order $m \times n$, $\mathscr{V}$ and $\mathscr{U}$ be subspaces in $\mathscr{E}^n$ and $\mathscr{E}^m$, respectively. In what follows we shall impose constraints of two different types to define a constrained inverse $\mathbf{G}$ of $\mathbf{A}$.

Constraints of Type 1

c: $\mathbf{G}$ maps vectors of $\mathscr{E}^m$ into $\mathscr{V}$

r: $\mathbf{G}^*$ maps vectors of $\mathscr{E}^n$ into $\mathscr{U}$

Constraints of Type 2

C: $\mathbf{GA}$ is an identity in $\mathscr{V}$

R: $(\mathbf{AG})^*$ is an identity in $\mathscr{U}$.

Inverses obtained by choosing various combinations of these constraints are listed below in Table 4.1 along with n.s. conditions for existence, and explicit forms, where $\mathbf{F}$ and $\mathbf{E}$ are matrices such that $\mathscr{V} = \mathscr{M}(\mathbf{E})$ and $\mathscr{U} = \mathscr{M}(\mathbf{F}^*)$.

TABLE 4.1
CONSTRAINED INVERSES OF VARIOUS TYPES
($\mathbf{V}$ AND $\mathbf{U}$ ARE ARBITRARY MATRICES)

Notation	N.S. Condition for Existence	Algebraic Expression	Reference to Theorem
$\mathbf{A}_{cC}$	$R(\mathbf{AE}) = R(\mathbf{E})$	$\mathbf{E}(\mathbf{AE})^{-}$	4.11.1
A_{rR}	$R(\mathbf{FA}) = R(\mathbf{F})$	$(\mathbf{FA})^{-}\mathbf{F}$	4.11.3
$\mathbf{A}_{cR}$	$R(\mathbf{FAE}) = R(\mathbf{F})$	$\mathbf{E}(\mathbf{FAE})^{-}\mathbf{F} + \mathbf{E}[\mathbf{I} - (\mathbf{FAE})^{-}\mathbf{FAE}]\mathbf{U}$	4.11.5
$\mathbf{A}_{rC}$	$R(\mathbf{FAE}) = R(\mathbf{E})$	$\mathbf{E}(\mathbf{FAE})^{-}\mathbf{F} + \mathbf{V}[\mathbf{I} - \mathbf{FAE}(\mathbf{FAE})^{-}]\mathbf{F}$	4.11.6
$\mathbf{A}_{crCR}$	$R(\mathbf{FAE}) = R(\mathbf{F}) = R(\mathbf{E})$	$\mathbf{E}(\mathbf{FAE})^{-}\mathbf{F}$	4.11.7

THEOREM 4.11.1 *A_{cC} exists iff $R(\mathbf{AE}) = R(\mathbf{E})$. In such a case $\mathbf{A}_{cC}$ is of the form $\mathbf{E}(\mathbf{AE})^-$.*

Proof: Using the condition c, $\mathbf{G} = \mathbf{EX}$ for some matrix $\mathbf{X}$. Then condition C gives

$$\mathbf{EXAE} = \mathbf{E}. \tag{4.11.2}$$

Equation (4.11.2) is solvable only if $R(\mathbf{AE}) = R(\mathbf{E})$ in which case, (4.11.2) $\Leftrightarrow$ $\mathbf{AEXAE} = \mathbf{AE}$, or

$$\mathbf{X} = (\mathbf{AE})^- \Rightarrow \mathbf{G} = \mathbf{E}(\mathbf{AE})^-. \tag{4.11.3}$$

The "if" part is trivial.

THEOREM 4.11.2 *$\mathbf{A}$ is a g-inverse of $\mathbf{A}_{cC}$, but not necessarily the other way. $\mathbf{A}_{cC}$ is a g-inverse of $\mathbf{A}$ iff $R(\mathbf{AE}) = R(\mathbf{A})$.*

Proof: Theorem 4.11.2 follows from Theorem 4.11.1 and Lemma 2.5.1. Theorems 4.11.3 and 4.11.4 follow along similar lines.

THEOREM 4.11.3 *$\mathbf{A}_{rR}$ exists iff $R(\mathbf{FA}) = R(\mathbf{F})$. In such a case $\mathbf{A}_{rR}$ is of the form $(\mathbf{FA})^-\mathbf{F}$.*

THEOREM 4.11.4 *$\mathbf{A}$ is a g-inverse of $\mathbf{A}_{rR}$ but not necessarily the other way. $\mathbf{A}_{rR}$ is a g-inverse of $\mathbf{A}$ iff $R(\mathbf{FA}) = R(\mathbf{A})$.*

THEOREM 4.11.5 *$\mathbf{A}_{cR}$ exists iff $R(\mathbf{FAE}) = R(\mathbf{F})$. In such a case A_{cR} is of the form*

$$\mathbf{E}(\mathbf{FAE})^-\mathbf{F} + \mathbf{E}[\mathbf{I} - (\mathbf{FAE})^-\mathbf{FAE}]\mathbf{U}, \tag{4.11.4}$$

where $\mathbf{U}$ is arbitrary.

Proof: Using the condition c, $\mathbf{G} = \mathbf{EX}$ for some matrix $\mathbf{X}$. Then condition R gives

$$\mathbf{FAEX} = \mathbf{F}. \tag{4.11.5}$$

Equation (4.11.5) is solvable if $R(\mathbf{FAE}) = R(\mathbf{F})$, in which case a general solution is given by

$$\mathbf{X} = (\mathbf{FAE})^-\mathbf{F} + [\mathbf{I} - (\mathbf{FAE})^-\mathbf{FAE}]\mathbf{U}, \tag{4.11.6}$$

where $\mathbf{U}$ is arbitrary. The 'if' part is easy. Thus Theorem 4.11.5 is established. Theorem 4.11.6 can be proved along similar lines.

THEOREM 4.11.6 *$\mathbf{A}_{rC}$ exists iff $R(\mathbf{FAE}) = R(\mathbf{E})$. In such a case $\mathbf{A}_{rC}$ is of the form*

$$\mathbf{E}(\mathbf{FAE})^-\mathbf{F} + \mathbf{V}[\mathbf{I} - \mathbf{FAE}(\mathbf{FAE})^-]\mathbf{F}, \tag{4.11.7}$$

where $\mathbf{V}$ is arbitrary.

THEOREM 4.11.7 $\mathbf{A}_{crCR}$ *exists iff* $R(\mathbf{FAE}) = R(\mathbf{F}) = R(\mathbf{E})$. *In such a case* $\mathbf{A}_{crCR}$ *is unique and is given by the expression* $\mathbf{E}(\mathbf{FAE})^{-}\mathbf{F}$.

Proof: The 'if' part is trivial. The necessity of the rank condition follows as in Theorems 4.11.5 and 4.11.6. The uniqueness follows, since under the condition $R(\mathbf{FAE}) = R(\mathbf{F}) = R(\mathbf{E})$ both $\mathbf{A}_{cR}$ and $\mathbf{A}_{rC}$ are uniquely determined by the expression $\mathbf{E}(\mathbf{FAE})^{-}\mathbf{F}$. Look for example at the expression (4.11.4), for $\mathbf{A}_{cR}$ and check that when $R(\mathbf{FAE}) = R(\mathbf{E})$.

$$\mathbf{FAE}[\mathbf{I} - (\mathbf{FAE})^{-}\mathbf{FAE}] = \mathbf{0} \Rightarrow \mathbf{E}[\mathbf{I} - (\mathbf{FAE})^{-}\mathbf{FAE}] = \mathbf{0}.$$

Note 1. Let $\mathbf{E}_1$ and $\mathbf{F}_1$ be matrices such that $\mathscr{M}(\mathbf{E}_1) = \mathscr{M}(\mathbf{E})$ and $\mathscr{M}(\mathbf{F}_1) = \mathscr{M}(\mathbf{F})$, where $\mathbf{F}$ and $\mathbf{E}$ are as defined in Theorem 4.11.1. Then

$$R(\mathbf{AE}) = R(\mathbf{E}) \Rightarrow R(\mathbf{AE}_1) = R(\mathbf{E}_1)$$

$$R(\mathbf{FAE}) = R(\mathbf{E}) \Rightarrow R(\mathbf{F}_1\mathbf{AE}_1) = R(\mathbf{E}_1)$$

$$\mathbf{E}_1(\mathbf{F}_1\mathbf{AE}_1)^{-}\mathbf{F}_1 = \mathbf{E}(\mathbf{FAE})^{-}\mathbf{F}.$$

so that $\mathbf{A}_{crCR}$ is unique for any choice of the matrices generating the subspaces $\mathscr{V}$ and $\mathscr{U}$.

Note 2. In particular let $\mathbf{P}$ and $\mathbf{Q}$ be projection operators onto $\mathscr{V}$ and $\mathscr{U}$ respectively. Then

$$\mathbf{A}_{crCR} = \mathbf{P}(\mathbf{QAP})^{-}\mathbf{Q}. \qquad (4.11.8)$$

Note 3. $\mathbf{A}$ is g-inverse of $\mathbf{A}_{crCR}$ but the converse is true only under the additional condition $R(\mathbf{FAE}) = R(\mathbf{A})$.

Note 4. When $\mathscr{V} = \mathscr{M}(\mathbf{A}^*)$ and $\mathscr{U} = \mathscr{M}(\mathbf{A})$, $\mathbf{A}_{crCR}$ coincides with $\mathbf{A}^{+}$. It may be of some historical interest to observe that Moore (1920, 1935) introduced his general reciprocal of a matrix as a constrained inverse of the type we are considering in this section. Now we consider the special case where $\mathbf{A}$ is $m \times m$ (square) matrix and the subspaces $\mathscr{V}$ and $\mathscr{U}$ are the same and discuss it in some detail. The constrained inverse $\mathbf{G}$ in such a case may be defined by the following conditions:

(i) $\mathbf{G}^*$ maps vectors of $\mathscr{E}^m$ into the subspace $\mathscr{V} \subset \mathscr{E}^m$.
(ii) $\mathbf{GA}$ is an identity in $\mathscr{V}$.

This is a special case of $\mathbf{A}_{rC}$, but we shall represent a matrix $\mathbf{G}$ satisfying the above two conditions by $\mathbf{T}$, following the notation used by Bott and Duffin. [In the condition (i) Bott and Duffin used $\mathbf{G}$ instead of $\mathbf{G}^*$ which does not characterize the matrix $\mathbf{T}$ used by them. Their definition leads to an inverse of the type $\mathbf{A}_{cC}$ which is not unique, etc.]

THEOREM 4.11.8 *Let* $\mathbf{E}$ *be a matrix such that* $\mathscr{V} = \mathscr{M}(\mathbf{E})$. *Then* $\mathbf{T}$ *exists iff* $R(\mathbf{E}^*\mathbf{AE}) = R(\mathbf{E})$, *in which case it is unique, and is of the form*

$$\mathbf{T} = \mathbf{E}(\mathbf{E}^*\mathbf{AE})^{-}\mathbf{E}^*. \tag{4.11.9}$$

Further $\mathbf{T}$ *is independent of the choice of* $\mathbf{E}$.

Proof is on the same lines as in Theorem 4.11.7.

THEOREM 4.11.9 *Let* $\mathbf{P}$ *be the projection operator onto* $\mathscr{V}$ *and* $R(\mathbf{PAP}) = R(\mathbf{P})$. *Then*

$$\mathbf{T} = \mathbf{P}(\mathbf{PAP})^{-}\mathbf{P} = \mathbf{P}(\mathbf{AP} + \mathbf{I} - \mathbf{P})^{-1}. \tag{4.11.10}$$

Proof: The first part of Equation (4.11.10) follows from Theorem 4.11.8, as we can choose $\mathbf{E}$ to be $\mathbf{P}$. For the second part, it is easy to see that $(\mathbf{AP} + \mathbf{I} - \mathbf{P})$ is nonsingular and admits a regular inverse when $R(\mathbf{PAP}) = R(\mathbf{P})$. Further

$$[\mathbf{P}(\mathbf{PAP})^{-}\mathbf{P} - \mathbf{P}(\mathbf{AP} + \mathbf{I} - \mathbf{P})^{-1}](\mathbf{AP} + \mathbf{I} - \mathbf{P}) = \mathbf{0}$$

giving $\mathbf{P}(\mathbf{PAP})^{-}\mathbf{P} = \mathbf{P}(\mathbf{AP} + \mathbf{I} - \mathbf{P})^{-1}$, which is the expression used by Bott and Duffin.

THEOREM 4.11.10 *Let* $\mathbf{A}$ *be* $m \times m$ *matrix and* $\mathbf{T}$ *be the constrained inverse as obtained in Theorem* 4.11.8. *Then*

(i) *Any arbitrary vector* $\mathbf{h}$ *admits a unique decomposition*

$$\mathbf{h} = \mathbf{Au} + \mathbf{w}, \ \mathbf{u} \in \mathscr{V} \text{ and } \mathbf{w} \in \mathscr{V}^{\perp}.$$

(ii) *The quadratic function* $Q = (\mathbf{v} - \mathbf{e})^*\mathbf{A}(\mathbf{v} - \mathbf{e}) - 2\mathbf{f}^*\mathbf{v}$, *where* $\mathbf{e}$ *and* $\mathbf{f}$ *are given vectors, attains a stationary value for variations of* $\mathbf{v}$ *in* $\mathscr{V}$. *If* $\mathbf{A}$ *is an n.n.d. matrix* $\mathbf{Q}$ *attains the minimum.*

Proof of (i): Let $\mathbf{h} = \mathbf{Au} + \mathbf{w}$. Multiplying by $\mathbf{T}$ on both sides $\mathbf{Th} = \mathbf{TAu} + \mathbf{Tw} = \mathbf{u}$. Then $\mathbf{w} = \mathbf{h} - \mathbf{ATh}$. It is easily checked that $\mathbf{Th} \in \mathscr{V}$ and $\mathbf{h} - \mathbf{ATh} \in \mathscr{V}^{\perp}$. Further, if $\mathbf{Au}_1 + \mathbf{w}_1$ is another decomposition, $\mathbf{0} = \mathbf{A}(\mathbf{u} - \mathbf{u}_1) + \mathbf{w} - \mathbf{w}_1$. Multiplying both sides by $\mathbf{T}$, $\mathbf{u} - \mathbf{u}_1 = \mathbf{0}$ and, hence $\mathbf{w} - \mathbf{w}_1 = \mathbf{0}$, so that the decomposition is unique.

Proof of (ii): Substituting $\mathbf{v} = \mathbf{v}_0 + \boldsymbol{\delta}$, $\mathbf{v}_0 \in \mathscr{V}$, $\boldsymbol{\delta} \in \mathscr{V}$, and retaining only linear terms in $\boldsymbol{\delta}$, the quadratic form becomes

$$(\mathbf{v}_0 - \mathbf{e})^*\mathbf{A}(\mathbf{v}_0 - \mathbf{e}) - 2\mathbf{f}^*\mathbf{v}_0 - 2\boldsymbol{\delta}^*(\mathbf{f} + \mathbf{Ae} - \mathbf{Av}_0).$$

Then $\mathbf{v}_0$ is a stationary point if $\boldsymbol{\delta}^*(\mathbf{Av}_0 - \mathbf{f} - \mathbf{Ae}) = \mathbf{0}$ or

$$\mathbf{Av}_0 + \mathbf{w} = \mathbf{Ae} + \mathbf{f} = \mathbf{h} \text{ (say)},$$

where $\mathbf{w} \in \mathscr{V}^{\perp}$. Applying the result (1) of the Theorem, $\mathbf{v}_0$ exists and has the value $\mathbf{v}_0 = \mathbf{Th} = \mathbf{T}(\mathbf{Ae} + \mathbf{f})$.

To show that Q attains a minimum at $\mathbf{v}_0$ when $\mathbf{A}$ is n.n.d., let us observe that for any $\mathbf{v} \in \mathscr{V}$,

$$(\mathbf{v}-\mathbf{e})^*\mathbf{A}(\mathbf{v}-\mathbf{e}) - 2\mathbf{f}^*\mathbf{v} = (\mathbf{v}_0-\mathbf{e})^*\mathbf{A}(\mathbf{v}_0-\mathbf{e}) - 2\mathbf{f}^*\mathbf{v}_0 + (\mathbf{v}-\mathbf{v}_0)^*\mathbf{A}(\mathbf{v}-\mathbf{v}_0).$$

This completes the proof of Theorem 4.11.10.

We consider a mechanical example to which the results of Theorem 4.11.10 can be applied. An elastic shaft with torsional rigidity constant g_i is held fixed at one end at the angle e_i. If the other end is twisted by angle v_i, then the elastic potential energy is $(1/2)g_i(v_i - e_i)^2$. Let us have a system with n such shafts whose free ends are connected by a chain of gears. This gives rise to linear constraints on the angles v_i. The gear chain employs differential gear boxes to perform the operation of addition and subtraction. Thus any constraint of the type $\Sigma c_i' v_i = 0$ can be enforced. A string is wound around each shaft and attached to a weight giving rise to a constant torque e_i'. The total potential energy of the system of shafts and weights is

$$Q = \tfrac{1}{2}\sum g_i(v_i - e_i)^2 - \sum e_i' v_i.$$

The potential energy is a function of the co-ordinates, so by a general principle of static mechanics, equilibrium is attained when Q is stationary subject to given constraints.

COMPLEMENTS

1. A g-inverse $\mathbf{G}$ of $\mathbf{A}$ is said to have property R if the reciprocals of nonzero eigenvalues of $\mathbf{A}$ are eigenvalues of $\mathbf{G}$, and conversely. $\mathbf{G}$ has property V if $\mathbf{x}$ is an eigenvector of $\mathbf{A}$ with eigenvalue $\lambda \neq 0$ implies that $\mathbf{x}$ is an eigenvector of $\mathbf{G}$ with eigenvalue λ^{-1}, and vice versa. Obviously, property R is weaker than property V.

(a) If $\mathbf{A}$ is hermitian, then $\mathbf{A}_{lr}^-$ has property R.

(b) If $\mathbf{A}$ and $\mathbf{A}_r^-$ commute, then $\mathbf{A}_r^-$ has property V.

(c) If $\mathbf{A}$ is normal, then $\mathbf{A}^+$ has property V. Conversely, if $\mathbf{A}$ and $\mathbf{G}$ are normal and $\mathbf{G}$ has property V, then $\mathbf{G} = \mathbf{A}^+$. (Rohde, 1966.)

2. For a positive integer k, $(\mathbf{A}_\chi^-)^k$ is one choice of $(\mathbf{A}^k)_\chi^-$, $(\mathbf{A}_\rho^-)^k$ is one choice of $(\mathbf{A}^k)_\rho^-$ and $(\mathbf{A}_{\rho\chi}^-)^k = (\mathbf{A}^k)_{\rho\chi}^-$.

3. Let $\mathbf{A}$ and $\mathbf{\Lambda}$ be square matrices of the same order and $\mathbf{\Lambda}$ be nonsingular. A g-inverse $\mathbf{G}$ of $\mathbf{A}$, such that $\mathscr{M}(\mathbf{G}) \subset \mathscr{M}(\mathbf{\Lambda A})$ is denoted by $\mathbf{A}_{\chi(\mathbf{\Lambda})}^-$. A g-inverse $\mathbf{G}$ of $\mathbf{A}$ such that $\mathscr{M}(\mathbf{G}^*) \subset \mathscr{M}(\mathbf{\Lambda}^*\mathbf{A}^*)$ is denoted by $\mathbf{A}_{\rho(\mathbf{\Lambda})}^-$. If $\mathbf{G}$ is both $\mathbf{A}_{\rho(\mathbf{\Lambda})}^-$ and $\mathbf{A}_{\chi(\mathbf{\Lambda})}^-$, it is denoted by $\mathbf{A}_{\rho\chi(\mathbf{\Lambda})}^-$.

(i) $\mathbf{A}_{\rho(\mathbf{\Lambda})}^-$, $\mathbf{A}_{\chi(\mathbf{\Lambda})}^-$ and $\mathbf{A}_{\rho\chi(\mathbf{\Lambda})}^-$ exist iff $R(\mathbf{A\Lambda A}) = R(\mathbf{A})$.

(ii) $\mathbf{A}_{\rho\chi(\mathbf{\Lambda})}^-$, if it exists, is unique and is given by $\mathbf{\Lambda A}(\mathbf{A\Lambda A\Lambda A})^-\mathbf{A\Lambda}$.

(iii) If $\mathbf{G}$ is $\mathbf{A}_{\chi((\mathbf{\Lambda})}^-$, $\mathbf{A}$ is one choice of $\mathbf{G}_{\chi(\mathbf{\Lambda}^{-1})}^-$.
If $\mathbf{G}$ is $\mathbf{A}_{\rho(\mathbf{\Lambda})}^-$, $\mathbf{A}$ is one choice of $\mathbf{G}_{\rho(\mathbf{\Lambda}^{-1})}^-$.
If $\mathbf{G}$ is $\mathbf{A}_{\rho\chi(\mathbf{\Lambda})}^-$, $\mathbf{A} = \mathbf{G}_{\rho\chi(\mathbf{\Lambda}^{-1})}^-$.

(iv) If $\mathbf{G}$ is $\mathbf{A}^-_{\chi(\Lambda)}$, $\mathbf{G}^*$ is one choice of $(\mathbf{A}^*)^-_{\rho(\Lambda^*)}$.
If $\mathbf{G}$ is $\mathbf{A}^-_{\rho(\Lambda)}$, $\mathbf{G}^*$ is one choice of $(\mathbf{A}^*)^-_{\chi(\Lambda^*)}$.

(v) If λ is a nonnull eigenvalue of $\mathbf{A}$ w.r.t. $\mathbf{\Lambda}^{-1}$ and $\mathbf{x}$ is the corresponding (right) eigenvector, that is, if $(\mathbf{A} - \lambda\mathbf{\Lambda}^{-1})\mathbf{x} = \mathbf{0}$, then λ^{-1} is an eigenvalue of $\mathbf{A}^-_{\chi(\Lambda)}$ w.r.t. $\mathbf{\Lambda}$ and the corresponding right eigenvector is given by $\mathbf{\Lambda}^{-1}\mathbf{x}$.

(vi) $\mathbf{A}^-_{\rho\chi(\Lambda)}\mathbf{A}\mathbf{\Lambda} = \mathbf{\Lambda}\mathbf{A}\mathbf{A}^-_{\rho\chi(\Lambda)}$.

(vii) For a g-inverse $\mathbf{G}$ of $\mathbf{A}$ to exist satisfying the condition $\mathbf{G}\mathbf{A}\mathbf{\Lambda} = \mathbf{\Lambda}\mathbf{A}\mathbf{G}$, it is both necessary and sufficient that $R(\mathbf{A}\mathbf{\Lambda}\mathbf{A}) = R(\mathbf{A})$.

(viii) The most general solution to a g-inverse of $\mathbf{A}$ satisfying the condition $\mathbf{G}\mathbf{A}\mathbf{\Lambda} = \mathbf{\Lambda}\mathbf{A}\mathbf{G}$ is given by

$$\mathbf{G} = \mathbf{A}^-_{\rho\chi(\Lambda)} + (\mathbf{I} - \mathbf{A}^-_{\rho\chi(\Lambda)}\mathbf{A})\mathbf{D}(\mathbf{I} - \mathbf{A}\mathbf{A}^-_{\rho\chi(\Lambda)}),$$

where $\mathbf{D}$ is abritrary.

4. If $R(\mathbf{A}) = R(\mathbf{A}^2)$, $R(\mathbf{B}) = R(\mathbf{B}^2)$, $R(\mathbf{A} + \mathbf{B}) = R(\mathbf{A}) + R(\mathbf{B})$, $\mathbf{AB} = \mathbf{BA} = \mathbf{0}$, then

(a) $R(\mathbf{A} + \mathbf{B})^2 = R(\mathbf{A} + \mathbf{B})$.

(b) $(\mathbf{A} + \mathbf{B})^-_{\rho\chi} = \mathbf{A}^-_{\rho\chi} + \mathbf{B}^-_{\rho\chi}$.

5. If $R(\mathbf{AB}) = R(\mathbf{A}) = R(\mathbf{B})$, $\mathscr{M}(\mathbf{A}) = \mathscr{M}(\mathbf{B})$, $\mathscr{M}(\mathbf{A}^*) = \mathscr{M}(\mathbf{B}^*)$ then $(\mathbf{AB})^-_{\rho\chi} = \mathbf{B}^-_{\rho\chi}\mathbf{A}^-_{\rho\chi}$.

6. (alternative expressions for principal idempotents of $\mathbf{A}$)

(a) The principal idempotents K_λ of $\mathbf{A}$ as defined in Section 2.9 are given by

$$K_\lambda = \mathbf{I} - (\mathbf{A} - \lambda\mathbf{I})^m[(\mathbf{A} - \lambda\mathbf{I})^m]^-_{\rho\chi},$$

where m is a positive integer sufficiently large.

(b) $\mathbf{K}_\lambda$ as well as the nilpotent $(\mathbf{A} - \lambda\mathbf{I})\mathbf{K}_\lambda = \mathbf{N}_\lambda$ are polynomials of finite degree in $\mathbf{A}$.

(c) $\mathbf{N}_\lambda = \mathbf{0}$ if $(\mathbf{A} - \lambda\mathbf{I})$ is of index 1, that is, $R(\mathbf{A} - \lambda\mathbf{I}) = R(\mathbf{A} - \lambda\mathbf{I})^2$, in which case the corresponding m in (a) can be taken to be equal to 1.

(d) $\mathbf{A} = \Sigma(\lambda\mathbf{K}_\lambda + \mathbf{N}_\lambda) = \Sigma\lambda\mathbf{K}_\lambda$ iff $\mathbf{A}$ is semisimple, where the summation extends over all the eigenvalues of $\mathbf{A}$.

(e) Matrices $\mathbf{A}$ and $\mathbf{B}$ commute iff their principal idempotents and nilpotents do so.

7. Consider the consistent equations $\mathbf{Ax} = \mathbf{y}$. A g-inverse $\mathbf{G}$ such that the solution $\mathbf{x} = \mathbf{Gy} \in \mathscr{M}(\mathbf{J})\ \forall\ \mathbf{y} \in \mathscr{M}(\mathbf{A})$ is called a restricted g-inverse and denoted by $\mathbf{A}^-_{(\mathbf{J})}$. Subclasses $\mathbf{A}^-_{r(\mathbf{J})}$, $\mathbf{A}^-_{m(\mathbf{N},\mathbf{J})}$, $\mathbf{A}^-_{l(\mathbf{M},\mathbf{J})}$, $\mathbf{A}^+_{\mathbf{MN}(\mathbf{J})}$ are defined as in sections 2.5, 3.1, 3.2 and 3.3 respectively.

(a) $\mathbf{A}^-_{(\mathbf{J})}$ exists iff $R(\mathbf{AJ}) = R(\mathbf{A})$ in which case the most general form of $\mathbf{A}^-_{(\mathbf{J})}$ is given by

$$\mathbf{A}^-_{(\mathbf{J})} = \mathbf{J}(\mathbf{AJ})^- + \mathbf{U}(\mathbf{I} - \mathbf{AA}^-)$$

where $\mathbf{U}$ is arbitrary and $(\mathbf{AJ})^-$ is an arbitrary g-inverse of $\mathbf{AJ}$.

(b) $\mathbf{G}$ is $\mathbf{A}^-_{r(\mathbf{J})}$ iff $\mathbf{G} = \mathbf{J}(\mathbf{AJ})^-_r$ or equivalently $\mathbf{G} = \mathbf{A}^-_{(\mathbf{J})}\mathbf{AA}^-$.

(c) $\mathbf{G}$ is $\mathbf{A}^-_{m(\mathbf{N},\mathbf{J})}$ iff

$$\mathbf{G} = \mathbf{J}(\mathbf{AJ})^-_{m(\mathbf{K})} + \mathbf{U}(\mathbf{I} - \mathbf{AA}^-)$$

where $\mathbf{K} = \mathbf{J}^*\mathbf{NJ}$, and $\mathbf{U}$ is arbitrary.

(d) $\mathbf{G}$ is $\mathbf{A}^-_{l(\mathbf{M},\mathbf{J})}$ iff $\mathbf{G} = \mathbf{J}(\mathbf{AJ})^-_{l(\mathbf{M})}$. Further such a $\mathbf{G}$ is a g-inverse of $\mathbf{A}$ if $\mathbf{M}$ is p.d. It is unique if $R(\mathbf{J}) = R(\mathbf{AJ})$.

(e) $A^+_{\mathbf{MN}(\mathbf{J})} = \mathbf{J}(\mathbf{AJ})^+_{\mathbf{MK}}$ where $\mathbf{K} = \mathbf{J}^*\mathbf{NJ}$.

8. Let **A** be a $m \times n$ matrix. A n.s. condition for **G** to be $\mathbf{A}^-$ is that there exists a matrix **B** such that **GA** is an identity in $\mathscr{M}(\mathbf{B})$ and

$$\mathscr{M}(\mathbf{B}) + \mathscr{N}(\mathbf{A}) = \mathscr{E}^n$$

9. Let **B** and **C** be such that

$$\mathscr{M}(\mathbf{A}) + \mathscr{M}(\mathbf{C}) = \mathscr{E}^m, \text{ and}$$

$$\mathscr{M}(\mathbf{B}) + \mathscr{N}(\mathbf{A}) = \mathscr{E}^n$$

then there exists an unique g-inverse such that **GA** is an identity in $\mathscr{M}(\mathbf{B})$ and $\mathbf{Gw} = \mathbf{0}\ \forall\ \mathbf{w} \in \mathscr{M}(\mathbf{C})$. Such a **G** is called an oblique pseudoinverse by Milne (1968). The oblique pseudoinverse is indeed a special case of the constrained inverse $\mathbf{A}_{rC}$, defined in section 4.11. Note that $\mathbf{Gw} = \mathbf{0}\ \forall\ \mathbf{w} \in \mathscr{M}(\mathbf{C}) \Rightarrow \mathbf{G}^*$ maps vectors in $\mathscr{E}^n$ into $\mathscr{N}(\mathbf{C}^*)$, which is condition r and **GA** is an identity in $\mathscr{M}(\mathbf{B})$ is condition C.

10. The pseudoinverse of Scroggs and Odell and naturally $\mathbf{A}^-_{\rho\chi}$ are special cases of the constrained inverse.

11. $\mathbf{A}_{\text{com}}$ is a special case of the constrained inverse.

CHAPTER 5

Projectors, Idempotent Matrices and Partial Isometry

5.1 PROJECTORS AND THEIR PROPERTIES

In this chapter we develop the theory of a class of operators called projectors and obtain explicit expressions for them. We provide a very general definition of projectors without explicitly using the concept of inner product or norm, and also study the closely related idempotent operators.

Decomposition of a Linear Space

The subspaces $\mathscr{S}_1, \mathscr{S}_2$ of a linear space $\mathscr{S}$ are said to be a decomposition of $\mathscr{S}$ if (a) $\mathscr{S}_1$ and $\mathscr{S}_2$ are virtually disjoint, that is, $\mathscr{S}_1 \cap \mathscr{S}_2$ consists only of the null vector and (b) every vector $\mathbf{x} \in \mathscr{S}$ can be expressed as $\mathbf{x} = \mathbf{x}_1 + \mathbf{x}_2$, such that $\mathbf{x}_1 \in \mathscr{S}_1$ and $\mathbf{x}_2 \in \mathscr{S}_2$. In such a case $\mathscr{S}$ is called the direct sum of $\mathscr{S}_1$ and $\mathscr{S}_2$. In symbols, this is expressed as $\mathscr{S} = \mathscr{S}_1 \oplus \mathscr{S}_2$. If $\mathscr{S} = \mathscr{S}_1 \oplus \mathscr{S}_2$ then $\mathbf{x}_1$ and $\mathbf{x}_2$ are unique.

Example. Let $\mathbf{A}$ and $\mathbf{B}$ be matrices of order $n \times p$ and $n \times q$ such that

$$R(\mathbf{A}) + R(\mathbf{B}) = R(\mathbf{A} \,\vdots\, \mathbf{B}) = n \tag{5.1.1}$$

then $\mathscr{M}(\mathbf{A})$ and $\mathscr{M}(\mathbf{B})$ provide a decomposition of E^n.

Projector. Consider an arbitrary vector $\mathbf{x} \in \mathscr{S} = \mathscr{S}_1 \oplus \mathscr{S}_2$ and express $\mathbf{x} = \mathbf{x}_1 + \mathbf{x}_2$ such that $\mathbf{x}_1 \in \mathscr{S}_1$ and $\mathbf{x}_2 \in \mathscr{S}_2$, where $\mathbf{x}_1$ and $\mathbf{x}_2$ are unique. The mapping $\mathbf{P}: \mathbf{x} \to \mathbf{x}_1$ is called the projector on $\mathscr{S}_1$ along $\mathscr{S}_2$.

LEMMA 5.1.1 *The projector* $\mathbf{P}$ *is a linear homogeneous operator.*

Lemma 5.1.1 is easy to establish and it suggests that in a finite dimensional vector space the projector can always be identified with a square matrix of a suitable order.

THEOREM 5.1.1 *A homogeneous linear operator* **P** *is a projector if and only if* **P** *is idempotent, that is,*

$$\mathbf{P}^2 = \mathbf{P}. \tag{5.1.2}$$

Proof: Let **P** be a projector on $\mathscr{S}_1$ along $\mathscr{S}_2$. It is seen that if $\mathbf{u} \in \mathscr{S}_1$, the projection of **u** on $\mathscr{S}_1$ along $\mathscr{S}_2$ is **u** itself. Hence

$$\mathbf{P}^2\mathbf{x} = \mathbf{PPx} = \mathbf{Px}\ \forall\, \mathbf{x} \in \mathscr{S} \Rightarrow \mathbf{P}^2 = \mathbf{P}.$$

Conversely, let **P** be an idempotent homogeneous linear operator. Define $\mathscr{S}_1$ to be the set of all vectors **v** such that $\mathbf{v} = \mathbf{Pu}$ for some $\mathbf{u} \in \mathscr{S}$ and let $\mathscr{S}_2$ be the set of all vectors **u** such that $\mathbf{Pu} = \mathbf{0}$. Clearly, $\mathscr{S}_1$ and $\mathscr{S}_2$ are subspaces of $\mathscr{S}$ and further $\mathscr{S} = \mathscr{S}_1 \oplus \mathscr{S}_2$.

Remark 1. If **P** is the projector on $\mathscr{S}_1$ along $\mathscr{S}_2$, then $\mathbf{I} - \mathbf{P}$ is the projector on $\mathscr{S}_2$ along $\mathscr{S}_1$.

Remark 2. It is instructive to identify $\mathscr{S}_1$ with $\mathscr{R}$, the range of **P** and $\mathscr{S}_2$ with $\mathscr{N}$, the null space of **P**.

THEOREM 5.1.2 *If* $\mathbf{P}_1$ *is the projector on* $\mathscr{R}_1$ *along* $\mathscr{N}_1$, $\mathbf{P}_2$ *the projector on* $\mathscr{R}_2$ *along* $\mathscr{N}_2$, *then* $\mathbf{P} = \mathbf{P}_1 + \mathbf{P}_2$ *is a projector if and only if*

$$\mathbf{P}_1\mathbf{P}_2 = \mathbf{P}_2\mathbf{P}_1 = \mathbf{0} \tag{5.1.3}$$

in which case **P** *is the projector on* $\mathscr{R} = \mathscr{R}_1 \oplus \mathscr{R}_2$ *along* $\mathscr{N} = \mathscr{N}_1 \cap \mathscr{N}_2$.

Proof: The 'if' part of Theorem 5.1.2 is trivial. To establish the 'only if' part observe that

$$\mathbf{P}_1^2 = \mathbf{P}_1, \qquad \mathbf{P}_2^2 = \mathbf{P}_2,$$

$$\mathbf{P}^2 = \mathbf{P} \Rightarrow \mathbf{P}_1\mathbf{P}_2 + \mathbf{P}_2\mathbf{P}_1 = \mathbf{0} \Rightarrow \mathbf{P}_1(\mathbf{P}_1\mathbf{P}_2 + \mathbf{P}_2\mathbf{P}_1) = \mathbf{P}_1\mathbf{P}_2 + \mathbf{P}_1\mathbf{P}_2\mathbf{P}_1$$
$$= \mathbf{0} \Rightarrow (\mathbf{P}_1\mathbf{P}_2 + \mathbf{P}_1\mathbf{P}_2\mathbf{P}_1)\mathbf{P}_1 = 2\mathbf{P}_1\mathbf{P}_2\mathbf{P}_1 = \mathbf{0}.$$

Back substitution in preceding steps gives $\mathbf{P}_1\mathbf{P}_2 = \mathbf{P}_2\mathbf{P}_1 = \mathbf{0}$.

To show $\mathscr{R} = \mathscr{R}_1 \oplus \mathscr{R}_2$, the required verifications are: (i) the subspaces $\mathscr{R}_1$ and $\mathscr{R}_2$ of $\mathscr{S}$ are subspaces of $\mathscr{R}$, for which check that if $\mathbf{u} \in \mathscr{R}_i$, then $\mathbf{u} = \mathbf{P}_i\mathbf{u}$. Hence $\mathbf{Pu} = \mathbf{PP}_i\mathbf{u} = \mathbf{P}_i^2\mathbf{u} = \mathbf{P}_i\mathbf{u} = \mathbf{u} \Rightarrow \mathbf{u} \in \mathscr{R} \Rightarrow \mathscr{R}_i \subset \mathscr{R}$, (ii) the only vector common to $\mathscr{R}_1$ and $\mathscr{R}_2$ is the nullvector, for which check that if $\mathbf{u} \in \mathscr{R}_1 \cap \mathscr{R}_2$, then

$$\mathbf{u} = \mathbf{P}_1\mathbf{u} = \mathbf{P}_2\mathbf{u} = \mathbf{P}_2\mathbf{P}_1\mathbf{u} = \mathbf{0}.$$

(iii) every vector $\mathbf{u} \in \mathscr{R}$ could be resolved along $\mathscr{R}_1$ and $\mathscr{R}_2$ for which observe that $\mathbf{u} = \mathbf{Pu} = \mathbf{P}_1\mathbf{u} + \mathbf{P}_2\mathbf{u}$ is the desired resolution.

Let $\mathbf{u} \in \mathscr{N}$, then $\mathbf{Pu} = \mathbf{0} \Rightarrow \mathbf{P}_1\mathbf{Pu} = \mathbf{P}_1^2\mathbf{u} = \mathbf{P}_1\mathbf{u} = \mathbf{0} \Rightarrow \mathbf{u} \in \mathscr{N}_1$. Similarly, it can be seen that $\mathbf{u} \in \mathscr{N}_2$. Hence, $\mathscr{N} \subset \mathscr{N}_1 \cap \mathscr{N}_2$. Conversely, if $\mathbf{u} \in \mathscr{N}_1 \cap \mathscr{N}_2$,

$\mathbf{Pu} = \mathbf{P}_1\mathbf{u} + \mathbf{P}_2\mathbf{u} = \mathbf{0} + \mathbf{0} \Rightarrow \mathbf{u} \in \mathcal{N}$. Hence, $\mathcal{N}_1 \cap \mathcal{N}_2 \subset \mathcal{N}$ implying consequently that $\mathcal{N} = \mathcal{N}_1 \cap \mathcal{N}_2$.

THEOREM 5.1.3 *If* $\mathbf{P}_1$ *is the projector on* $\mathcal{R}_1$ *along* $\mathcal{N}_1$ *and* $\mathbf{P}_2$ *is the projector on* $\mathcal{R}_2$ *along* $\mathcal{N}_2$, *then* $\mathbf{P} = \mathbf{P}_1 - \mathbf{P}_2$ *is a projector if and only if*

$$\mathbf{P}_1\mathbf{P}_2 = \mathbf{P}_2\mathbf{P}_1 = \mathbf{P}_2 \tag{5.1.4}$$

in which case $\mathbf{P}$ *is the projector on* $\mathcal{R} = \mathcal{R}_1 \cap \mathcal{N}_2$ *along* $\mathcal{N} = \mathcal{N}_1 \oplus \mathcal{R}_2$.

Proof: (5.1.4) $\Rightarrow (\mathbf{P}_1 - \mathbf{P}_2)^2 = \mathbf{P}_1^2 - \mathbf{P}_1\mathbf{P}_2 - \mathbf{P}_2\mathbf{P}_1 + \mathbf{P}_2^2 = \mathbf{P}_1 - \mathbf{P}_2 \Rightarrow \mathbf{P}$ is a projector. If $\mathbf{P}$ is a projector, $\mathbf{I} - \mathbf{P} = (\mathbf{I} - \mathbf{P}_1) + \mathbf{P}_2$ expresses the projector $\mathbf{I} - \mathbf{P}$ as the sum of two projectors $\mathbf{I} - \mathbf{P}_1$ and $\mathbf{P}_2$ for which, according to Theorem 5.1.2, it is necessary that

$$(\mathbf{I} - \mathbf{P}_1)\mathbf{P}_2 = \mathbf{P}_2(\mathbf{I} - \mathbf{P}_1) = \mathbf{0} \Rightarrow (5.1.4).$$

Also by Theorem 5.1.2 the range $\mathcal{N}$ of $\mathbf{I} - \mathbf{P}$ is given by the direct sum of the range $\mathcal{N}_1$ of $\mathbf{I} - \mathbf{P}_1$ and the range $\mathcal{R}_2$ of $\mathbf{P}_2$, and the null space $\mathcal{R}$ of $\mathbf{I} - \mathbf{P}$ is the intersection of the null space $\mathcal{R}_1$ of $\mathbf{I} - \mathbf{P}_1$ and the null space $\mathcal{N}_2$ of $\mathbf{P}_2$.

THEOREM 5.1.4 *If* $\mathbf{P}_1$ *is the projector on* $\mathcal{R}_1$ *along* $\mathcal{N}_1$ *and* $\mathbf{P}_2$ *is the projector on* $\mathcal{R}_2$ *along* $\mathcal{N}_2$, *then* $\mathbf{P} = \mathbf{P}_1\mathbf{P}_2$ *is a projector if*

$$\mathbf{P}_1\mathbf{P}_2 = \mathbf{P}_2\mathbf{P}_1 \tag{5.1.5}$$

in which case $\mathbf{P}$ *is the projector on* $\mathcal{R} = \mathcal{R}_1 \cap \mathcal{R}_2$ *along* $\mathcal{N} = \mathcal{N}_1 + \mathcal{N}_2$.

Proof: The first part is straightforward. $\mathcal{R} \subset \mathcal{R}_1 \cap \mathcal{R}_2$: This is seen as follows. If $\mathbf{u} \in \mathcal{R}$, then $\mathbf{u} = \mathbf{P}_1\mathbf{P}_2\mathbf{u} \Rightarrow \mathbf{P}_1\mathbf{u} = \mathbf{P}_1^2\mathbf{P}_2\mathbf{u} = \mathbf{P}_1\mathbf{P}_2\mathbf{u} = \mathbf{u} \Rightarrow \mathbf{u} \in \mathcal{R}_1$. Similarly it is seen that $\mathbf{u} \in \mathcal{R} \Rightarrow \mathbf{u} \in \mathcal{R}_2$. Hence, $\mathcal{R} \subset \mathcal{R}_1 \cap \mathcal{R}_2$. $\mathcal{R}_1 \cap \mathcal{R}_2 \subset \mathcal{R}$: For this, check that if $\mathbf{u} \in \mathcal{R}_1 \cap \mathcal{R}_2$, then $\mathbf{u} = \mathbf{P}_1\mathbf{u} = \mathbf{P}_2\mathbf{u} = \mathbf{P}_1\mathbf{P}_2\mathbf{u} = \mathbf{Pu} \Rightarrow \mathbf{u} \in \mathcal{R}$. Hence, $\mathcal{R}_1 \cap \mathcal{R}_2 \subset \mathcal{R}$. Inclusion both ways implies $\mathcal{R} = \mathcal{R}_1 \cap \mathcal{R}_2$.

Let $\mathbf{u} \in \mathcal{N}$, then $\mathbf{P}_1\mathbf{P}_2\mathbf{u} = \mathbf{0} \Rightarrow \mathbf{P}_2\mathbf{u} \in \mathcal{N}_1$. Trivially, $(\mathbf{I} - \mathbf{P}_2)\mathbf{u} \in \mathcal{N}_2$. Therefore, $\mathbf{u} = \mathbf{P}_2\mathbf{u} + (\mathbf{I} - \mathbf{P}_2)\mathbf{u} \Rightarrow \mathbf{u} \in \mathcal{N}_1 + \mathcal{N}_2$. Conversely, if $\mathbf{u} \in \mathcal{N}_1 + \mathcal{N}_2$, $\mathbf{u}$ can be expressed as $\mathbf{u} = \mathbf{u}_1 + \mathbf{u}_2$, such that $\mathbf{u}_1 \in \mathcal{N}_1$ and $\mathbf{u}_2 \in \mathcal{N}_2$. Now,

$$\mathbf{P}_1\mathbf{P}_2\mathbf{u} = \mathbf{P}_1\mathbf{P}_2\mathbf{u}_1 = \mathbf{P}_2\mathbf{P}_1\mathbf{u}_1 = \mathbf{0} \Rightarrow \mathbf{u} \in \mathcal{N}.$$

We have, therefore, $\mathcal{N} = \mathcal{N}_1 + \mathcal{N}_2$.

5.2 ORTHOGONAL PROJECTOR

Let $\mathcal{S}$ be a vector space over the field of complex numbers with an inner product $(\mathbf{u}, \mathbf{v})$ defined for each pair of vectors $\mathbf{u}, \mathbf{v}$ in $\mathcal{S}$.

Definition: The projector on $\mathcal{S}_1$ along $\mathcal{S}_2$ is called the orthogonal projector on $\mathcal{S}_1$, if $\mathcal{S}_2$ is the orthogonal complement of $\mathcal{S}_1$ in $\mathcal{S}$.

THEOREM 5.2.1 *A homogeneous linear operator* $\mathbf{P}$ *is an orthogonal projector if and only if (a)* $\mathbf{P}^2 = \mathbf{P}$ *and (b)* $\mathbf{P}^\# = \mathbf{P}$, *where* $\mathbf{P}^\#$ *is the adjoint of* $\mathbf{P}$, *defined by the condition* $(\mathbf{x}, \mathbf{Py}) = (\mathbf{P}^\#\mathbf{x}, \mathbf{y})$ *for each* $\mathbf{x}, \mathbf{y}$ *in* $\mathcal{S}$.

Proof: Since (a) is covered by Theorem 5.1.1, we confine our attention to (b). Observe that $\mathbf{Px} \in \mathcal{S}_1$, $(\mathbf{I} - \mathbf{P})\mathbf{y} \in \mathcal{S}_2 \ \forall\, \mathbf{x}, \mathbf{y} \in \mathcal{S}$. Hence if $\mathbf{P}$ is an orthogonal projector,

$$\begin{aligned}((\mathbf{I} - \mathbf{P})\mathbf{y}, \mathbf{Px}) = \mathbf{0} \ \forall\, \mathbf{x}, \mathbf{y} \in \mathcal{S} &\Rightarrow (\mathbf{P}^\#(\mathbf{I} - \mathbf{P})\mathbf{y}, \mathbf{x}) = \mathbf{0} \ \forall\, \mathbf{x}, \mathbf{y} \in \mathcal{S} \\ &\Rightarrow \mathbf{P}^\#(\mathbf{I} - \mathbf{P})\mathbf{y} = \mathbf{0} \ \forall\, \mathbf{y} \in \mathcal{S} \Rightarrow \mathbf{P}^\#(\mathbf{I} - \mathbf{P}) = \mathbf{0} \Rightarrow \mathbf{P}^\# \\ &= \mathbf{P}^\#\mathbf{P} \Rightarrow \mathbf{P} = (\mathbf{P}^\#)^\# = (\mathbf{P}^\#\mathbf{P})^\# = \mathbf{P}^\#\mathbf{P} \Rightarrow \mathbf{P}^\# = \mathbf{P}.\end{aligned}$$

The sufficiency of (a) and (b) is likewise established by defining $\mathcal{S}_1$ and $\mathcal{S}_2$ as in Theorem 5.1.1 and showing that $\mathcal{S}_2$ is the orthogonal complement of $\mathcal{S}_1$.

If $\mathcal{S}$ is $\mathcal{E}^n$ with the unitary inner product $(\mathbf{x}, \mathbf{y}) = \mathbf{y}^*\mathbf{x}$, observe that $\mathbf{P}^\# = \mathbf{P}^*$, the hermitian conjugate of $\mathbf{P}$. If the inner product relation is defined by the matrix $\mathbf{\Lambda}$ (where $\mathbf{\Lambda}$ is a hermitian p.d. matrix of order n), $\mathbf{P}^\# = \mathbf{\Lambda}^{-1}\mathbf{P}^*\mathbf{\Lambda}$.

In Section 1.1 of Chapter 1 a projector on the space generated by a matrix $\mathbf{A}$ is defined as a matrix $\mathbf{P}$, such that $\mathbf{Px} \in \mathcal{M}(\mathbf{A})$, and $\|\mathbf{x} - \mathbf{Px}\|$ is a minimum where $\|\cdot\|$ denotes a norm or a seminorm. If $\|\mathbf{x}\| = (\mathbf{x}^*\mathbf{Mx})^{1/2}$, where $\mathbf{M}$ is p.s.d., it is shown that $\mathbf{P}$ satisfied the conditions (1) $\mathbf{P}^*\mathbf{MP} = \mathbf{MP}$, (2) $\mathbf{MPA} = \mathbf{MA}$, (3) $R(\mathbf{P}) = R(\mathbf{A})$. It is seen that when $\mathbf{M}$ is p.d., we obtain the same conditions as in Theorem 5.2.1.

5.3 EXPLICIT REPRESENTATION OF PROJECTORS

We prove some lemmas which are of general interest and which will be used in obtaining explicit representation of projectors.

LEMMA 5.3.1 *Let* $\mathbf{A}_i$ *be a matrix of order* $n \times m_i$ *of rank* r_i *for* $i = 1, \ldots, k$ *such that for each* i, $\mathcal{M}(\mathbf{A}_i)$ *and* $\mathcal{M}(\mathbf{A}_i \vdots \cdots \vdots \mathbf{A}_{i-1} \vdots \mathbf{A}_{i+1} \vdots \cdots \vdots \mathbf{A}_k)$ *are virtually disjoint and* $\Sigma r_i = n$. *Then there exists a p.d. matrix* $\mathbf{\Lambda}$ *such that*

$$\mathbf{A}_i^*\mathbf{\Lambda}\mathbf{A}_j = \mathbf{0} \ \forall\, i \neq j. \tag{5.3.1}$$

The most general form of $\mathbf{\Lambda}$ *is*

$$\mathbf{\Lambda} = \sum \mathbf{B}_i\mathbf{\Lambda}_i\mathbf{B}_i^* \tag{5.3.2}$$

where $\mathbf{\Lambda}_i$ *are arbitrary hermitian matrices subject only to* $\mathbf{\Lambda}$ *being p.d., and*

$(\mathbf{B}_1 \vdots \cdots \vdots \mathbf{B}_k)^*$ *is the true inverse of* $(\mathbf{A}_{1f} \vdots \cdots \vdots \mathbf{A}_{kf})$ *where* $\mathbf{A}_{if}$ *denotes the matrix obtained from* $\mathbf{A}_i$ *by retaining only the independent columns.*

Proof: Let us observe that by construction

$$\mathbf{B}_j^*\mathbf{A}_{if} = \begin{cases} \mathbf{0} & \text{if} \quad i \neq j, \\ \mathbf{I}_{r_i} & \text{if} \quad i = j. \end{cases} \tag{5.3.3}$$

Since $\mathbf{\Lambda}$ is p.d. matrix, $\mathbf{\Lambda} = \mathbf{CC}^*$ where $\mathbf{C}$ can be written as $\mathbf{C} = \mathbf{B}_1\mathbf{E}_1 + \cdots + \mathbf{B}_k\mathbf{E}_k$, so that

$$\mathbf{\Lambda} = \sum \mathbf{B}_i\mathbf{E}_i\mathbf{E}_j^*\mathbf{B}_j^*. \tag{5.3.4}$$

Since $\mathbf{A}_i^*\mathbf{\Lambda}\mathbf{A}_j = \mathbf{0} \Rightarrow \mathbf{A}_{if}^*\mathbf{\Lambda}\mathbf{A}_{jf} = \mathbf{0}$, we have by multiplying (5.3.4) by $\mathbf{A}_{if}^*$ from the left and $\mathbf{A}_{jf}$ from the right and using (5.3.3)

$$\mathbf{E}_i\mathbf{E}_j^* = \mathbf{0}, \qquad i \neq j, \tag{5.3.5}$$

so that we arrive at the desired form (5.3.2) by choosing $\mathbf{\Lambda}_i = \mathbf{E}_i\mathbf{E}_i^*$. Thus the necessity is established. Sufficiency can be easily verified.

LEMMA 5.3.2 *Let* $\mathbf{A}_i$, $i = 1, \ldots, k$, *be as defined in Lemma 5.3.1 and* $\mathbf{\Lambda}$ *be any p.d. matrix such that (5.3.1) holds. Then the following identities hold:*

(i) $\mathbf{I} = \sum \mathbf{A}_i(\mathbf{A}_i^*\mathbf{\Lambda}\mathbf{A}_i)^-\mathbf{A}_i^*\mathbf{\Lambda}$, (5.3.6)

(ii) $\mathbf{I} = \sum \mathbf{A}_i(\mathbf{C}_i^*\mathbf{A}_i)^-\mathbf{C}_i^*$, (5.3.7)

where $\mathbf{C}_i$ *is a matrix whose columns generate the space orthogonal to the space generated by the columns of* $\mathbf{A}_1, \ldots, \mathbf{A}_{i-1}, \mathbf{A}_{i+1}, \ldots, \mathbf{A}_k$, *and*

(iii) $\mathbf{I} = \sum \mathbf{A}_i\mathbf{D}_i^*$, (5.3.8)

where $(\mathbf{D}_1 \vdots \cdots \vdots \mathbf{D}_k)^*$ *is any g-inverse of* $(\mathbf{A}_1 \vdots \cdots \vdots \mathbf{A}_k)$.

Further $\mathbf{A}_i(\mathbf{A}_i^*\mathbf{\Lambda}\mathbf{A}_i)^-\mathbf{A}_i^*\mathbf{\Lambda}$, $\mathbf{A}_i(\mathbf{C}_i^*\mathbf{A}_i)^-\mathbf{C}_i^*$ and $\mathbf{A}_i\mathbf{D}_i^*$ *are unique and equal for any choice of the g-inverse involved and any choice of* $\mathbf{\Lambda}$, $\mathbf{C}_i$ *and* $\mathbf{D}_i$ *subject to the conditions imposed on them.*

Proof: Since for each i the spaces $\mathcal{M}(\mathbf{A}_i)$ and $\mathcal{M}(\mathbf{A}_1 \vdots \cdots \vdots \mathbf{A}_{i-1} \vdots \mathbf{A}_{i+1} \vdots \cdots \vdots \mathbf{A}_k)$ are virtually disjoint and $\Sigma\, R(\mathbf{A}_i) = n$, the identity matrix can be written as

$$\mathbf{I} = \mathbf{A}_1\mathbf{F}_1 + \cdots + \mathbf{A}_k\mathbf{F}_k, \tag{5.3.9}$$

where $\mathbf{A}_i\mathbf{F}_i$ is unique (though not $\mathbf{F}_i$), $i = 1, \ldots, k$. Multiplying (5.3.9) by $\mathbf{A}_i^*\mathbf{\Lambda}$,

$$\mathbf{A}_i^*\mathbf{\Lambda} = \mathbf{A}_i^*\mathbf{\Lambda}\mathbf{A}_i\mathbf{F}_i, \qquad i = 1, \ldots, k, \tag{5.3.10}$$

and hence one choice of $\mathbf{F}_i$ is

$$(\mathbf{A}_i^*\mathbf{\Lambda}\mathbf{A}_i)^-\mathbf{A}_i^*\mathbf{\Lambda}, \tag{5.3.11}$$

so that $\mathbf{A}_i\mathbf{F}_i = \mathbf{A}_i(\mathbf{A}_i^*\mathbf{\Lambda}\mathbf{A}_i)^-\mathbf{A}_i^*\mathbf{\Lambda}$ and the identity (5.3.6) is established. Since $\mathbf{A}_i\mathbf{F}_i$ is unique, $\mathbf{A}_i(\mathbf{A}_i^*\mathbf{\Lambda}\mathbf{A}_i)^-\mathbf{A}_i^*\mathbf{\Lambda}$ is unique both with respect to choice of the g-inverse as well as of $\mathbf{\Lambda}$ subject to (5.3.2).

The identity (5.3.7) and uniqueness of the components in it are similarly established by multiplying both sides of (5.3.9) by $\mathbf{C}_i^*$.

The identity (5.3.8) is established by noting that $\mathbf{F}$, given by $\mathbf{F}^* = (\mathbf{F}_1^* \vdots \mathbf{F}_2^* \vdots \cdots \vdots \mathbf{F}_k^*)$, is, according to (5.3.9), a right inverse of $(\mathbf{A}_1 \vdots \mathbf{A}_2 \vdots \cdots \vdots \mathbf{A}_k)$.

LEMMA 5.3.3 *Let the spaces generated by $\mathbf{A}_1 \ldots, \mathbf{A}_k$ be virtually disjoint in the sense of Lemma 5.3.1 and $\Sigma\, R(\mathbf{A}_i) = n$. Then the projectors on $\mathscr{M}(\mathbf{A}_i)$ along the space generated by the rest of the matrices has the explicit representations*

$$\mathbf{A}_i(\mathbf{A}_i^*\mathbf{\Lambda}\mathbf{A}_i)^-\mathbf{A}_i^*\mathbf{\Lambda}, \qquad \mathbf{A}_i(\mathbf{C}_i^*\mathbf{A}_i)^-\mathbf{C}_i^*, \qquad \mathbf{A}_i\mathbf{D}_i^*,$$

where $\mathbf{\Lambda}$, $\mathbf{C}_i$ and $\mathbf{D}_i$ are as defined in the identities (5.3.6), (5.3.7), and (5.3.8).

Proof: The result follows by choosing any one of the identities and multiplying both sides by an arbitrary vector $\mathbf{y}$ and observing that the ith component on the right-hand side belongs to $\mathscr{M}(\mathbf{A}_i)$.

Representations of the orthogonal projector on the subspace $\mathscr{M}(\mathbf{A}) \subset \mathscr{E}^m$ are as follows:

(i) $\mathbf{P} = \mathbf{A}\mathbf{A}_l^-$, where $\mathbf{A}_l^-$ is a least squares g-inverse under a given norm,
(ii) $\mathbf{P} = \mathbf{A}(\mathbf{A}^*\mathbf{A})^-\mathbf{A}^*$, when $\|\mathbf{x}\| = (\mathbf{x}^*\mathbf{x})^{\frac{1}{2}}$,
(iii) $\mathbf{P} = \mathbf{A}(\mathbf{A}^*\mathbf{\Lambda}\mathbf{A})^-\mathbf{A}^*\mathbf{\Lambda}$, when $\|\mathbf{x}\| = (\mathbf{x}^*\mathbf{\Lambda}\mathbf{x})^{\frac{1}{2}}$ and $\mathbf{\Lambda}$ is p.d.

5.4 IDEMPOTENT MATRICES

An interesting property of idempotent matrices was obtained in Lemma 2.7.1. In view of the close relationship between the idempotent matrices and the projectors on one hand and between idempotent matrices and g-inverses on the other, we include here some further results on idempotent matrices which are also of independent interest (Khatri, 1968).

LEMMA 5.4.1 *Let $\mathbf{H}_i$, $i = 1, \ldots, k$, be square matrices of order n and $\mathbf{H} = \Sigma\,\mathbf{H}_i$. Consider the following statements*

(a) $\mathbf{H}_i^2 = \mathbf{H}_i\ \forall\, i$
(b) $\mathbf{H}_i\mathbf{H}_j = \mathbf{0}\ \forall\, i \neq j,\quad R(\mathbf{H}_i^2) = R(\mathbf{H}_i)\ \forall\, i$
(c) $\mathbf{H}^2 = \mathbf{H}$
(d) $R(\mathbf{H}) = \Sigma\, R(\mathbf{H}_i)$.

Then any two of (a), (b) and (c) imply all four. (c) and (d) imply (a) and (b).

Proof:

(i) (a) and (b) $\Rightarrow$ (c): This is straightforward.

(ii) (a) and (c) $\Rightarrow$ (d): This follows since rank of an idempotent matrix is same as its trace. Therefore, (a) and (b) $\Rightarrow$ (c) and (d).

(iii) (b) and (c) $\Rightarrow$ (a): Observe (b) and (c) $\Rightarrow \mathbf{H}_i^2 = \mathbf{H}\mathbf{H}_i = \mathbf{H}^2\mathbf{H}_i = \mathbf{H}_i^3$, that is, $\mathbf{H}_i^2(\mathbf{I} - \mathbf{H}_i) = \mathbf{0}$. Also $R(\mathbf{H}_i) = R(\mathbf{H}_i^2) \Rightarrow \mathcal{M}(\mathbf{H}_i') = \mathcal{M}(\mathbf{H}_i^{2\prime})$, that is, $\mathbf{H}_i = \mathbf{D}\mathbf{H}_i^2$ for some $\mathbf{D}$. Therefore, $\mathbf{H}_i^2(\mathbf{I} - \mathbf{H}_i) = \mathbf{0} \Rightarrow \mathbf{H}_i(\mathbf{I} - \mathbf{H}_i) = \mathbf{0}$. (iii) together with (ii) shows (b) and (c) $\Rightarrow$ (a) and (d).

(iv) (c) and (d) $\Rightarrow$ (a) and (b): Define $\mathbf{H}_0 = \mathbf{I} - \mathbf{H}$. Observe that (c) and (d) $\Rightarrow \Sigma_0^k \mathbf{H}_i = \mathbf{I}$, $\Sigma_0^k R(\mathbf{H}_i) = n$. Observe also that $R(\mathbf{I} - \mathbf{H}_i) = R(\Sigma_{j \neq i} \mathbf{H}_j) \leq \Sigma_{j \neq i} R(\mathbf{H}_j) = n - R(\mathbf{H}_i)$. However, $R(\mathbf{H}_i) + R(\mathbf{I} - \mathbf{H}_i) \geq n$. Hence $R(\mathbf{H}_i) + R(\mathbf{I} - \mathbf{H}_i) = n$. From Lemma 2.2.8, we see that this implies $\mathbf{H}_i^2 = \mathbf{H}_i \; \forall \, i$. Further

$$(\mathbf{H}_i + \mathbf{H}_j)^2 = \mathbf{H}_i + \mathbf{H}_j, \qquad \mathbf{H}_i^2 = \mathbf{H}_i,$$

$$\mathbf{H}_j^2 = \mathbf{H}_j \Rightarrow \mathbf{H}_i\mathbf{H}_j + \mathbf{H}_j\mathbf{H}_i = \mathbf{0}$$

$$\Rightarrow (\mathbf{H}_i\mathbf{H}_j + \mathbf{H}_j\mathbf{H}_i)\mathbf{H}_j = \mathbf{H}_i\mathbf{H}_j + \mathbf{H}_j\mathbf{H}_i\mathbf{H}_j = \mathbf{0}$$

$$\Rightarrow \mathbf{H}_j(\mathbf{H}_i\mathbf{H}_j + \mathbf{H}_j\mathbf{H}_i\mathbf{H}_j) = 2\mathbf{H}_j\mathbf{H}_i\mathbf{H}_j = \mathbf{0}.$$

Back substitution in the preceding step gives $\mathbf{H}_i\mathbf{H}_j = \mathbf{0} \; \forall \, i \neq j$. Thus (c) and (d) $\Rightarrow$ (a) and (b).

Remark. The following example shows that the rank condition in (b) is essential and cannot even be replaced by the condition $tr(\mathbf{H}_i^2) = tr(\mathbf{H}_i) \; \forall \, i$

$$\mathbf{H}_1 = \begin{pmatrix} 1 & 0 & 0 & 0 \\ 0 & 0 & 0 & 0 \\ 0 & 0 & 0 & 0 \\ 0 & 0 & 1 & 0 \end{pmatrix}, \quad \mathbf{H}_2 = \begin{pmatrix} 0 & 0 & 0 & 0 \\ 0 & 1 & 0 & 0 \\ 0 & 0 & 0 & 0 \\ 0 & 0 & -1 & 0 \end{pmatrix}.$$

Observe here that $\mathbf{H}_1 + \mathbf{H}_2$ is idempotent, $\mathbf{H}_1\mathbf{H}_2 = \mathbf{0}$, $\mathbf{H}_2\mathbf{H}_1 = \mathbf{0}$, $tr\,\mathbf{H}_i^2 = tr\,\mathbf{H}_i \; \forall \, i$. Yet $\mathbf{H}_1$ and $\mathbf{H}_2$ are not idempotent.

5.5 REPRESENTATION OF IDEMPOTENT MATRICES

Let $\mathbf{P}$ be an idempotent matrix, that is, $\mathbf{P}^2 = \mathbf{P}$ in which case, as shown in Theorem 5.1.1, $\mathbf{P}$ is a projector onto $\mathcal{M}(\mathbf{P})$ along the space $\mathcal{M}(\mathbf{I} - \mathbf{P})$. Then the representations of projector discussed in Lemma 5.3.3 apply. Thus we have the following lemma.

LEMMA 5.5.1 *Let* $\mathbf{P}$ *be an* $n \times m$ *idempotent matrix of rank* r. *Then the following representations hold*:

(i) $\mathbf{P} = \mathbf{A}(\mathbf{A}^*\mathbf{\Lambda}\mathbf{A})^-\mathbf{A}^*\mathbf{\Lambda}$, *where* $\mathbf{\Lambda}$ *is a p.d. matrix or a nonsingular matrix such that* $R(\mathbf{A}^*\mathbf{\Lambda A}) = R(\mathbf{A})$.

(ii) $\mathbf{P} = \mathbf{A}(\mathbf{C}^*\mathbf{A})^-\mathbf{C}^*$,

where $\mathbf{C}^* = \mathbf{A}^*\mathbf{E}$ *for some* $\mathbf{E}$ *and* $R(\mathbf{C}^*\mathbf{A}) = R(\mathbf{A})$.

(iii) $\mathbf{P} = \mathbf{AD}$,

where $\mathbf{D}$ *is a g-inverse of* $\mathbf{A}$.

All the results are easy to establish.

5.6 TRIPOTENT MATRICES

A square matrix is said to be tripotent if $\mathbf{A}^3 = \mathbf{A}$. We state a number of lemmas which are easy to prove.

LEMMA 5.6.1 *If* $\mathbf{A}$ *is tripotent, then*:

(a) $\mathbf{A}^2$ *is idempotent,*
(b) $-\mathbf{A}$ *is tripotent,*
(c) *the eigenvalues of* $\mathbf{A}$ *consist only of* $-1, 0, +1$,
(d) $R(\mathbf{A}) = \text{trace}\,(\mathbf{A}^2)$.

LEMMA 5.6.2 $\mathbf{A} = \mathbf{A}^-$ *iff* $\mathbf{A}$ *is tripotent.*

LEMMA 5.6.3 $R(\mathbf{A}) = R(\mathbf{A}^2)$ *and* $\mathbf{A}^2$ *idempotent* $\Leftrightarrow \mathbf{A}$ *is tripotent.*

Proof: If $\mathbf{A}$ is tripotent then $\mathbf{A}^3 = \mathbf{A} \Rightarrow \mathbf{A}^4 = \mathbf{A}^2$ and $R(\mathbf{A}) = R(\mathbf{A}^2)$, which proves the 'only if' part. To prove the 'if' part, let us observe that $R(\mathbf{A}) = R(\mathbf{A}^2) \Rightarrow \mathbf{A} = \mathbf{A}^2\mathbf{D}$ for some matrix $\mathbf{D}$. Since $\mathbf{A}^2$ is idempotent $\mathbf{A}^4 = \mathbf{A}^2 \Rightarrow \mathbf{A}^2\mathbf{A}^2\mathbf{D} = \mathbf{A}^2\mathbf{D} \Rightarrow \mathbf{A}^2\mathbf{A} = \mathbf{A}$ which is the required result.

LEMMA 5.6.4 *Let* $\mathbf{A}$ *be semisimple. Then a n.s. condition that* $\mathbf{A}$ *is tripotent is that the characteristic roots of* $\mathbf{A}$ *consist only of* $-1, 0, +1$.

Proof: Necessity is noted in Lemma 5.6.1. To prove sufficiency consider the decomposition $\mathbf{A} = \mathbf{P}^{-1}\mathbf{\Delta P}$, where $\mathbf{P}$ is nonsingular and $\mathbf{\Delta}$ is the diagonal matrix of characteristic roots of $\mathbf{A}$. If $\mathbf{\Delta}$ consists of only $-1, 0, 1$, we have $\mathbf{\Delta}^3 = \mathbf{\Delta}$. Then

$$\mathbf{A}^3 = \mathbf{P}^{-1}\mathbf{\Delta}^3\mathbf{P} = \mathbf{P}^{-1}\mathbf{\Delta P} = \mathbf{A}.$$

LEMMA 5.6.5 *Let* $\mathbf{A}$ *be tripotent with* n_1 *characteristic roots equal to* 1, n_2 *to* -1 *and* n_3 *to* 0. *Then*

$$\tfrac{1}{2}\,tr(\mathbf{A}^2 + \mathbf{A}) = n_1, \qquad \tfrac{1}{2}\,tr(\mathbf{A}^2 - \mathbf{A}) = n_2, \qquad tr(\mathbf{I} - \mathbf{A}^2) = n_3.$$

Proof: The results follow since $tr\,\mathbf{A} = n_1 - n_2$, $tr\,\mathbf{I} = n_1 + n_2 + n_3$, and $tr\,\mathbf{A}^2 = R(\mathbf{A}) = n_1 + n_2$ by (d) of Lemma 5.6.1.

LEMMA 5.6.6 *Let* $\mathbf{A}$ *be* $n \times n$ *matrix. For* $\mathbf{A}$ *to be tripotent, a n.s. condition is that there exist two idempotent and disjoint matrices* $\mathbf{C}$ *and* $\mathbf{D}$ *such that* $\mathbf{A} = \mathbf{C} - \mathbf{D}$. $\mathbf{C}$ *and* $\mathbf{D}$ *are unique, and* $\mathbf{C} = 2^{-1}(\mathbf{A}^2 + \mathbf{A})$ *and* $\mathbf{D} = 2^{-1}(\mathbf{A}^2 - \mathbf{A})$.

Proof: If $\mathbf{C}$ and $\mathbf{D}$ are idempotent and disjoint, (i.e., $\mathbf{DC} = \mathbf{0}$, $\mathbf{CD} = \mathbf{0}$), then $\mathbf{A}^3 = (\mathbf{C} - \mathbf{D})^3 = (\mathbf{C} - \mathbf{D})(\mathbf{C} - \mathbf{D})^2 = (\mathbf{C} - \mathbf{D})(\mathbf{C} + \mathbf{D}) = \mathbf{C}^2 - \mathbf{D}^2 = \mathbf{C} - \mathbf{D}$, which proves sufficiency. Necessity is proved by choosing $\mathbf{C}$ and $\mathbf{D}$, as given in the lemma.

To prove uniqueness let $\mathbf{A} = \mathbf{E} - \mathbf{F} = \mathbf{C} - \mathbf{D}$ with $\mathbf{E}$ and $\mathbf{F}$ satisfying the conditions of the lemma. Then $(\mathbf{E} - \mathbf{F})^2 = (\mathbf{C} - \mathbf{D})^2 \Rightarrow \mathbf{E} + \mathbf{F} = \mathbf{C} + \mathbf{D}$. Hence $\mathbf{E} = \mathbf{C}$ and $\mathbf{F} = \mathbf{D}$.

5.7 PARTIAL ISOMETRY (SUBUNITARY TRANSFORMATION)

Let us consider two finite dimensional vector spaces $\mathscr{E}^m$ and $\mathscr{E}^n$ furnished with inner products $(\cdot,\cdot)_m$, $(\cdot,\cdot)_n$ and associated norms $\|\cdot\|_m$ and $\|\cdot\|_n$. We define the adjoint of an $m \times n$ matrix $\mathbf{A}$, as in (1.1.8),

$$(\mathbf{Ax}, \mathbf{y})_m = (\mathbf{x}, \mathbf{A}^{\#}\mathbf{y})_n. \tag{5.7.1}$$

Let $\mathbf{A}$ be $m \times m$ (square matrix) such that

$$\|\mathbf{y}_1 - \mathbf{y}_2\|_m = \|\mathbf{Ay}_1 - \mathbf{Ay}_2\|_m \; \forall\, \mathbf{y}_1, \mathbf{y}_2 \in \mathscr{E}^m. \tag{5.7.2}$$

Then it is known that $\mathbf{A}^{-1} = \mathbf{A}^{\#}$ and conversely. In such a case the square matrix $\mathbf{A}$ is said to be unitary (or a unitary transformation). We shall extend this concept to linear transformations from $\mathscr{E}^n$ to $\mathscr{E}^m$ defined by a $m \times n$ matrix $\mathbf{A}$ ($\mathbf{y} = \mathbf{Ax}$, $\mathbf{x} \in \mathscr{E}^n$, $\mathbf{y} \in \mathscr{E}^m$).

Definition: An $m \times n$ matrix $\mathbf{A}$ is said to be a partial isometry (subunitary transformation) if

$$\|\mathbf{x}_1 - \mathbf{x}_2\|_n = \|\mathbf{Ax}_1 - \mathbf{Ax}_2\|_m \; \forall\, \mathbf{x}_1, \mathbf{x}_2 \in \mathscr{M}(\mathbf{A}^{\#}) \tag{5.7.3}$$

which is equivalent to

$$(\mathbf{x}_1, \mathbf{x}_2)_n = (\mathbf{Ax}_1, \mathbf{Ax}_2)_m \; \forall\, \mathbf{x}_1, \mathbf{x}_2 \in \mathscr{M}(\mathbf{A}^{\#}), \tag{5.7.4}$$

where $\mathbf{A}^{\#}$ is the adjoint of $\mathbf{A}$ as defined in (5.7.1).

It may be noted that $\mathscr{M}(\mathbf{A}^{\#})$ is the same as the subspace in $\mathscr{E}^n$ which is orthogonal to a null space of $\mathbf{A}$ or the set of all vectors $\mathbf{x}$ in $\mathscr{E}^n$ such that $\mathbf{Ax} = \mathbf{0}$. It is clear that a relation such as (5.7.3) cannot hold for all $\mathbf{x}_1, \mathbf{x}_2 \in \mathscr{E}^n$ but only for a suitable subset. The concept of partial isometry is thus a natural generalization of a unitary transformation (Erdelyi, 1966a).

LEMMA 5.7.1 $\mathbf{A}$ *is a partial isometry iff*

$$\mathbf{AA}^{\#}\mathbf{AA}^{\#} = \mathbf{AA}^{\#}, \tag{5.7.5}$$

or equivalently $\mathbf{A}^{\#}$ *is a Moore–Penrose inverse of* $\mathbf{A}$ *for appropriate norms in* $\mathscr{E}^m$ *and* $\mathscr{E}^n$.

Proof: Writing $\mathbf{x}_1 = \mathbf{A}^{\#}\mathbf{y}_1$ and $\mathbf{x}_2 = \mathbf{A}^{\#}\mathbf{y}_2$ and using (5.7.3) we have

$$(\mathbf{A}^{\#}\mathbf{y}_1, \mathbf{A}^{\#}\mathbf{y}_2) = (\mathbf{A}\mathbf{A}^{\#}\mathbf{y}_1, \mathbf{A}\mathbf{A}^{\#}\mathbf{y}_2)\ \forall\, \mathbf{y}_1, \mathbf{y}_2$$

$$\Rightarrow (\mathbf{y}_1, \mathbf{A}\mathbf{A}^{\#}\mathbf{y}_2) = (\mathbf{y}_1, \mathbf{A}\mathbf{A}^{\#}\mathbf{A}\mathbf{A}^{\#}\mathbf{y}_2)\ \forall\, \mathbf{y}_1, \mathbf{y}_2$$

$$\Rightarrow \mathbf{A}\mathbf{A}^{\#} = \mathbf{A}\mathbf{A}^{\#}\mathbf{A}\mathbf{A}^{\#}$$

which proves (5.7.5).

It easily follows from (5.7.5) that

$$\mathbf{A}\mathbf{A}^{\#} = \mathbf{P}_{\mathbf{A}} \qquad \text{and} \qquad \mathbf{A}^{\#}\mathbf{A} = \mathbf{P}_{\mathbf{A}^{\#}}, \tag{5.7.6}$$

so that $\mathbf{A}^{\#}$ is the Moore–Penrose inverse.

In particular if the inner products in $\mathscr{E}^m$ and $\mathscr{E}^n$ are of the form $(\mathbf{x}, \mathbf{y}) = \mathbf{y}^*\mathbf{x}$, then a n.s. condition for $\mathbf{A}$ to be subunitary is

$$\mathbf{A}^* = \mathbf{A}^+. \tag{5.7.7}$$

LEMMA 5.7.2 *If* $\mathbf{A}$ *is a partial isometry* $\|\mathbf{x}\|_n = \|\mathbf{A}\mathbf{x}\|_m$ *if and only if* $\mathbf{x} \in \mathscr{M}(\mathbf{A}^{\#})$.

Proof: The 'if' part follows from the definition of isometry. To prove the 'only if' part consider the orthogonal decomposition $\mathbf{x} = \mathbf{x}_1 + \mathbf{x}_2$ where $\mathbf{x}_1 \in \mathscr{N}(\mathbf{A})$, the null space of $\mathbf{A}$, and $\mathbf{x}_2 \in \mathscr{M}(\mathbf{A}^{\#})$. Then $\mathbf{A}\mathbf{x} = \mathbf{A}\mathbf{x}_2$ and $\|\mathbf{A}\mathbf{x}\| = \|\mathbf{A}\mathbf{x}_2\| = \|\mathbf{x}_2\|$, since $\mathbf{A}$ is a partial isometry. But $\|\mathbf{A}\mathbf{x}\| = \|\mathbf{x}\| = (\|\mathbf{x}_1\|^2 + \|\mathbf{x}_2\|^2)^{1/2}$. Then $\|\mathbf{x}_1\| = 0$ or $\mathbf{x}_1 = \mathbf{0}$.

In the rest of the propositions we shall assume that $(\mathbf{x}_1, \mathbf{x}_2)_n = \mathbf{x}_2^*\mathbf{x}_1$ and $(\mathbf{y}_1, \mathbf{y}_2)_m = \mathbf{y}_2^*\mathbf{y}_1$, without loss of generality.

LEMMA 5.7.3 *Let* $\mathbf{P}$ *and* $\mathbf{Q}$ *be orthogonal projectors of the same rank* r *onto subspaces of* $\mathscr{E}^m$ *and* $\mathscr{E}^n$, *respectively. Then there exists a partial isometry* $\mathbf{X}$ *of order* $m \times n$ *such that*

$$\mathbf{X}\mathbf{X}^{\#} = \mathbf{P}, \qquad \mathbf{X}^{\#}\mathbf{X} = \mathbf{Q}, \qquad \mathbf{P}\mathbf{X} = \mathbf{X}, \qquad \mathbf{Q}\mathbf{X}^{\#} = \mathbf{X}^{\#}. \tag{5.7.8}$$

Proof: Note that $\mathbf{P}$ and $\mathbf{Q}$ are selfadjoint idempotents in which case

$$\mathbf{P} = \mathbf{L}\mathbf{L}^{\#}, \qquad \mathbf{Q} = \mathbf{R}\mathbf{R}^{\#}, \tag{5.7.9}$$

where $\mathbf{L}$ is an $m \times r$ matrix and $\mathbf{R}$ is an $n \times r$ matrix, each of rank r such that $\mathbf{L}^{\#}\mathbf{L} = \mathbf{R}^{\#}\mathbf{R} = \mathbf{I}_r$. Then it is easily seen that $\mathbf{X} = \mathbf{L}\mathbf{R}^{\#}$ satisfies (5.7.8) although the solution may not be unique and this is so whatever proper norm is taken in $\mathscr{E}^r$ for the definition of $\mathbf{L}^{\#}$ and $\mathbf{R}^{\#}$. Indeed $\mathbf{X}^{\#} = \mathbf{X}^+$ so that $\mathbf{X}$ is a partial isometry.

Remark. With $\mathbf{L}$ and $\mathbf{R}$ as defined in (5.7.9) the most general solution for the equations

$$\mathbf{XX}^{+} = \mathbf{P}, \qquad \mathbf{X}^{+}\mathbf{X} = \mathbf{Q}, \qquad \mathbf{PX} = \mathbf{X}, \qquad \mathbf{QX}^{\#} = \mathbf{X}^{\#}$$

is given by

$$\mathbf{X} = \mathbf{LER}, \tag{5.7.10}$$

where $\mathbf{E}$ is an arbitrary nonsingular matrix.

Semiunitary Matrix and Supplements of a Partial Isometry. A partial isometry of full rank is called semiunitary. Notice that if the matrix $\mathbf{A}$ of order $m \times n$ is semiunitary, then either $\mathbf{AA}^{\#} = \mathbf{I}_m$ or $\mathbf{A}^{\#}\mathbf{A} = \mathbf{I}_n$ according as $\min(m, n)$ is m or n.

Lemma 5.7.4 *If* $\mathbf{A}$ *is a partial isometry (subunitary matrix) of order* $m \times n$ *and rank* $r < \min(m, n)$, *then there exists a partial isometry of the same order such that* $\mathbf{A} + \mathbf{B}$ *is semiunitary and* $\mathbf{AB}^{\#} = \mathbf{0}$ *or* $\mathbf{B}^{\#}\mathbf{A} = \mathbf{0}$ *according as* $\min(m, n)$ *is* m *or* n *(such a matrix is called a supplement of the partial isometry* $\mathbf{A}$*).*

Proof: Suppose $m = \min(m, n)$. Write $\mathbf{P} = \mathbf{I}_m - \mathbf{AA}^{\#}$ and let $\mathbf{Q}$ be an orthogonal projector of order $n \times n$ and rank $n - r$, such that $\mathbf{AQ} = \mathbf{0}$. Determine $\mathbf{B}$ as in Lemma 5.7.3 to satisfy

$$\mathbf{BB}^{\#} = \mathbf{P}, \qquad \mathbf{B}^{\#}\mathbf{B} = \mathbf{Q}, \qquad \mathbf{PB} = \mathbf{B}, \qquad \mathbf{QB}^{\#} = \mathbf{B}^{\#}.$$

$\mathbf{AQ} = \mathbf{0} \Rightarrow \mathbf{AB}^{\#} = \mathbf{0} \Rightarrow (\mathbf{A} + \mathbf{B})(\mathbf{A} + \mathbf{B})^{\#} = \mathbf{I}$. Hence, $\mathbf{A} + \mathbf{B}$ is semiunitary. The case $n = \min(m, n)$ can be proven along similar lines.

The following Lemma gives an interesting characterization of partial isometries in terms of semiunitary matrices.

Lemma 5.7.5 *Let* $\mathbf{A}$ *be a matrix of order* $m \times n$ $(m < n)$. *Then* $\mathbf{A}$ *is a partial isometry iff* $\mathbf{A} = \mathbf{UQ}$, *where* $\mathbf{U}$ *is semiunitary and* $\mathbf{Q}$ *is the orthogonal projector onto the range of* $\mathbf{A}^{\#}$.

Proof: The 'if' part is trivial. To prove the 'only if' part assume that $\mathbf{A}$ is a partial isometry, obtain $\mathbf{B}$ as in Lemma 5.7.4 and check that

$$\mathbf{A} = (\mathbf{A} + \mathbf{B})\mathbf{A}^{\#}\mathbf{A}.$$

Eigen Values and Vectors

Lemma 5.7.6 *The eigenvalues of a square partial isometry have absolute magnitudes on the closed interval* $[0, 1]$.

Proof. Let $\mathbf{A}$ be a partial isometry and $\mathbf{x}$ an eigenvector of $\mathbf{A}$ corresponding to its eigenvalue λ, then

$$\lambda\bar{\lambda}(\mathbf{x}, \mathbf{x}) = (\mathbf{Ax}, \mathbf{Ax}) = (\mathbf{x}, \mathbf{A}^{\#}\mathbf{Ax}) = (\mathbf{x}, \mathbf{Px})$$
$$= (\mathbf{x}, \mathbf{P}^{\#}\mathbf{Px}) = (\mathbf{Px}, \mathbf{Px}),$$

where $\mathbf{P} = \mathbf{A}^{\#}\mathbf{A}$ is the orthogonal projector onto $\mathcal{M}(\mathbf{A}^{\#})$. Hence

$$|\lambda|\ \|\mathbf{x}\| = \|\mathbf{Px}\| \Rightarrow |\lambda| = \frac{\|\mathbf{Px}\|}{\|\mathbf{x}\|} \leq 1.$$

LEMMA 5.7.7 *Let λ, $\mathbf{x}$ be the eigenvalue and the corresponding eigenvector of a partial isometry $\mathbf{A}$ of order $n \times n$. Then*

(a) $\lambda = 0$ *iff* $\mathbf{x} \in \mathcal{N}(\mathbf{A})$
(b) $|\lambda| = 1$ *iff* $\mathbf{x} \in [\mathcal{N}(\mathbf{A})]^{\perp}$
(c) $0 < \lambda < 1$ *iff* $\mathbf{x} \notin \mathcal{N}(\mathbf{A}) \cup [\mathcal{N}(\mathbf{A})]^{\perp}$.

Proof: (a) is easily proved. To prove (b) observe that $|\lambda| = 1 \Leftrightarrow \mathbf{A}^{\#}\mathbf{Ax} = \mathbf{x} \Leftrightarrow \mathbf{x} \in \mathcal{M}(\mathbf{A}^{\#}) = [\mathcal{N}(\mathbf{A})]^{\perp}$. (c) follows from (a), (b) and Lemma 5.7.6.

COROLLARY *If the partial isometry* $\mathbf{A}$ *is* EPr *and λ is a nonnull eigenvalue of* $\mathbf{A}$, *then* $|\lambda| = 1$.

Proof: Observe that the corresponding eigenvector $\mathbf{x} \in \mathcal{M}(\mathbf{A}) = \mathcal{M}(\mathbf{A}^{\#})$, since $\mathbf{A}$ is EPr. Apply Lemma 5.7.7.

LEMMA 5.7.8 *Let* $\mathbf{A}$ *be a partial isometry of order $m \times n$ and* $\mathbf{P}, \mathbf{Q}$ *be unitary matrices of order $m \times m$ and $n \times n$, respectively. Then* $\mathbf{B} = \mathbf{PAQ}$ *is a partial isometry.*

Proof: Lemma 5.7.8 follows from the identities

$$\mathbf{PP}^{\#} = \mathbf{P}^{\#}\mathbf{P} = \mathbf{I}_m, \quad \mathbf{QQ}^{\#} = \mathbf{Q}^{\#}\mathbf{Q} = \mathbf{I}_n.$$

LEMMA 5.7.9 *Let* $\mathbf{A}, \mathbf{B}$ *be partial isometries of order $m \times n$ and $n \times p$, respectively, then* $\mathbf{C} = \mathbf{AB}$ *is a partial isometry iff*

$$\mathbf{A}^{\#}\mathbf{AB} = \mathbf{B}. \tag{5.7.11}$$

Proof: $(5.7.11) \Rightarrow (\mathbf{ABx}, \mathbf{ABx}) = (\mathbf{Bx}, \mathbf{A}^{\#}\mathbf{ABx})$
$$= (\mathbf{Bx}, \mathbf{Bx}) = (\mathbf{x}, \mathbf{x})\ \forall\ \mathbf{x} \in \mathcal{M}(\mathbf{C}^{\#}) \subset \mathcal{M}(\mathbf{B}^{\#}).$$

Conversely, if $\mathbf{Bx} \notin \mathcal{M}(\mathbf{A}^{\#})$

$$(\mathbf{ABx}, \mathbf{ABx}) < (\mathbf{Bx}, \mathbf{Bx}) = (\mathbf{x}, \mathbf{Px}) = (\mathbf{Px}, \mathbf{Px})$$
$$\leq (\mathbf{x}, \mathbf{x}),$$

where $\mathbf{P}$ is the orthogonal projector $\mathbf{B}^{\#}\mathbf{B}$.

COMPLEMENTS

1. If $\mathbf{A},\mathbf{B}$ are matrices of order $n \times p$ and $n \times q$ such that $R(\mathbf{A} \vdots \mathbf{B}) = R(\mathbf{A}) + R(\mathbf{B}) = n$ then

$$(\mathbf{AA}^* + \mathbf{BB}^*)^{-1}$$

is a p.d. matrix and $\mathbf{A}^*(\mathbf{AA}^* + \mathbf{BB}^*)^{-1}\mathbf{B} = \mathbf{0}$.

2. Let $\mathbf{A}(n \times p)$, $\mathbf{B}(n \times q)$, $\mathbf{C}(n \times s)$ be matrices such that $R(\mathbf{A}) + R(\mathbf{B}) + R(\mathbf{C}) = R(\mathbf{A} \vdots \mathbf{B} \vdots \mathbf{C}) = n$, then $\mathbf{A}^*(\mathbf{AA}^* + \mathbf{BB}^* + \mathbf{CC}^*)^{-1}\mathbf{B} = \mathbf{0}$.

3. If $\mathbf{A}(n \times p)$, $\mathbf{B}(n \times q)$ are matrices such that $R(\mathbf{A}) + R(\mathbf{B}) = R(\mathbf{A} \vdots \mathbf{B})$, then a general p.d. solution of the equation

$$\mathbf{A}^*\mathbf{\Lambda B} = \mathbf{0}$$

is given by $\mathbf{\Lambda} = \mathbf{\Lambda}_0\mathbf{A}\mathbf{\Sigma}_1\mathbf{A}^*\mathbf{\Lambda}_0 + \mathbf{\Lambda}_0\mathbf{B}\mathbf{\Sigma}_2\mathbf{B}^*\mathbf{\Lambda}_0$, where $\mathbf{\Lambda}_0 = (\mathbf{AA}^* + \mathbf{BB}^*)^{-1}$ and $\mathbf{\Sigma}_1\mathbf{\Sigma}_2$ are arbitrary matrices subject to $\mathbf{\Lambda}$ being p.d.

4. If $\mathbf{H}$ is an idempotent matrix and $\lambda \neq 0, -1$, then the matrix $\mathbf{H} + \lambda\mathbf{I}$ is nonsingular. Obtain the inverse of $(\mathbf{H} + \lambda\mathbf{I})$.

5. If $\mathbf{H}$ is idempotent, and $\mathbf{G} = 2\mathbf{H} - \mathbf{I}$, then the matrix $\mathbf{G}$ is an involution, that is, $\mathbf{G}^2 = \mathbf{I}$. Conversely, if $\mathbf{G}$ is an involution, then the matrix $2^{-1}(\mathbf{G} + \mathbf{I})$ is idempotent.

6. If $\mathbf{A}$ is $m \times n$ matrix and the columns of $\mathbf{B}$ are an orthonormal basis of $\mathcal{M}(\mathbf{A})$, then $\mathbf{BB}^* = \mathbf{A}(\mathbf{A}^*\mathbf{A})^-\mathbf{A}^*$ for any choice of $(\mathbf{A}^*\mathbf{A})^-$.

7(a). Let $\mathbf{A}$ and $\mathbf{B}$ be matrices of order $n \times m$ and $n \times p$, respectively, and the matrix $\mathbf{C}$ be such that $\mathcal{M}(\mathbf{C}) = \mathcal{M}(\mathbf{A}) \cap \mathcal{M}(\mathbf{B})$. Then $\mathbf{C}$ has the representation $\mathbf{C} = \mathbf{AF}$, where $\mathbf{F} = \mathbf{W}^\perp$, $\mathbf{W} = \mathbf{A}^*\mathbf{B}^\perp$ and $R(\mathbf{C}) = R(\mathbf{A}) + R(\mathbf{B}) - R(\mathbf{A} \vdots \mathbf{B})$.

7(b). If the matrix $\mathbf{C}$ be such that $\mathcal{M}(\mathbf{C}) = \mathcal{M}(\mathbf{A}) \cap \mathcal{O}(\mathbf{B})$, then $\mathbf{C}$ has the representation

$$\mathbf{C} = \mathbf{A}(\mathbf{A}^*\mathbf{B})^\perp$$

(To prove 7(a) observe $(\mathbf{B}^\perp)^*\mathbf{C} = \mathbf{0}$. Further

$$\begin{aligned} R(\mathbf{C}) &= R(\mathbf{A}^* \vdots \mathbf{W}) - R(\mathbf{W}) = R(\mathbf{A}^*) - R(\mathbf{W}) \\ &= R(\mathbf{A}) - [R(\mathbf{A} \vdots \mathbf{B}) - R(\mathbf{B})]. \end{aligned}$$

Refer to Lemma 7.1.2 in the connection. Proof of 7(b) is similar).

8. The projector $\mathbf{A}_i(\mathbf{A}_i^*\mathbf{\Lambda A}_i)^-\mathbf{A}_i^*\mathbf{\Lambda}$ can be expressed as a polynomial of finite degree in $\mathbf{A}_i\mathbf{A}_i^*\mathbf{\Lambda}$

[To prove 8, note that $\mathbf{A}_i(\mathbf{A}_i^*\mathbf{\Lambda A}_i)^-\mathbf{A}_i^*\mathbf{\Lambda} = \mathbf{A}_i(\mathbf{A}_i^*\mathbf{\Lambda A}_i)^-_{\rho\chi}\mathbf{A}_i^*\mathbf{\Lambda}$. An appeal to Lemma 4.6.1 therefore establishes the result.]

9. If $\mathbf{A}$ is semisimple such that $\mathbf{A}^t = \mathbf{A}^{t+1}$ for some positive integer t, then $\mathbf{A}$ is idempotent.

10. For a symmetric idempotent matrix, the ith diagonal element is equal to sum of squares of elements in the ith row.

11. Let $\mathbf{A}_i$ be hermitian, $i = 1, \ldots, k$ such that $\mathbf{A}_i\mathbf{A}_j = \mathbf{0}$, $i \neq j$. Then there exist scalars $\alpha_i > 0$, $i = 0, 1, \ldots, k$ such that $\mathbf{V} = \alpha_o\mathbf{I} + \Sigma\alpha_i\mathbf{A}_i$ is p.d.

12. If $\mathbf{B}$ is idempotent and $\mathbf{A}$ is such that $\mathcal{M}(\mathbf{A}) \subset \mathcal{M}(\mathbf{B})$, $\mathcal{M}(\mathbf{A}^*) \subset \mathcal{M}(\mathbf{B}^*)$, then $\mathbf{BA} = \mathbf{A}$, $\mathbf{AB} = \mathbf{A}$.

13. Conversely if $\mathbf{A}$ and $\mathbf{B}$ are such that $R(\mathbf{A}) = R(\mathbf{B})$ and $\mathbf{BA} = \mathbf{A}$ or $\mathbf{AB} = \mathbf{A}$, then $\mathbf{B}$ is idempotent.

14. An idempotent matrix is uniquely determined by its row space and column space.

15. Let $\mathbf{A}$ and $\mathbf{B}$ be hermitian matrices such that $\mathbf{C} = \mathbf{A} + \mathbf{B}$, is a projector. Then:

(a) $\mathbf{A}$ and $\mathbf{B}$ are projectors iff $\mathbf{AB} = \mathbf{0}$.

(b) If $\mathbf{A}$ is a projector and $\mathbf{B}$ is p.s.d., then $\mathbf{B}$ is a projector.

(c) If $R(\mathbf{A}) + R(\mathbf{B}) \leq R(\mathbf{C})$, then $\mathbf{A}$ and $\mathbf{B}$ are projectors.

16. $\mathbf{P}_{\mathbf{A}\otimes\mathbf{B}} = \mathbf{P}_{\mathbf{A}} \otimes \mathbf{P}_{\mathbf{B}}$.

17. For a subspace $\mathcal{S}$ of $\mathcal{E}^n$ and a specified vector $\mathbf{y}$ is $\mathcal{E}^n$, let $\{\mathbf{y} + \mathcal{S}\}$ denote the set of all vectors $\mathbf{z}$ in $\mathcal{E}^n$, such that $\mathbf{z} = \mathbf{y} + \mathbf{x}$ for arbitrary $\mathbf{x} \in \mathcal{S}$. Let $\mathbf{P}$ and $\mathbf{Q}$ denote arbitrary orthogonal projectors in $\mathcal{E}^n$, with respect to the inner product definition $(\mathbf{x}, \mathbf{y})_n = \mathbf{y}^*\mathbf{x}$. Then, if $\mathbf{y} \in \mathcal{M}(\mathbf{PQ})$, $(\mathbf{PQ})^+\mathbf{y}$ is the only vector which is common to both $\mathbf{y} + \mathcal{N}(\mathbf{QP})$ and the subspace $\mathcal{M}(\mathbf{QP})$, where $\mathcal{N}(\mathbf{QP})$ is the null space of $\mathbf{QP}$ (i.e. the set of all vectors $\mathbf{x} \in \mathcal{E}^n$ such that $\mathbf{QPx} = \mathbf{0}$). (Ben-Israel, 1967a).

18. Let $\mathbf{A}$ be a partial isometry of order $n \times n$, with respect to the inner product definition $(\mathbf{x}, \mathbf{y})_n = \mathbf{y}^*\mathbf{x}$, then

(a) $\mathbf{A}$ is normal $\Leftrightarrow$ $\mathbf{A}$ is EPr.

(b) if λ is an eigenvalue of $\mathbf{A}$ and $\mathbf{x}$ is a corresponding eigenvector, the following relation is true

$$|\lambda| = \frac{\|\mathbf{P}_{\mathbf{A}^*}\mathbf{x}\|}{\|\mathbf{x}\|}$$

(c) if $\mathbf{A}$ is normal, $|\lambda| = 1$ is true for each nonnull eigenvalue λ of $\mathbf{A}$.

19. Let $\mathbf{L}$ and $\mathbf{H} = \mathbf{H}^*$ be complex matrices of order $n \times n$, such that $\mathcal{M}(\mathbf{H}) \subset \mathcal{M}(\mathbf{L})$; and $\mathbf{P}_{\mathbf{L}}$ denote the orthogonal projector $\mathbf{L}(\mathbf{L}^*\mathbf{L})^-\mathbf{L}^*$. Then

$$(\mathbf{P}_{\mathbf{L}} + i\mathbf{H})(\mathbf{P}_{\mathbf{L}} - i\mathbf{H})^+ = \mathbf{U}$$

is a normal partial isometry with respect to the inner product definition $(\mathbf{x}, \mathbf{y})_n = \mathbf{y}^*\mathbf{x}$. Further $\mathcal{M}(\mathbf{U}) = \mathcal{M}(\mathbf{L})$ and $|\mathbf{U} + \mathbf{I}| \neq 0$. Conversely, let $\mathbf{U}$ be a normal partial isometry such that $\mathcal{M}(\mathbf{U}) = \mathcal{M}(\mathbf{L})$ and let the spectrum of $\mathbf{U}$ exclude -1 then

$$\mathbf{H} = i(\mathbf{P}_{\mathbf{L}} - \mathbf{U})(\mathbf{P}_{\mathbf{L}} + \mathbf{U})^+$$

is a hermitian matrix and $\mathcal{M}(\mathbf{H}) \subset \mathcal{M}(\mathbf{L})$. (Ben-Israel, 1966c.)

CHAPTER 6

Simultaneous Reduction of a Pair of Hermitian Forms

6.1 INTRODUCTION

Given two hermitian forms $Q_1 = \mathbf{x}^*\mathbf{A}\mathbf{x}$ and $Q_2 = \mathbf{x}^*\mathbf{B}\mathbf{x}$ of which (say) Q_2 is p.d., it is well known that both are simultaneously reducible to diagonal forms

$$\sum a_i z_i \bar{z}_i \quad \text{and} \quad \sum b_i z_i \bar{z}_i \tag{6.1.1}$$

by a nonsingular transformation $\mathbf{x} = \mathbf{T}\mathbf{z}$ and also by contragredient transformations. In this chapter, necessary and sufficient conditions are obtained for these results to hold in respect of a pair of hermitian forms, none of which may be p.d. The problem studied here has therefore a strong resemblance with Weirstrass's study of strict equivalence and the canonical form for regular pencils of matrices and Kronecker's generalizations dealing with singular pencils. The approach in this chapter is, however, radically different and in many respects more elementary. The concept of proper eigenvalues of a hermitian matrix $\mathbf{A}$ with respect to a n.n.d. matrix $\mathbf{B}$ is developed in Section 6.3. It will be seen that proper eigenvectors of $\mathbf{A}$ with respect to $\mathbf{B}$ determine the transformation for simultaneous reduction of the associated hermitian forms.

Let us write $\mathbf{N}$ for $\mathbf{B}^{\perp}$. If $\mathbf{B}$ is of order n and $R(\mathbf{B}) = r$, then $\mathbf{N}$ is of order $n \times (n - r)$ and rank $(n - r)$, such that $\mathbf{N}^*\mathbf{B} = \mathbf{0}$.

The results obtained in this chapter on simultaneous reducibility of hermitian forms Q_1 and Q_2 may be summarized as follows:

Q_1	Q_2	N.S. Conditions for Reducibility: Single Transformation	Contragredient Transformations
arbitrary	n.n.d.	$R(\mathbf{AN}) = R(\mathbf{N^*AN})$ (Theorem 6.2.2)	$R(\mathbf{AB}) = R(\mathbf{BAB})$ (Theorem 6.2.4)
n.n.d.	n.n.d.	always possible (Theorem 6.2.3)	always possible (Theorem 6.2.5)
arbitrary	nonsingular	$\mathbf{AB}^{-1}$ is semisimple with real eigenvalues (Theorem 6.4.1)	$\mathbf{AB}$ is semisimple with real eigenvalues (Theorem 6.4.4)
arbitrary	arbitrary	$R(\mathbf{AN}) = R(\mathbf{N^*AN})$, $[(\mathbf{A} - \mathbf{AN}(\mathbf{N^*AN})^{-}\mathbf{N^*A}]\mathbf{B}^{-}$ is semisimple with real eigenvalues (Theorem 6.4.2)	$R(\mathbf{AB}) = R(\mathbf{BAB})$, $\mathbf{AB}$ is semisimple with real eigenvalues (Theorem 6.4.5)

6.2 A PAIR OF HERMITIAN FORMS, ONE OF WHICH IS N.N.D.

THEOREM 6.2.1 *Let* $\mathbf{A}$ *and* $\mathbf{B}$ *be hermitian matrices of order n, where* $\mathbf{B}$ *is* n.n.d. *and* $R(\mathbf{B}) = r \leq n$. *Then there exists a matrix* $\mathbf{L}$, *such that* $\mathbf{L^*BL} = \mathbf{I}_r$ *and* $\mathbf{L^*AL}$ *is diagonal.*

Proof: Since $\mathbf{B}$ is n.n.d., there exists a $r \times n$ matrix $\mathbf{C}$, such that $\mathbf{B} = \mathbf{C^*C}$. Let $\mathbf{D}$ be a right inverse of $\mathbf{C}$ and $\mathbf{\Lambda}$ be the diagonal matrix of eigenvalues of $\mathbf{D^*AD}$. There exists a unitary matrix $\mathbf{M}$ such that $\mathbf{M^*D^*ADM} = \mathbf{\Lambda}$. If $\mathbf{L} = \mathbf{DM}$, then $\mathbf{L^*AL} = \mathbf{\Lambda}$ and $\mathbf{L^*BL} = \mathbf{M^*D^*C^*CDM} = \mathbf{M^*M} = \mathbf{I}_r$, which proves Theorem 6.2.1.

THEOREM 6.2.2 *Let* $\mathbf{A}$ *and* $\mathbf{B}$ *be as in Theorem 6.2.1. Then a necessary and sufficient condition that there exists a nonsingular transformation* $\mathbf{T}$, *such that* $\mathbf{T^*AT}$ *and* $\mathbf{T^*BT}$ *are both diagonal is*

$$R(\mathbf{N^*A}) = R(\mathbf{N^*AN}), \tag{6.2.1}$$

where $\mathbf{N}$ *is written for* $\mathbf{B}^{\perp}$ *for simplicity of notation.*

Proof: Necessity is easy to establish. To prove sufficiency consider $\mathbf{S} = (\mathbf{L} - \mathbf{NG} \vdots \mathbf{N})$ where $\mathbf{L}$ is as defined in Theorem 6.2.1 and $\mathbf{G} = (\mathbf{N^*AN})^{-}\mathbf{N^*AL}$. Observe that from (6.2.1) and Lemma 2.2.4(i) we have

$$\mathbf{N^*AL} - \mathbf{N^*ANG} = \mathbf{N^*AL} - \mathbf{N^*AN}(\mathbf{N^*AN})^{-}\mathbf{N^*AL} = \mathbf{0}.$$

Hence

$$\mathbf{S^*AS} = \begin{pmatrix} \mathbf{E} & \mathbf{0} \\ \mathbf{0} & \mathbf{F} \end{pmatrix},$$

where $\mathbf{E} = (\mathbf{L} - \mathbf{NG})^*\mathbf{A}(\mathbf{L} - \mathbf{NG})$ and $\mathbf{F} = \mathbf{N^*AN}$ are hermitian matrices and

$$\mathbf{S^*BS} = \begin{pmatrix} \mathbf{I}_r & \mathbf{0} \\ \mathbf{0} & \mathbf{0} \end{pmatrix}.$$

Let $\mathbf{M}$ and $\mathbf{Q}$ be unitary matrices such that $\mathbf{M^*EM} = \mathbf{\Lambda}_1$ and $\mathbf{Q^*FQ} = \mathbf{\Lambda}_2$ are diagonal and choose

$$\mathbf{T} = \mathbf{S}\begin{pmatrix} \mathbf{M} & \mathbf{0} \\ \mathbf{0} & \mathbf{Q} \end{pmatrix}.$$

Then it is easy to verify that

$$\mathbf{T^*AT} = \begin{pmatrix} \mathbf{\Lambda}_1 & \mathbf{0} \\ \mathbf{0} & \mathbf{\Lambda}_2 \end{pmatrix}, \qquad \mathbf{T^*BT} = \begin{pmatrix} \mathbf{I}_r & \mathbf{0} \\ \mathbf{0} & \mathbf{0} \end{pmatrix}.$$

Observe that $(\mathbf{L} \,\vdots\, \mathbf{N})\begin{pmatrix} \mathbf{a} \\ \mathbf{b} \end{pmatrix} = \mathbf{La} + \mathbf{Nb} = \mathbf{0} \Rightarrow \mathbf{a^*L^*BLa} = \mathbf{b^*N^*BNb} = 0 \Rightarrow \mathbf{a^*a} = 0 \Rightarrow \mathbf{a} = \mathbf{0} \Rightarrow \mathbf{Nb} = \mathbf{0} \Rightarrow \mathbf{b} = \mathbf{0}$, since the columns of $\mathbf{N}$ are linear independent. Hence $(\mathbf{L} \,\vdots\, \mathbf{N})$ is nonsingular. Since $R(\mathbf{S}) = R(\mathbf{L} \,\vdots\, \mathbf{N})$, this implies $\mathbf{S}$ is nonsingular and in turn $\mathbf{T}$ is seen to be nonsingular. Thus Theorem 6.2.2 is proved.

THEOREM 6.2.3 *Let* $\mathbf{A}$ *and* $\mathbf{B}$ *be both* n.n.d. *matrices of order n. Then there always exists a nonsingular transformation* $\mathbf{T}$ *such that* $\mathbf{T^*AT}$ *and* $\mathbf{T^*BT}$ *are both diagonal.*

Proof: When $\mathbf{A}$ and $\mathbf{B}$ are both n.n.d., it is easy to show that $\mathcal{M}(\mathbf{N^*A}) = \mathcal{M}(\mathbf{N^*AN})$ so that the condition (6.2.1) is automatically satisfied.

THEOREM 6.2.4 *Let* $\mathbf{B}$ *be* n.n.d. *Then a* n.s. *condition for the existence of a nonsingular matrix* $\mathbf{T}$ *such that* $\mathbf{T^*BT}$ *and* $\mathbf{T}^{-1}\mathbf{A}(\mathbf{T}^{-1})^*$ *are both diagonal (i.e.* $\mathbf{A}$ *and* $\mathbf{B}$ *are reducible by contragredient transformations) is*

$$R(\mathbf{BA}) = R(\mathbf{BAB}). \tag{6.2.2}$$

Proof: Necessity is easy to establish. To prove sufficiency we need Lemma 6.2.1.

LEMMA 6.2.1 *Let* $\mathbf{A}$ *be hermitian of order n,* $\mathbf{C}$ *of order* $n \times r$ *and rank r and* $\mathbf{X}$ *of order* $r \times n$ *be such that* $\mathbf{XC} = \mathbf{I}_r$. *Then* $\mathcal{M}(\mathbf{AC}) \subset \mathcal{M}(\mathbf{X}^*) \Rightarrow$

$R(\mathbf{C^*A}) = R(\mathbf{C^*AC})$. *Conversely if* $R(\mathbf{C^*A}) = R(\mathbf{C^*AC})$, *then there exists a left inverse* $\mathbf{X}$ *of* $\mathbf{C}$ *such that* $\mathcal{M}(\mathbf{AC}) \subset \mathcal{M}(\mathbf{X^*})$.

Proof of Lemma 6.2.1: The $\Rightarrow$ part is trivial. To prove the converse we note that it suffices to show that the rank condition implies the existence of a $r \times n$ matrix $\mathbf{Y}$ such that $\mathcal{M}(\mathbf{AC}) \subset \mathcal{M}(\mathbf{Y^*})$ and $\mathbf{YC}$ is nonsingular for $\mathbf{X} = (\mathbf{YC})^{-1}\mathbf{Y}$ is a left inverse of $\mathbf{C}$ and $\mathcal{M}(\mathbf{X^*}) = \mathcal{M}(\mathbf{Y^*})$.

If $R(\mathbf{C^*A}) = p$, choose p linearly independent rows of $\mathbf{C^*A}$. Let these form the rows of a matrix $\mathbf{Y}_1$. Take $\mathbf{Y^*} = (\mathbf{Y}_1^* : \mathbf{Y}_2^*)$, where $\mathbf{Y}_2$ is of order $(r - p) \times n$ and is so chosen that $\mathbf{YC}$ is nonsingular. For example, let $\mathbf{C^*}(\mathbf{Y}_1^* : \mathbf{I}_n) = \mathbf{W} = (\mathbf{W}_1^* : \mathbf{C^*})$, where clearly $R(\mathbf{W}) = r$. Since $R(\mathbf{C^*A}) = R(\mathbf{C^*AC}) = p \Rightarrow R(\mathbf{W}_1) = p$ one can choose $r - p$ independent rows of $\mathbf{C}$ such that these rows together with the rows of $\mathbf{W}_1$ determine a nonsingular matrix of order r. The corresponding $r - p$ rows of $\mathbf{I}_n$ may be taken to constitute the rows of $\mathbf{Y}_2$. Thus Lemma 6.2.1 is established.

To complete the proof of Theorem 6.2.4, write $\mathbf{B} = \mathbf{CC^*}$ where $\mathbf{C}$ is of order $n \times r$ and $r = R(\mathbf{B})$. Observe that $(6.2.2) \Rightarrow R(\mathbf{C^*A}) = R(\mathbf{C^*AC})$. Hence by Lemma 6.2.1, a left inverse $\mathbf{L^*}$ of $\mathbf{C}$ exists such that $\mathcal{M}(\mathbf{AC}) \subset \mathcal{M}(\mathbf{L})$. Consider the partitioned matrix, $\mathbf{S} = (\mathbf{L} : \mathbf{N})$. $\mathbf{S}$ is clearly nonsingular. For $\mathbf{La} + \mathbf{Nb} = \mathbf{0} \Rightarrow \mathbf{a} = \mathbf{C^*La} = -\mathbf{C^*Nb} = \mathbf{0}$, since $\mathbf{BN} = \mathbf{0} \Rightarrow \mathbf{N^*BN} = \mathbf{0} \Rightarrow \mathbf{N^*CC^*N} = \mathbf{0} \Rightarrow \mathbf{C^*N} = \mathbf{0}$. Then $\mathbf{a} = \mathbf{0}$ and $\mathbf{La} + \mathbf{Nb} = \mathbf{0} \Rightarrow \mathbf{Nb} = \mathbf{0} \Rightarrow \mathbf{b} = \mathbf{0}$ since columns of $\mathbf{N}$ are linearly independent. Let us write $\mathbf{S}^{-1} = \begin{pmatrix} \mathbf{C^*} \\ \mathbf{F^*} \end{pmatrix}$ so that $\mathbf{F^*L} = \mathbf{0}$. Since $\mathcal{M}(\mathbf{AC}) \subset \mathcal{M}(\mathbf{L})$, $\mathbf{F^*L} = \mathbf{0} \Rightarrow \mathbf{F^*AC} = \mathbf{0}$. Observe that

$$\mathbf{S^*BS} = \begin{pmatrix} \mathbf{I}_r & \mathbf{0} \\ \mathbf{0} & \mathbf{0} \end{pmatrix} \quad \text{and} \quad \mathbf{S}^{-1}\mathbf{A}(\mathbf{S}^{-1})^* = \begin{pmatrix} \mathbf{C^*AC} & \mathbf{0} \\ \mathbf{0} & \mathbf{F^*AF} \end{pmatrix}.$$

Let $\mathbf{M}$ and $\mathbf{Q}$ be unitary matrices such that $\mathbf{M^*C^*ACM}$ and $\mathbf{Q^*F^*AFQ}$ are diagonal. Then the required transformation is

$$\mathbf{T} = \mathbf{S}\begin{pmatrix} \mathbf{M} & \mathbf{0} \\ \mathbf{0} & \mathbf{Q} \end{pmatrix}.$$

It is a simple matter to check that $\mathbf{T}$ satisfies Theorem 6.2.4.

Theorem 6.2.5 is easy to establish.

THEOREM 6.2.5 *Let* $\mathbf{A}$ *and* $\mathbf{B}$ *be n.n.d. matrices. Then there always exists a nonsingular matrix* $\mathbf{T}$ *such that both* $\mathbf{T^*BT}$ *and* $\mathbf{T}^{-1}\mathbf{A}(\mathbf{T}^{-1})^*$ *are diagonal.*

THEOREM 6.2.6 *Whatever may be the choice of transformation* **T** *in Theorem 6.2.5 satisfying the conditions*

$$\mathbf{T}^*\mathbf{B}\mathbf{T} = \begin{pmatrix} \mathbf{I}_r & \mathbf{0} \\ \mathbf{0} & \mathbf{0} \end{pmatrix}, \mathbf{T}^{-1}\mathbf{A}(\mathbf{T}^{-1})^* = \begin{pmatrix} \mathbf{\Lambda}_1 & \mathbf{0} \\ \mathbf{0} & \mathbf{\Lambda}_2 \end{pmatrix}$$

the nonzero elements of $\mathbf{\Lambda}_1$ *are precisely the nonzero eigenvalues of the matrix* **AB**.

Proof: The result is established by considering the product

$$\mathbf{T}^{-1}\mathbf{A}(\mathbf{T}^{-1})^*\mathbf{T}^*\mathbf{B}\mathbf{T} = \mathbf{T}^{-1}\mathbf{A}\mathbf{B}\mathbf{T}$$

THEOREM 6.2.7 *Let* **A** *and* **B** *be hermitian matrices of order n. Then there exists a unitary matrix* **U** *such that* **U*AU** *and* **U*BU** *are diagonal if and only if* **A** *commutes with* **B**. *In the latter case* **AB** *is hermitian and* **U*ABU** *is also diagonal.*

Proof: The 'only if' part is trivial. To prove the 'if' part let us assume **AB** = **BA**. Let $\mathbf{q}_1$ be an eigenvector of **B**, corresponding to its eigenvalue $\lambda_1 . \mathbf{B}\mathbf{q}_1 = \lambda_1 \mathbf{q}_1 \Rightarrow \mathbf{B}\mathbf{A}\mathbf{q}_1 = \mathbf{A}\mathbf{B}\mathbf{q}_1 = \lambda_1 \mathbf{A}\mathbf{q}_1$. This shows $\mathbf{A}\mathbf{q}_1$ is also an eigenvector of **B** corresponding to the same eigenvalue λ_1. Repetition of this argument shows that each nonnull vector in the linear space $\mathscr{M}(\mathbf{q}_1, \mathbf{A}\mathbf{q}_1, \mathbf{A}^2\mathbf{q}_1, \ldots)$ spanned by $\mathbf{q}_1, \mathbf{A}\mathbf{q}_1, \mathbf{A}^2\mathbf{q}_1, \ldots$ is an eigenvector of **B** for the same eigenvalue λ_1. From Lemma 4.7.7 it is seen that $\mathscr{M}(\mathbf{q}_1, \mathbf{A}\mathbf{q}_1, \mathbf{A}^2\mathbf{q}_1, \ldots)$ contains an eigenvector $\mathbf{p}_1$ of **A** for some eigenvalue μ_1 of **A**. $\mathbf{p}_1$ is thus an eigenvector of both **A** and **B**. Let now $\mathbf{q}_2$ be an eigenvector of **B** orthogonal to $\mathbf{p}_1$ corresponding to some eigenvalue λ_2 of **B**. Consider the linear space $\mathscr{M}(\mathbf{q}_2, \mathbf{A}\mathbf{q}_2, \mathbf{A}^2\mathbf{q}_2, \ldots)$ and proceeding in the same way as above, determine a nonnull vector $\mathbf{p}_2$ in $\mathscr{M}(\mathbf{q}_2, \mathbf{A}\mathbf{q}_2, \mathbf{A}^2\mathbf{q}_2, \ldots)$ which is an eigenvector of both **A** and **B**. Since for every positive integer *m*, $\mathbf{p}_1^*\mathbf{A}^m\mathbf{q}_2 = (\bar{\mu}_1)^m \mathbf{p}_1^*\mathbf{q}_2 = 0$, $\mathbf{p}_1$ is clearly orthogonal to $\mathbf{p}_2$. Repetition of the algorithm just described leads to a set of *n* mutually orthogonal vectors $\mathbf{p}_1, \mathbf{p}_2, \ldots, \mathbf{p}_n$, each of which is an eigenvector of both **A** and **B**. Let $\mathbf{p}_i/\sqrt{\mathbf{p}_i^*\mathbf{p}_i}$ be the *i*-th column of a matrix **U**. It is easily seen that **U** satisfies the conditions of Theorem 6.2.7.

THEOREM 6.2.8 *Let* $\mathbf{A}_i (i = 1, 2, \ldots, k)$ *be hermitian matrices of the same order. There exists a unitary matrix* **U**, *such that* $\mathbf{U}^*\mathbf{A}_i\mathbf{U}$ *is diagonal for each* $i = 1, 2, \ldots, k$, *if and only if* $\mathbf{A}_i$ *commutes with* $\mathbf{A}_j$ *for each i, j.*

6.3 EIGENVALUES AND VECTORS OF A MATRIX WITH RESPECT TO N.N.D. MATRIX

Let **A** and **B** be two hermitian matrices of which **B** is n.n.d. Let λ be a constant and **w** a vector such that

$$\mathbf{A}\mathbf{w} = \lambda \mathbf{B}\mathbf{w}, \qquad \mathbf{B}\mathbf{w} \neq \mathbf{0}. \tag{6.3.1}$$

Then λ is called a proper eigenvalue and $\mathbf{w}$ a proper eigenvector of $\mathbf{A}$ with respect to $\mathbf{B}$. It is seen that there may exist a vector $\mathbf{w}$ such that $\mathbf{Aw} = \mathbf{Bw} = \mathbf{0}$, in which case the equation $\mathbf{Aw} = \lambda\mathbf{Bw}$ is satisfied for arbitrary λ. Such a vector $\mathbf{w}$ is called an improper eigenvector. The space of improper eigenvectors is precisely $\mathcal{O}(\mathbf{A}^*) \cap \mathcal{O}(\mathbf{B}^*)$. We shall now study the proper eigenvalues and vectors.

THEOREM 6.3.1 *Let $R(\mathbf{B}) = r$ and $R(\mathbf{N}^*\mathbf{A}) = R(\mathbf{N}^*\mathbf{AN})$, where $\mathbf{N}$ is written for $\mathbf{B}^\perp$. Then there are precisely r proper eigenvalues of $\mathbf{A}$ with respect to $\mathbf{B}$, $\lambda_1, \lambda_2, \ldots, \lambda_r$ some of which may be repeated or null. Also $\mathbf{w}_1, \mathbf{w}_2, \ldots, \mathbf{w}_r$, the corresponding eigenvectors, can be so chosen that if $\mathbf{W}$ be the $n \times r$ matrix with $\mathbf{w}_i$ as its ith column then*

$$\mathbf{W}^*\mathbf{BW} = \mathbf{I}_r, \qquad \mathbf{W}^*\mathbf{AW} = \mathbf{\Lambda}_1, \qquad \mathbf{W}^*\mathbf{AN} = \mathbf{0}, \tag{6.3.2}$$

where $\mathbf{\Lambda}_1$ is a diagonal matrix of eigenvalues $\lambda_1, \lambda_2, \ldots, \lambda_r$.

Proof: We make use of Theorem 6.2.2 which establishes under the condition $R(\mathbf{N}^*\mathbf{A}) = R(\mathbf{N}^*\mathbf{AN})$, the existence of a nonsingular matrix $\mathbf{T}$ such that

$$\mathbf{T}^*\mathbf{AT} = \begin{pmatrix} \mathbf{\Lambda}_1 & \mathbf{0} \\ \mathbf{0} & \mathbf{\Lambda}_2 \end{pmatrix} = \mathbf{U}_1, \qquad \mathbf{T}^*\mathbf{BT} = \begin{pmatrix} \mathbf{I}_r & \mathbf{0} \\ \mathbf{0} & \mathbf{0} \end{pmatrix} = \mathbf{U}_2. \tag{6.3.3}$$

Let us consider the matrices $\mathbf{U}_1$ and $\mathbf{U}_2$. It is easy to see that the proper eigenvalues of $\mathbf{U}_1$ with respect to $\mathbf{U}_2$ are the diagonal values of $\mathbf{\Lambda}_1$, which are r in number and the corresponding proper eigenvectors can be chosen as the columns of $\begin{pmatrix} \mathbf{I}_r \\ \mathbf{0} \end{pmatrix}$. Further there are no improper eigenvalues when $\mathbf{\Lambda}_2$ is of full rank. We choose

$$\mathbf{W} = \mathbf{T}\begin{pmatrix} \mathbf{I}_r \\ \mathbf{0} \end{pmatrix} = (\mathbf{T}_1 : \mathbf{T}_2)\begin{pmatrix} \mathbf{I}_r \\ \mathbf{0} \end{pmatrix} = \mathbf{T}_1$$

so that $\mathbf{W}^*\mathbf{AW} = \mathbf{\Lambda}_1$, $\mathbf{W}^*\mathbf{BW} = \mathbf{I}_r$, and $\mathbf{W}^*\mathbf{AN} = \mathbf{0}$. Now,

$$\mathbf{U}_1\begin{pmatrix} \mathbf{I}_r \\ \mathbf{0} \end{pmatrix} = \mathbf{U}_2\begin{pmatrix} \mathbf{I}_r \\ \mathbf{0} \end{pmatrix}\mathbf{\Lambda}_1 \Rightarrow \mathbf{T}^*\mathbf{AT}\begin{pmatrix} \mathbf{I}_r \\ \mathbf{0} \end{pmatrix} = \mathbf{T}^*\mathbf{BT}\begin{pmatrix} \mathbf{I}_r \\ \mathbf{0} \end{pmatrix}\mathbf{\Lambda}_1$$

$$\Rightarrow \mathbf{T}^*\mathbf{AW} = \mathbf{T}^*\mathbf{BW\Lambda}_1 \Rightarrow \mathbf{AW} = \mathbf{BW\Lambda}_1,$$

which shows that $\mathbf{\Lambda}_1$ is the matrix of proper eigenvalues of $\mathbf{A}$ with respect to $\mathbf{B}$. Note that each column of $\mathbf{BW}$ is nonnull, since $\mathbf{W}^*\mathbf{BW} = \mathbf{I}_r$ and the theorem is established.

THEOREM 6.3.2 *Let the condition $R(\mathbf{N}^*\mathbf{A}) = R(\mathbf{N}^*\mathbf{AN})$ be satisfied. Then the nonnull proper eigenvalues of $\mathbf{A}$ with respect to $\mathbf{B}$ are same as the* *nonnull eigenvalues of* $(\mathbf{A} - \mathbf{AN}(\mathbf{N}^*\mathbf{AN})^-\mathbf{N}^*\mathbf{A})\mathbf{B}^-$ *and vice versa for any choice of the g-inverses involved.*

Proof: Let $\mathbf{H} = \mathbf{I} - \mathbf{AN}(\mathbf{N}^*\mathbf{AN})^-\mathbf{N}^*$. Then $\mathbf{HAN} = \mathbf{0} \Rightarrow \mathbf{HA} = \mathbf{DB}$ for some matrix $\mathbf{D}$. Also $\mathbf{DB} = \mathbf{BD}^*$ since $\mathbf{HA}$ is hermitian by Lemma 2.2.6(d).

Let λ be a nonnull eigenvalue of $\mathbf{HAB}^-$. Then $\mathbf{HAB}^-\mathbf{x} = \lambda\mathbf{x} \Rightarrow \mathbf{N}^*\mathbf{HAB}^-\mathbf{x} = \lambda\mathbf{N}^*\mathbf{x} = \mathbf{0} \Rightarrow \mathbf{x} = \mathbf{Bz}$ for some $\mathbf{z}$. Choosing $\mathbf{w} = \mathbf{H}^*\mathbf{B}^-\mathbf{x}$, we have

$$\mathbf{Bw} = \mathbf{BH}^*\mathbf{B}^-\mathbf{x} = \mathbf{BB}^-\mathbf{x} = \mathbf{BB}^-\mathbf{Bz} = \mathbf{Bz} = \mathbf{x},$$

$$\mathbf{Aw} = \mathbf{AH}^*\mathbf{B}^-\mathbf{x} = \mathbf{HAB}^-\mathbf{x} = \lambda\mathbf{x}, = \lambda\mathbf{Bw} \neq \mathbf{0},$$

which shows λ is a proper eigenvalue of $\mathbf{A}$ with respect to $\mathbf{B}$. To prove the converse, let λ be an eigenvalue of $\mathbf{A}$ with respect to $\mathbf{B}$. Then $\mathbf{Aw} = \lambda\mathbf{Bw}$ for some vector $\mathbf{w}$, and

$$\begin{aligned}\mathbf{HAB}^-\mathbf{Bw} &= \mathbf{DBB}^-\mathbf{Bw} = \mathbf{DBw} = \mathbf{HAw}\\ &= \mathbf{Aw} - \mathbf{AN}(\mathbf{N}^*\mathbf{AN})^-\mathbf{N}^*(\lambda\mathbf{Bw}) = \mathbf{Aw} = \lambda\mathbf{Bw},\end{aligned}$$

which shows that λ is an eigenvalue of $\mathbf{HAB}^-$ with the corresponding eigenvector $\mathbf{Bw}$.

Note that

$$[\mathbf{A} - \mathbf{AN}(\mathbf{N}^*\mathbf{AN})^-\mathbf{N}^*\mathbf{A}]\mathbf{B}^- = [\mathbf{A} - \mathbf{AP}(\mathbf{PAP})^-\mathbf{PA}]\mathbf{B}^-,$$

where $\mathbf{P}$ is the projection operator $[\mathbf{P} = \mathbf{I} - \mathbf{B}(\mathbf{B}^2)^-\mathbf{B}]$ so that we could have used $\mathbf{P}$ instead of $\mathbf{N}$ in the statement of Theorem 6.3.2.

THEOREM 6.3.3 *A necessary and sufficient condition that there are no improper eigenvectors is that rank of* $\mathbf{N}^*\mathbf{AN}$ *is full.*

The result follows since there are no improper eigenvalues when $\mathbf{\Lambda}_2$ in (6.3.3) is nonsingular.

THEOREM 6.3.4 *Let* $R(\mathbf{N}^*\mathbf{A}) = R(\mathbf{N}^*\mathbf{AN})$, $\mathbf{W}$ *be the matrix of standardized mutually orthogonal proper eigenvectors of* $\mathbf{A}$ *with respect to* $\mathbf{B}$, *and* $\mathbf{\Lambda}_1$ *be the corresponding diagonal matrix of proper eigenvalues. Then the transformation*

$$(\mathbf{W} \vdots \mathbf{N})\begin{pmatrix}\mathbf{I}_r & \mathbf{0}\\ \mathbf{0} & \mathbf{M}\end{pmatrix},$$

where $\mathbf{M}$ *is a unitary matrix such that* $\mathbf{M}^*\mathbf{N}^*\mathbf{ANM}$ *is diagonal provides the simultaneous reduction of* $\mathbf{A}$ *and* $\mathbf{B}$

$$\mathbf{T}^*\mathbf{AT} = \begin{pmatrix}\mathbf{\Lambda}_1 & \mathbf{0}\\ \mathbf{0} & \mathbf{\Lambda}_2\end{pmatrix}, \quad \mathbf{T}^*\mathbf{BT} = \begin{pmatrix}\mathbf{I}_r & \mathbf{0}\\ \mathbf{0} & \mathbf{0}\end{pmatrix},$$

where $\mathbf{\Lambda}_2$ *is the matrix of eigenvalues of* $\mathbf{N}^*\mathbf{AN}$.

Theorem 6.3.4 which is simply a restatement of the results in Theorems 6.3.1 and 6.2.2 establishes the relationship between the transformation **T** and the proper eigenvectors of **A** with respect to **B**.

6.4 A PAIR OF ARBITRARY HERMITIAN FORMS

In this section we establish the condition under which two arbitrary hermitian forms can be simultaneously reduced by cogredient and contragredient transformations.

THEOREM 6.4.1 *Let* **A** *and* **B** *be a pair of hermitian matrices and* **B** *be nonsingular. Then there exists a nonsingular matrix* **T** *such that* $\mathbf{T^*AT}$ *and* $\mathbf{T^*BT}$ *are diagonal if and only if* $\mathbf{AB}^{-1}$ *is semisimple with real eigenvalues.*

Proof: The 'only if' part is trivial. To prove the 'if' part we proceed as follows. Since $\mathbf{AB}^{-1}$ is semisimple, there exists a nonsingular matrix **L** such that $\mathbf{LAB}^{-1}\mathbf{L}^{-1} = \mathbf{D}_1$ is diagonal. Write $\mathbf{LA} = \mathbf{D}_1\mathbf{LB}$ and observe $\mathbf{LAL^*} = \mathbf{D}_1\mathbf{LBL^*}$ is hermitian. Therefore, $\mathbf{D}_1\mathbf{LBL^*} = (\mathbf{D}_1\mathbf{LBL^*})^* = \mathbf{LBL^*D}_1$, since $\mathbf{D}_1$ and $\mathbf{LBL^*}$ are hermitian. The matrices $\mathbf{LBL^*}$ and $\mathbf{D}_1$ commute. Hence by Theorem 6.2.7, there exists a unitary matrix **U**, such that $\mathbf{U^*LBL^*U}$ and $\mathbf{U^*D}_1\mathbf{U}$ are diagonal. Clearly,

$$\mathbf{U^*LAL^*U} = \mathbf{U^*D}_1\mathbf{LBL^*U} = \mathbf{U^*D}_1\mathbf{UU^*LBL^*U}$$

is also diagonal. Thus $\mathbf{T} = \mathbf{L^*U}$ satisfies Theorem 6.4.1.

THEOREM 6.4.2 *Let* **A** *and* **B** *be a pair of hermitian matrices. A necessary and sufficient condition that there exists a nonsingular transformation* **T** *such that* $\mathbf{T^*AT}$ *and* $\mathbf{T^*BT}$ *are both diagonal is*

(i) $R(\mathbf{N^*A}) = R(\mathbf{N^*AN})$ (6.4.1)

(ii) $[\mathbf{A} - \mathbf{AN}(\mathbf{N^*AN})^-\mathbf{N^*A}]\mathbf{B}^-$ *is semisimple with real eigenvalues.* (6.4.2)

Proof: While proving Theorem 6.3.2, we have noted that if $\mathbf{H} = \mathbf{I} - \mathbf{AN}(\mathbf{N^*AN})^-\mathbf{N^*}$ then $\mathbf{HA} = \mathbf{DB}$ for some matrix **D**. Hence

$$\mathbf{HAB}^- = \mathbf{DB}(\mathbf{B}^-)^*\mathbf{BB}^- = \mathbf{HAB}_r^-$$

writing $\mathbf{B}(\mathbf{B}^-)^*\mathbf{B}$ for **B** which is obviously true, since **B** is hermitian, and putting $\mathbf{B}_r^-$ for $(\mathbf{B}^-)^*\mathbf{BB}^-$. $\mathbf{B}_r^-$ is clearly a hermitian reflexive g-inverse of **B**. We may thus assume without any loss of generality that (6.4.2) is true for some such g-inverse $\mathbf{B}_r^-$ of **B**. Since **B** is hermitian, we can express **B** as

$$\mathbf{B} = \mathbf{CDC^*}, \tag{6.4.3}$$

where $\mathbf{C^*C} = \mathbf{I}_r$ and **D** is diagonal.

Let us write $\mathbf{B}_r^- = \mathbf{Y}^*\mathbf{\Lambda}\mathbf{Y}$, where $\mathbf{Y}\mathbf{Y}^* = \mathbf{I}_r$ and $\mathbf{\Lambda}$ is diagonal. $\mathbf{B}\mathbf{B}_r^-\mathbf{B} = \mathbf{B} \Rightarrow \mathbf{C}\mathbf{D}\mathbf{C}^*\mathbf{Y}^*\mathbf{\Lambda}\mathbf{Y}\mathbf{C}\mathbf{D}\mathbf{C}^* = \mathbf{C}\mathbf{D}\mathbf{C}^* \Rightarrow \mathbf{D}(\mathbf{C}^*\mathbf{Y}^*)\mathbf{\Lambda}(\mathbf{Y}\mathbf{C})\mathbf{D} = \mathbf{D} \Rightarrow (\mathbf{C}^*\mathbf{Y}^*)\mathbf{\Lambda}(\mathbf{Y}\mathbf{C}) = \mathbf{D}^{-1} \Rightarrow \mathbf{\Lambda} = (\mathbf{C}^*\mathbf{Y}^*)^{-1}\mathbf{D}^{-1}(\mathbf{Y}\mathbf{C})^{-1} \Rightarrow \mathbf{B}_r^- = \mathbf{Y}^*(\mathbf{C}^*\mathbf{Y}^*)^{-1}\mathbf{D}^{-1}(\mathbf{Y}\mathbf{C})^{-1}\mathbf{Y} = \mathbf{Z}\mathbf{D}^{-1}\mathbf{Z}^*$, where

$$\mathbf{Z} = \mathbf{Y}^*(\mathbf{C}^*\mathbf{Y}^*)^{-1} \quad \text{and} \quad \mathbf{C}^*\mathbf{Z} = \mathbf{I}_r.$$

Now consider $\mathbf{S} = (\mathbf{Z} - \mathbf{N}\mathbf{G} \vdots \mathbf{N})$, where $\mathbf{G} = (\mathbf{N}^*\mathbf{A}\mathbf{N})^-\mathbf{N}^*\mathbf{A}\mathbf{Z}$. Observe that if (6.4.1) is satisfied

$$\mathbf{S}^*\mathbf{A}\mathbf{S} = \begin{pmatrix} \mathbf{E} & \mathbf{0} \\ \mathbf{0} & \mathbf{F} \end{pmatrix},$$

where $\mathbf{E} = \mathbf{Z}^*\mathbf{H}\mathbf{A}\mathbf{Z}$ and $\mathbf{F} = \mathbf{N}^*\mathbf{A}\mathbf{N}$ are hermitian matrices, while

$$\mathbf{S}^*\mathbf{B}\mathbf{S} = \begin{pmatrix} \mathbf{D} & \mathbf{0} \\ \mathbf{0} & \mathbf{0} \end{pmatrix}.$$

If $\mathbf{E}\mathbf{D}^{-1}$ is semisimple, with real eigenvalues, then by Theorem 6.4.1 there exists a nonsingular matrix $\mathbf{K}$, such that $\mathbf{K}^*\mathbf{D}\mathbf{K}$ and $\mathbf{K}^*\mathbf{E}\mathbf{K}$ are diagonal. Let $\mathbf{Q}$ be a unitary matrix, such that $\mathbf{Q}^*\mathbf{F}\mathbf{Q}$ is diagonal. Choose

$$\mathbf{T} = \mathbf{S}\begin{pmatrix} \mathbf{K} & \mathbf{0} \\ \mathbf{0} & \mathbf{Q} \end{pmatrix},$$

and check that $\mathbf{T}^*\mathbf{A}\mathbf{T}$ and $\mathbf{T}^*\mathbf{B}\mathbf{T}$ are diagonal. To complete the proof of Theorem 6.4.2, it remains therefore to show that (6.4.2) implies the semisimplicity of $\mathbf{E}\mathbf{D}^{-1}$, which is true since, if $\mathbf{H}\mathbf{A}\mathbf{Z}\mathbf{D}^{-1}\mathbf{Z}^*$ is semisimple

$$\begin{pmatrix} \mathbf{Z}^* \\ \mathbf{N}^* \end{pmatrix}\mathbf{H}\mathbf{A}\mathbf{Z}\mathbf{D}^{-1}\mathbf{Z}^*\begin{pmatrix} \mathbf{Z}^* \\ \mathbf{N}^* \end{pmatrix}^{-1} = \begin{pmatrix} \mathbf{Z}^* \\ \mathbf{0} \end{pmatrix}\mathbf{H}\mathbf{A}\mathbf{Z}\mathbf{D}^{-1}(\mathbf{I}_r \vdots \mathbf{0}) = \begin{pmatrix} \mathbf{Z}^*\mathbf{H}\mathbf{A}\mathbf{Z}\mathbf{D}^{-1} & \mathbf{0} \\ \mathbf{0} & \mathbf{0} \end{pmatrix}$$

is obviously semisimple.

An alternative set of necessary and sufficient conditions for simultaneous reducibility of arbitrary hermitian matrices $\mathbf{A}$ and $\mathbf{B}$ is due to Kronecker. A direct proof of Kronecker's theorem, stated below as Theorem 6.4.3, will be found in Section 6, Chapter II of Gantmacher (1959). We give below certain definitions which are needed in this context. For a pair of complex matrices $\mathbf{A}$, $\mathbf{B}$ of order n, consider the matrix $\mu\mathbf{A} - \lambda\mathbf{B}$ defined in terms of 'homogeneous' parameters λ, μ. Let $g_t(\lambda, \mu)$ be the greatest common divisor (g.c.d.) of all $t \times t$ subdeterminants of $\mu\mathbf{A} - \lambda\mathbf{B}$. The invariant polynomials are obtained by the wellknown formulae

$$f_t(\lambda, \mu) = \frac{g_t(\lambda, \mu)}{g_{t-1}(\lambda, \mu)}, \quad t = 1, 2, \ldots, n,$$

where $g_o(\lambda, \mu) = 1$ by convention. The elementary divisors of the pencil $\mu\mathbf{A} - \lambda\mathbf{B}$ are powers of homogeneous polynomials irreducible in the complex field, which appear as factors of the invariant polynomials. Note that in the complex field, these polynomials are necessarily of degree 1. If $e(\lambda, \mu)$ is an elementary divisor of the pencil $\mu\mathbf{A} - \lambda\mathbf{B}$, $e(\lambda, 1)$ is obviously an elementary divisor of the pencil $\mathbf{A} - \lambda\mathbf{B}$. Conversely, if $e(\lambda)$ is an elementary divisor of the pencil $\mathbf{A} - \lambda\mathbf{B}$ of degree m in λ, $e(\lambda, \mu) = \mu^m e(\lambda/\mu)$ is an elementary divisor of $\mu\mathbf{A} - \lambda\mathbf{B}$, and all elementary divisors of the pencil $\mu\mathbf{A} - \lambda\mathbf{B}$ are obtainable this way, with the exception of elementary divisors of the type μ^m. Elementary divisors of type μ^m exist only if $|\mathbf{B}| = 0$ and are referred to as infinite elementary divisors for $\mathbf{A} - \lambda\mathbf{B}$.

Consider the equation $(\mathbf{A} - \lambda\mathbf{B})\mathbf{x} = \mathbf{0}$, and its solutions $\mathbf{x}(\lambda)$, which are polynomials in λ. Of all such solutions let $\mathbf{x}_1(\lambda)$ be a nonzero solution of minimum degree d_1. Of all other solutions which are linearly independent of $\mathbf{x}_1(\lambda)$, let $\mathbf{x}_2(\lambda)$ be a nonzero solution of minimum degree d_2. Proceeding this way, we obtain a fundamental series of solutions $\mathbf{x}_1(\lambda), \mathbf{x}_2(\lambda), \ldots, \mathbf{x}_p(\lambda)$ of degrees $d_1 \le d_2 \le \cdots \le d_p$, where p is an integer obviously $\le n$. The integers $d_1, d_2, \ldots, d_p$ do not depend upon the choice of the fundamental series of solutions of the equation $(\mathbf{A} - \lambda\mathbf{B})\mathbf{x} = \mathbf{0}$. These integers are called the minimal indices for the columns of the pencil $\mathbf{A} - \lambda\mathbf{B}$. The minimal indices for the rows could be defined in a similar fashion.

THEOREM 6.4.3 *A pair of hermitian forms* $\mathbf{x}^*\mathbf{A}\mathbf{x}$ *and* $\mathbf{x}^*\mathbf{B}\mathbf{x}$ *can be simultaneously reduced to diagonal forms (6.1.1), if and only if (i) all the elementary divisors (finite and infinite) of the pencil of matrices* $\mathbf{A} - \lambda\mathbf{B}$ *are of the first degree and (ii) all the minimal indices are equal to zero.*

It may be of some interest to deduce the sufficiency part of Theorem 6.4.2 from Theorem 6.4.3. For this purpose we show that (6.4.1) $\Rightarrow$ (ii) and (6.4.1), (6.4.2) $\Rightarrow$ (i). If all the (column) minimal indices are not equal to zero, let s be a nonzero minimal index. Consider the corresponding solution $\mathbf{x}(\lambda)$ of degree s, included in the fundamental series, and write it as

$$\mathbf{x}(\lambda) = \mathbf{x}_0 + \lambda\mathbf{x}_1 + \cdots + \lambda^s\mathbf{x}_s.$$

Since $\mathbf{x}(\lambda)$ is linearly independent of all solutions of lower degree, it is clear that $(\mathbf{A} - \lambda\mathbf{B})\mathbf{x}_s$ is not equal to $\mathbf{0}$, identically in λ. We observe that

$$(\mathbf{A} - \lambda\mathbf{B})\mathbf{x}(\lambda) = \mathbf{0} \Rightarrow \mathbf{A}\mathbf{x}_0 = \mathbf{0},$$

$\mathbf{A}\mathbf{x}_1 = \mathbf{B}\mathbf{x}_0, \ldots, \mathbf{A}\mathbf{x}_s = \mathbf{B}\mathbf{x}_{s-1}$, $\mathbf{B}\mathbf{x}_s = \mathbf{0} \Rightarrow \mathbf{x}_s = \mathbf{N}\mathbf{y}$ for some $\mathbf{y}$. Further $\mathbf{B}\mathbf{x}_{s-1} = \mathbf{A}\mathbf{x}_s = \mathbf{A}\mathbf{N}\mathbf{y} \ne \mathbf{0}$. However $\mathbf{N}^*\mathbf{A}\mathbf{N}\mathbf{y} = \mathbf{N}^*\mathbf{B}\mathbf{x}_{s-1} = \mathbf{0}$. This shows $R(\mathbf{A}\mathbf{N}) > R(\mathbf{N}^*\mathbf{A}\mathbf{N})$. Hence, if (6.4.1) is true, all the minimal indices necessarily have to be zero. Let $h_t(\lambda, \mu)$ be the g.c.d. of all $t \times t$ subdeterminants of $\mu\mathbf{H}\mathbf{A}\mathbf{B}^- - \lambda\mathbf{I}$. The semisimplicity of $\mathbf{H}\mathbf{A}\mathbf{B}^-$ implies that the invariant polynomials of the pencil $\mu\mathbf{HAB}^- - \lambda\mathbf{I}$, that is, $h_t(\lambda, \mu)/h_{t-1}(\lambda, \mu)$ are produ of distinct linear factors. Consider $\mathbf{Z}$ as defined in the proof of Theorem 6. and observe that $\mathbf{Z^*C} = \mathbf{I}_r$ implies the existence of matrices $\mathbf{M}$ and $\mathbf{W}$ order $n \times (n - r)$ each, such that

$$\begin{pmatrix} \mathbf{Z}^* \\ \mathbf{M}^* \end{pmatrix} (\mathbf{C} \vdots \mathbf{W}) = \mathbf{I}_n.$$

Since $\begin{pmatrix} \mathbf{BB}^- \\ \mathbf{M}^* \end{pmatrix}$ and $(\mathbf{B} \vdots \mathbf{W})$ are matrices of order $(2n - r) \times n$ and $n \times (2n -$ respectively, and rank n each, it is clear that $h_t(\lambda, \mu)$, as defined above, is al the g.c.d. of all $t \times t$ subdeterminants of

$$\begin{pmatrix} \mathbf{BB}^- \\ \mathbf{M}^* \end{pmatrix} (\mu\mathbf{HAB}^- - \lambda\mathbf{I})(\mathbf{B} : \mathbf{W}) = \begin{pmatrix} \mu\mathbf{HA} - \lambda\mathbf{B} & \mathbf{0} \\ \mathbf{0} & -\lambda\mathbf{I}_{n-r} \end{pmatrix}$$

whenever (6.4.1) holds.

If $k_t(\lambda, \mu)$ is the g.c.d. of all $t \times t$ subdeterminants of $\mu\mathbf{HA} - \lambda\mathbf{B}$, t following relations follow almost straightforward from the definition, one merely notes that $h_t(\lambda, \mu) = 0$ for $t \geq n + 1$,

$$h_n(\lambda, \mu) = \lambda^{n-r} k_r(\lambda, \mu),$$

$$h_{n-1}(\lambda, \mu) = \lambda^{n-r} k_{r-1}(\lambda, \mu) \text{ if } k_r(\lambda, \mu) \text{ has } \lambda \text{ as a factor,}$$

$$h_{n-1}(\lambda, \mu) = \lambda^{n-r-1} k_{r-1}(\lambda, \mu) \text{ otherwise, and so on.} \qquad (6.4$$

Let now $g_t(\lambda, \mu)$ be the g.c.d. of all $t \times t$ subdeterminants of $\mu\mathbf{A} - \lambda$ since $\mathbf{H} = \mathbf{I} - \mathbf{AN}(\mathbf{N^*AN})^-\mathbf{N}^*$ and $\mathbf{F} = \mathbf{I} - \mathbf{N}(\mathbf{N^*AN})^-\mathbf{N^*A}$, it is seen th $(\mathbf{H}^* \vdots \mathbf{N})$ and $(\mathbf{F} \vdots \mathbf{N})$ are matrices of order $n \times (2n - r)$ and rank n ea Hence $g_t(\lambda, \mu)$ is also the g.c.d. of all $t \times t$ subdeterminants of

$$\begin{pmatrix} \mathbf{H} \\ \mathbf{N}^* \end{pmatrix} (\mu\mathbf{A} - \lambda\mathbf{B})(\mathbf{F} \vdots \mathbf{N}) = \begin{pmatrix} \mu\mathbf{HA} - \lambda\mathbf{B} & \mathbf{0} \\ \mathbf{0} & \mu\mathbf{N^*AN} \end{pmatrix}$$

If $s = R(\mathbf{N^*AN})$ we have therefore

$$g_t(\lambda, \mu) = 0 \text{ for } t \geq r + s + 1$$

$$g_{r+s}(\lambda, \mu) = \mu^s k_r(\lambda, \mu) \qquad (6.4$$

$$g_{r+s-1}(\lambda, \mu) = \mu^{s-1} k_{r-1}(\lambda, \mu) \text{ and so on}$$

(6.4.4) and (6.4.5) imply that, if (6.4.1) is true, and $h_t(\lambda, \mu)/h_{t-1}(\lambda, \mu)$ a products of distinct linear factors then so are $g_t(\lambda, \mu)/g_{t-1}(\lambda, \mu)$ and vi versa. Thus (6.4.1) and (6.4.2) $\Rightarrow$ (i).

THEOREM 6.4.4 *Let* **A** *and* **B** *be a pair of hermitian matrices and* **B** *be nonsingular. Then there exists a nonsingular matrix* **T** *such that* $\mathbf{T}^*\mathbf{B}\mathbf{T}$ *and* $\mathbf{T}^{-1}\mathbf{A}(\mathbf{T}^{-1})^*$ *are diagonal if and only if* **AB** *is semisimple, with real eigenvalues.*

Proof: Theorem 6.4.4 follows a simple consequence of Theorem 6.4.1.

THEOREM 6.4.5 *Let* **A** *and* **B** *be a pair of hermitian matrices. Then a necessary and sufficient condition that there exists a nonsingular matrix* **T** *such that* $\mathbf{T}^*\mathbf{B}\mathbf{T}$ *and* $\mathbf{T}^{-1}\mathbf{A}(\mathbf{T}^{-1})^*$ *are both diagonal is that*

$$\left.\begin{array}{l}\text{(i) } R(\mathbf{BA}) = R(\mathbf{BAB}) \\ \text{(ii) } \mathbf{AB} \text{ is semisimple with real eigenvalues.}\end{array}\right\} \tag{6.4.6}$$

Proof: Necessity is easy to establish. To prove sufficiency, determine **C** as in (6.4.3) and note that (6.4.6) $\Rightarrow R(\mathbf{C}^*\mathbf{A}) = R(\mathbf{C}^*\mathbf{AC})$. Hence, proceeding as in the proof of Theorem 6.2.4, one can determine **S** such that

$$\mathbf{S}^*\mathbf{BS} = \begin{pmatrix} \mathbf{D} & \mathbf{0} \\ \mathbf{0} & \mathbf{0} \end{pmatrix} \quad \text{and} \quad (\mathbf{S}^{-1})\mathbf{A}(\mathbf{S}^{-1})^* = \begin{pmatrix} \mathbf{C}^*\mathbf{AC} & \mathbf{0} \\ \mathbf{0} & \mathbf{F}^*\mathbf{AF} \end{pmatrix}.$$

Note that (6.4.6) implies

$$(\mathbf{S}^{-1})\mathbf{A}(\mathbf{S}^{-1})^*\mathbf{S}^*\mathbf{BS} = \begin{pmatrix} \mathbf{C}^*\mathbf{ACD} & \mathbf{0} \\ \mathbf{0} & \mathbf{0} \end{pmatrix}$$

is semisimple. Hence $\mathbf{C}^*\mathbf{ACD}$ is semisimple. Since **D** is nonsingular, by Theorem 6.4.4 there exists a nonsingular matrix **K** such that $\mathbf{K}^*\mathbf{DK}$ and $\mathbf{K}^{-1}\mathbf{C}^*\mathbf{AC}(\mathbf{K}^{-1})^*$ are both diagonal. Let **Q** be a unitary matrix such that $\mathbf{Q}^*\mathbf{F}^*\mathbf{AFQ}$ is diagonal. It is easy to check that

$$\mathbf{T} = \mathbf{S}\begin{pmatrix} \mathbf{K} & \mathbf{0} \\ \mathbf{0} & \mathbf{Q} \end{pmatrix}$$

satisfies Theorem 6.4.5.

6.5 SIMULTANEOUS REDUCTION OF SEVERAL HERMITIAN FORMS

We have already considered at least two situations in which several hermitian forms

$$Q_i = \mathbf{x}^*\mathbf{A}_i\mathbf{x}, \qquad i = 1, 2, \ldots, k$$

can be simultaneously diagonalized.

(i) If $\mathbf{A}_i\mathbf{A}_j = \mathbf{A}_j\mathbf{A}_i$ for all i, j, then there exists a unitary transformation $\mathbf{T}(\mathbf{TT}^* = \mathbf{I})$ such that $\mathbf{T}^*\mathbf{A}_i\mathbf{T}$ is diagonal for each i. (Theorem 6.2.8.)

(ii) If $\mathbf{A} = \Sigma \mathbf{A}_i$ is n.n.d. and $R(\mathbf{A}) = \Sigma R(\mathbf{A}_i)$, then there exists a matrix $\mathbf{T}$ such that $\mathbf{T}^*\mathbf{A}_i\mathbf{T}$ is diagonal for each i [(ii) follows from (i)].

In this section we consider a few other possibilities due to Bhimasankaram (1971).

THEOREM 6.5.1 *Let* $\mathbf{A}_1, \ldots, \mathbf{A}_k$ *be p.d. matrices. Then there exists a nonsingular matrix* $\mathbf{T}$*, such that* $\mathbf{T}^*\mathbf{A}_j\mathbf{T}$ *is diagonal for* $j = 1, \ldots, k$ *iff there exists an* i, $1 \le i \le k$ *such that* $\mathbf{A}_j^*\mathbf{A}_i^{-1}\mathbf{A}_m = \mathbf{A}_m^*\mathbf{A}_i^{-1}\mathbf{A}_j$ *for all* j, m.

Proof: Let $\mathbf{A}_i^{-1} = \mathbf{LL}^*$, then $\mathbf{L}^*\mathbf{A}_j\mathbf{LL}^*\mathbf{A}_m\mathbf{L} = \mathbf{L}^*\mathbf{A}_m\mathbf{LL}^*\mathbf{A}_j\mathbf{L}$. Hence using the result (i) above, there exists a unitary matrix $\mathbf{P}$ such that

$$\mathbf{P}^*\mathbf{L}^*\mathbf{A}_j\mathbf{LP} \text{ is diagonal for all } j.$$

Thus the 'if' part is proved by choosing $\mathbf{T} = \mathbf{LP}$.

To prove the 'only if' part, consider $\mathbf{T}^*\mathbf{A}_1\mathbf{T} = \mathbf{\Delta}$ (diagonal matrix with diagonal element $d_i > 0, i = 1, 2, \ldots$). Let $\mathbf{F} = \text{diag.}\,(1/\sqrt{d_1}, 1/\sqrt{d_2}, \ldots, 1/\sqrt{d_n})$. Then $\mathbf{F}^*\mathbf{T}^*\mathbf{A}_1\mathbf{TF} = \mathbf{I}$. Hence $\mathbf{A}_1^{-1} = \mathbf{TFF}^*\mathbf{T}^*$. Further, $\mathbf{F}^*\mathbf{T}^*\mathbf{A}_j\mathbf{TF}$ is diagonal for $j = 2, \ldots, k$. Hence

$$\mathbf{F}^*\mathbf{T}^*\mathbf{A}_j\mathbf{TFF}^*\mathbf{T}^*\mathbf{A}_m\mathbf{TF} = \mathbf{F}^*\mathbf{T}^*\mathbf{A}_m\mathbf{TFF}^*\mathbf{T}^*\mathbf{A}_j\mathbf{TF} \qquad \forall\, m \text{ and } j,$$

$$\Rightarrow \mathbf{A}_j\mathbf{A}_1^{-1}\mathbf{A}_m = \mathbf{A}_m\mathbf{A}_1^{-1}\mathbf{A}_j \qquad \forall\, m \text{ and } j.$$

COROLLARY: Let $\mathbf{A}_1, \mathbf{A}_2, \ldots, \mathbf{A}_k$ be p.d. matrices. If there exists $i : 1 \le i \le k$ such that $\mathbf{A}_m\mathbf{A}_i^{-1}\mathbf{A}_j = \mathbf{A}_j\mathbf{A}_i^{-1}\mathbf{A}_m \,\forall\, m$ and j, then for every positive integer p, $1 \le p \le k$ we have

$$\mathbf{A}_m\mathbf{A}_p^{-1}\mathbf{A}_j = \mathbf{A}_j\mathbf{A}_p^{-1}\mathbf{A}_m.$$

THEOREM 6.5.2 *Let* $\mathbf{A}_1, \mathbf{A}_2, \ldots, \mathbf{A}_k$ *be hermitian matrices of order* $n \times n$. *If* $\mathbf{B} = \Sigma_{i=1}^k \mathbf{A}_i$ *is n.n.d. and* $\mathcal{M}(\mathbf{A}_i) \subset \mathcal{M}(\mathbf{B})$ *for* $i = 1, 2, \ldots, k$, *then there exists a nonsingular matrix* $\mathbf{M}$ *such that* $\mathbf{M}^*\mathbf{A}_i\mathbf{M}$ *is diagonal for* $i = 1, 2, \ldots, k$ *if and only if*

$$\mathbf{A}_i\mathbf{B}^-\mathbf{A}_j = \mathbf{A}_j\mathbf{B}^-\mathbf{A}_i \quad \text{for } i, j = 1, 2, \ldots, k,$$

where $\mathbf{B}^-$ *is any g-inverse of* $\mathbf{B}$.

Proof: To prove the 'if part' observe that $\mathbf{A}_i\mathbf{B}^-\mathbf{A}_j$ is invariant under the choices of g-inverses of $\mathbf{B}$. So without loss of generality, we have

$$\mathbf{A}_i\mathbf{B}^+\mathbf{A}_j = \mathbf{A}_j\mathbf{B}^+\mathbf{A}_i \quad \text{for } i, j = 1, 2, \ldots, k.$$

Let $\mathbf{B}^+ = \mathbf{LL}^*$ be a rank factorization of $\mathbf{B}^+$. $\mathbf{L}$ is of order $n \times r$ and $r = R(\mathbf{B})$. Then it follows from the hypothesis that $\mathbf{L}^*\mathbf{A}_i\mathbf{L}$ commutes with $\mathbf{L}^*\mathbf{A}_j\mathbf{L}$ for all i and j. Hence there exists a unitary matrix $\mathbf{P}$ such that $\mathbf{P}^*\mathbf{L}^*\mathbf{A}_i\mathbf{LP}$ is diagonal for $i = 1, 2, \ldots, k$.

Let $\mathbf{N}$ be a $n \times (n - r)$ matrix of maximum rank such that $\mathbf{BN} = \mathbf{0}$. Since $\mathcal{M}(\mathbf{L}) = \mathcal{M}(\mathbf{B}^+) = \mathcal{M}(\mathbf{B})$, it follows that $\mathbf{L}^*\mathbf{N} = \mathbf{0}$. Choose $\mathbf{M} = (\mathbf{LP} \vdots \mathbf{N})$ and observe that, first, $\mathbf{M}$ is nonsingular, since $\mathbf{P}^*\mathbf{L}^*\mathbf{N} = \mathbf{0}$ and, secondly, $\mathbf{M}^*\mathbf{A}_i\mathbf{M}$ is diagonal for $i = 1, 2, \ldots, k$.

To prove the 'only if' part, let $\mathbf{M} = (\mathbf{M}_1 \vdots \mathbf{M}_2)$, where $\mathbf{M}_1$ is of order $n \times r$. Clearly, $\mathbf{M}_1^*\mathbf{B}\mathbf{M}_1$ is diagonal. Without loss of generality we can assume that $\mathbf{M}_1^*\mathbf{B}\mathbf{M}_1 = \mathbf{I}_r$. It immediately follows that $\mathbf{M}_1\mathbf{M}_1^*$ is a g-inverse of $\mathbf{B}$.

Now, $\mathbf{M}_1^*\mathbf{A}_i\mathbf{M}_1$, $i = 1, 2, \ldots, k$ are diagonal. Hence,

$$\begin{aligned} \mathbf{M}_1^*\mathbf{A}_i\mathbf{M}_1\mathbf{M}_1^*\mathbf{A}_j\mathbf{M}_1 &= \mathbf{M}_1^*\mathbf{A}_j\mathbf{M}_1\mathbf{M}_1^*\mathbf{A}_i\mathbf{M}_1 \qquad \forall\, i, j \\ \Rightarrow \mathbf{M}_1^*\mathbf{A}_i\mathbf{B}^-\mathbf{A}_j\mathbf{M}_1 &= \mathbf{M}_1^*\mathbf{A}_j\mathbf{B}^-\mathbf{A}_i\mathbf{M}_1 \qquad \forall\, i, j \\ \Rightarrow \mathbf{A}_i\mathbf{B}^-\mathbf{A}_j &= \mathbf{A}_j\mathbf{B}^-\mathbf{A}_i \qquad \forall\, i, j. \end{aligned}$$

Remark. If $\mathbf{A}_1, \mathbf{A}_2, \ldots, \mathbf{A}_k$ are n.n.d. then a n.s. condition for simultaneous diagonal reduction is

$$\mathbf{A}_i\mathbf{B}^-\mathbf{A}_j = \mathbf{A}_j\mathbf{B}^-\mathbf{A}_i \quad \text{for all } i, j,$$

where $\mathbf{B} = \mathbf{A}_1 + \cdots + \mathbf{A}_k$.

THEOREM 6.5.3 *Let $\mathbf{A}_1, \mathbf{A}_2, \ldots, \mathbf{A}_k$ be hermitian matrices of order $n \times n$, and let $\mathbf{A}_1$ be nonsingular. Then there exists a nonsingular matrix $\mathbf{M}$ such that $\mathbf{M}^*\mathbf{A}_i\mathbf{M}$ is diagonal for $i = 1, 2, \ldots, k$ iff*

(i) $\mathbf{A}_i\mathbf{A}_1^{-1}$ *is semisimple with real eigenvalues* $\forall i$ (6.5.1)

and

(ii) $\mathbf{A}_i\mathbf{A}_1^{-1}\mathbf{A}_j = \mathbf{A}_j\mathbf{A}_1^{-1}\mathbf{A}_i \;\forall\, i, j.$ (6.5.2)

Proof: The 'only if' part is trivial. For the 'if' part we need Lemma 6.5.1.

LEMMA 6.5.1 *Let $\mathbf{B}_1, \mathbf{B}_2, \ldots, \mathbf{B}_k$ be semisimple matrices of order $n \times n$. Then $\mathbf{B}_i$ and $\mathbf{B}_j$ commute $\forall\, i, j$ iff $\mathbf{B}_1, \mathbf{B}_2, \ldots, \mathbf{B}_k$ can be expressed as polynomials of a common semisimple matrix $\mathbf{B}$.*

Proof of Lemma: We prove the lemma for the case $k = 2$. Extension to the general case is straightforward. The 'if' part is easy. For the only part we write, as in example 6(d) of complements to Chapter 4

$$\mathbf{B}_1 = \sum \lambda_i \mathbf{K}_{\lambda_i}, \qquad \mathbf{B}_2 = \sum \mu_j \mathbf{L}_{\mu_j},$$

where $\lambda_i, \mathbf{K}_{\lambda_i}$ are, respectively, the distinct eigenvalues and the corresponding principal idempotents of $\mathbf{B}_1$ and $\mu_j, \mathbf{L}_{\mu_j}$ are defined analogously for $\mathbf{B}_2$. We put

$$\mathbf{E}_{ij} = \mathbf{K}_{\lambda_i}\mathbf{L}_{\mu_j}$$

and observe, using example 6(e) of complements to Chapter 4 that

$$\mathbf{E}_{ij}^2 = \mathbf{E}_{ij}, \mathbf{E}_{ij}\mathbf{E}_{i'j'} = \mathbf{0}, \text{ if } (i,j) \neq (i',j') \tag{6.5.3}$$

and

$$\sum_{i,j} \mathbf{E}_{ij} = \mathbf{I}.$$

Consider

$$\mathbf{B} = \sum_{i,j} c_{ij}\mathbf{E}_{ij},$$

where c_{ij} are arbitrary but distinct complex numbers. **B** is clearly semisimple, and c_{ij} is an eigenvalue of **B**, if $\mathbf{E}_{ij}$ is nonnull.

Let $p(\cdot)$ be a polynomial of finite degree with complex coefficients. (6.5.3) $\Rightarrow$

$$p(\mathbf{B}) = \sum_{i,j} p(c_{ij})\mathbf{E}_{ij}.$$

Since c_{ij} are distinct, one can choose polynomials $p(\cdot)$ and $q(\cdot)$ such that

$$p(c_{ij}) = \lambda_i, \qquad q(c_{ij}) = \mu_j \;\forall\, i, j.$$

Note that $\mathbf{B}_1 = p(\mathbf{B})$, $\mathbf{B}_2 = q(\mathbf{B})$. This completes the proof of Lemma 6.5.1.

In view of Lemma 6.5.1, (6.5.1) and (6.5.2), imply the existence of a nonsingular matrix **T** such that for each i, $\mathbf{T}^{-1}\mathbf{A}_i\mathbf{A}_1^{-1}\mathbf{T} = \mathbf{D}_i$, a diagonal matrix with real elements. Let us write $\mathbf{T}^{-1}\mathbf{A}_i(\mathbf{T}^{-1})^* = \mathbf{M}_i$ and observe that, since $\mathbf{M}_i$ is hermitian, $\mathbf{T}^{-1}\mathbf{A}_i\mathbf{A}_1^{-1}\mathbf{T} = \mathbf{D}_i \Rightarrow \mathbf{M}_i = \mathbf{D}_i\mathbf{M}_1 = \mathbf{M}_1\mathbf{D}_i$. Thus the matrices $\mathbf{M}_1, \mathbf{D}_1, \mathbf{D}_2, \ldots, \mathbf{D}_k$ commute and by applying Theorem 6.2.8, $\exists$ a unitary matrix **L** such that $\mathbf{L}^*\mathbf{M}_1\mathbf{L}$ is diagonal and so are $\mathbf{L}^*\mathbf{D}_i\mathbf{L} \;\forall\, i$. This implies $\mathbf{L}^*\mathbf{M}_i\mathbf{L} = \mathbf{L}^*\mathbf{D}_i\mathbf{L}\mathbf{L}^*\mathbf{M}_1\mathbf{L}$ is diagonal $\forall\, i$. Theorem 6.5.3 is thus true for $\mathbf{M} = (\mathbf{T}^{-1})^*\mathbf{L}$.

COROLLARY: *Let* $\mathbf{A}_1, \mathbf{A}_2, \ldots, \mathbf{A}_k$ *be hermitian matrices of order* $n \times n$ *such that* $\mathscr{M}(\mathbf{A}_i) \subset \mathscr{M}(\mathbf{A}_1) \;\forall\, i$. *Then there exists a nonsingular matrix* **M** *such that* $\mathbf{M}^*\mathbf{A}_i\mathbf{M}$ *is diagonal* $\forall\, i$, *iff for some g-inverse* $\mathbf{A}_1^-$ *of* $\mathbf{A}_1$

(i) $\mathbf{A}_i\mathbf{A}_1^-$ *is semisimple with real eigenvalues* $\forall\, i$ (6.5.5)

(ii) $\mathbf{A}_i\mathbf{A}_1^-\mathbf{A}_j = \mathbf{A}_j\mathbf{A}_1^-\mathbf{A}_i \;\forall\, i, j.$ (6.5.6)

The proof of the corollary is straightforward.

COMPLEMENTS

1. Let **A** and **B** be hermitian matrices of order n and **B** be n.n.d. Then

(a) the hermitian forms $\mathbf{x}^*\mathbf{B}\mathbf{x}$ and $\mathbf{x}^*\mathbf{B}\mathbf{A}\mathbf{B}\mathbf{x}$ are simultaneously reducible to diagonal forms by a cogredient transformation, and

(b) the hermitian forms $\mathbf{x}^*\mathbf{B}\mathbf{x}$ and $\mathbf{x}^*\mathbf{CAC}\mathbf{x}$ are simultaneously reducible to diagonal forms by contragredient transformations, where $\mathbf{C} = \mathbf{I} - \mathbf{P}_{\mathbf{B}}$.

2. If $\mathbf{M}$ be a nonsingular matrix of order n and $\mathbf{B}$ a hermitian matrix of order n, then for every positive integer k the hermitian forms $\mathbf{x}^*\mathbf{M}^*\mathbf{BM}\mathbf{x}$, $\mathbf{x}^*\mathbf{M}^*\mathbf{B}^2\mathbf{M}\mathbf{x}, \ldots,$ $\mathbf{x}^*\mathbf{M}^*\mathbf{B}^k\mathbf{M}\mathbf{x}$ are simultaneously reducible to diagonal forms by a cogredient transformation.

3. Let $\mathbf{A}$ be a p.d. matrix of order n and for a positive integer $m \le n$, the matrix $\mathbf{A}_m$ be obtained from $\mathbf{A}$ by replacing its trailing minor $\mathbf{A}_{22}$ of order m by $\mathbf{A}_{21}\mathbf{A}_{11}^{-1}\mathbf{A}_{12}$, where $\mathbf{A}_{11}, \mathbf{A}_{12}, \mathbf{A}_{21}$ are determined by the partitioned form of $\mathbf{A}$

$$\mathbf{A} = \begin{pmatrix} \mathbf{A}_{11} & \mathbf{A}_{12} \\ \mathbf{A}_{21} & \mathbf{A}_{22} \end{pmatrix}.$$

The hermitian forms $\mathbf{x}^*\mathbf{A}\mathbf{x}$ and $\mathbf{x}^*\mathbf{A}_i\mathbf{x}$, $i = 1, 2, \ldots, n$, are simultaneously reducible to diagonal forms by a cogredient transformation.

4. Let $\mathbf{x}^*\mathbf{A}\mathbf{x}$ and $\mathbf{x}^*\mathbf{B}\mathbf{x}$ be hermitian forms of which $\mathbf{x}^*\mathbf{B}\mathbf{x}$ is n.n.d. If $\mathbf{x}^*\mathbf{A}\mathbf{x}$ and $\mathbf{x}^*\mathbf{B}\mathbf{x}$ be simultaneously diagonable by (a) cogredient transformation or (b) contragredient transformations, then so are (i) $\mathbf{x}^*\mathbf{A}\mathbf{x}$ and $\mathbf{x}^*\mathbf{B}^k\mathbf{x}$ for any positive integer k, (ii) $\mathbf{x}^*\mathbf{A}\mathbf{x}$ and $\mathbf{x}^*\mathbf{B}^+\mathbf{x}$.

5. Let $\mathbf{A}$ and $\mathbf{B}$ be hermitian matrices of which $\mathbf{B}$ is nonsingular. If $\mathbf{x}^*\mathbf{A}\mathbf{x}$ and $\mathbf{x}^*\mathbf{B}\mathbf{x}$ be simultaneously diagonable by a cogredient transformation, then so are $\mathbf{x}^*\mathbf{AB}^{-1}\mathbf{A}\mathbf{x}$ and $\mathbf{x}^*\mathbf{B}\mathbf{x}$.

6. If $\mathbf{x}^*\mathbf{A}\mathbf{x}$ and $\mathbf{x}^*\mathbf{B}\mathbf{x}$ be simultaneously diagonable by contragredient transformations then so are $\mathbf{x}^*\mathbf{ABA}\mathbf{x}$ and $\mathbf{x}^*\mathbf{B}\mathbf{x}$.

7. $\mathbf{B}$ is a hermitian matrix. $\mathbf{x}$ is a proper eigenvector of $\mathbf{B}$ w.r.t. $\mathbf{B}^2$ iff $\mathbf{x}$ is the sum of two vectors, one an eigenvector of $\mathbf{B}$, corresponding to a nonnull eigenvalue, and the other is in the null space of $\mathbf{B}$.

8. (a) k hermitian forms $Q_1, Q_2, \ldots, Q_k$ in n variables are simultaneously diagonable by a single transformation, if $\exists\, n$ linearly independent hermitian forms $Q_1^0, Q_2^0, \ldots, Q_n^0$ each of rank 1, such that each Q_i is a linear combination of $Q_1^0, Q_2^0, \ldots, Q_n^0$.

(b) If $Q_1, Q_2, \ldots, Q_k$ contain n linearly independent forms, then each Q_i^0 in (a) can be expressed as a linear combination of $Q_1, Q_2, \ldots, Q_k$.

9. $(n + 1)$ linearly independent hermitian forms in n variables are not simultaneously diagonable by a single transformation.

CHAPTER 7

Estimation of Parameters in Linear Models

7.1 GAUSS-MARKOV MODEL

We consider a n-vector random variable $\mathbf{Y}$ such that

$$E(\mathbf{Y}) = \mathbf{X}\boldsymbol{\beta}, \qquad D(\mathbf{Y}) = \boldsymbol{\Sigma}, \tag{7.1.1}$$

where $\mathbf{X}$ is a given $n \times m$ matrix, $\boldsymbol{\beta}$ is a m-vector of unknown parameters and $\boldsymbol{\Sigma}$, the dispersion matrix of $\mathbf{Y}$, may be partly known. In addition, it may be known that $\boldsymbol{\beta}$ is subject to linear restrictions

$$\mathbf{R}\boldsymbol{\beta} = \mathbf{c}, \tag{7.1.2}$$

where $\mathbf{R}$, a $k \times m$ matrix and $\mathbf{c}$, a k-vector are known. In some situations $\mathbf{c}$ itself may be a random variable independent of $\mathbf{Y}$, in which case, instead of (7.1.2), we have

$$E(\mathbf{c}) = \mathbf{R}\boldsymbol{\beta}, \qquad D(\mathbf{c}) = \boldsymbol{\Theta}, \tag{7.1.3}$$

in addition to (7.1.1). In such a case the entire model (7.1.1), together with (7.1.3), can be written in the form (7.1.4)

$$E\begin{pmatrix}\mathbf{Y}\\ \mathbf{c}\end{pmatrix} = \begin{pmatrix}\mathbf{X}\\ \mathbf{R}\end{pmatrix}\boldsymbol{\beta}, \qquad D\begin{pmatrix}\mathbf{Y}\\ \mathbf{c}\end{pmatrix} = \begin{pmatrix}\boldsymbol{\Sigma} & \mathbf{0}\\ \mathbf{0} & \boldsymbol{\Theta}\end{pmatrix}. \tag{7.1.4}$$

It may be seen that the model (7.1.1) with the restrictions (7.1.2), is a limiting case of (7.1.4) as $\boldsymbol{\Theta} \to \mathbf{0}$. Indeed, we can work with a model of the type (7.1.4) and derive the results for a model with restrictions by taking appropriate limits. The model (7.1.4) is again a special case of the model (7.1.1) without restrictions, when we do not place any restriction on $\boldsymbol{\Sigma}$.

We shall represent the model (7.1.1) with the possible restrictions (7.1.2) as

$$(\mathbf{Y},\ \mathbf{X}\boldsymbol{\beta} | \mathbf{R}\boldsymbol{\beta} = \mathbf{c}, \boldsymbol{\Sigma}). \tag{7.1.5}$$

The problem we consider is the estimation of linear functions of $\boldsymbol{\beta}$ by means of linear functions of $\mathbf{Y}$. We use the following definitions and preliminary results.

Definition 1: A linear function $\mathbf{p}'\boldsymbol{\beta}$ is said to be unbiasedly estimable (u.e.) under model (7.1.5), if there exists a linear function $\mathbf{L}'\mathbf{Y} + d$ such that

$$E(\mathbf{L}'\mathbf{Y} + d) \equiv \mathbf{p}'\boldsymbol{\beta} \text{ given } \mathbf{R}\boldsymbol{\beta} = \mathbf{c}.$$

LEMMA 7.1.1 *For $\mathbf{p}'\boldsymbol{\beta}$ to be u.e. under the model $(\mathbf{Y}, \mathbf{X}\boldsymbol{\beta}|\mathbf{R}\boldsymbol{\beta} = \mathbf{c}, \boldsymbol{\Sigma})$, it is n.s. that there exist vectors $\mathbf{L}$ and $\boldsymbol{\lambda}$ such that*

$$\mathbf{X}'\mathbf{L} + \mathbf{R}'\boldsymbol{\lambda} = \mathbf{p}. \tag{7.1.6}$$

Proof: Let $\mathbf{L}'\mathbf{Y} + d$ be an unbiased estimator of $\mathbf{p}'\boldsymbol{\beta}$. Then

$$E(\mathbf{L}'\mathbf{Y} + d) = \mathbf{L}'\mathbf{X}\boldsymbol{\beta} + d = \mathbf{p}'\boldsymbol{\beta},$$

when $\boldsymbol{\beta}$ satisfies the equation $\mathbf{R}\boldsymbol{\beta} = \mathbf{c}$. This implies

$$(\mathbf{L}'\mathbf{X} - \mathbf{p}')\boldsymbol{\beta} + d \equiv \boldsymbol{\lambda}'(\mathbf{c} - \mathbf{R}\boldsymbol{\beta})$$

that is,

$$\mathbf{L}'\mathbf{X} + \boldsymbol{\lambda}'\mathbf{R} = \mathbf{p}' \quad \text{and} \quad d = \boldsymbol{\lambda}'\mathbf{c},$$

which proves the result (7.1.6).

COROLLARY 1: *If the model is $(\mathbf{Y}, \mathbf{X}\boldsymbol{\beta}, \boldsymbol{\Sigma})$ without any restriction on $\boldsymbol{\beta}$, then the condition for u.e. of $\mathbf{p}'\boldsymbol{\beta}$ is $\mathbf{p} \in \mathscr{M}(\mathbf{X}')$, that is, there exists a vector $\mathbf{L}$ such that*

$$\mathbf{X}'\mathbf{L} = \mathbf{p}. \tag{7.1.7}$$

LEMMA 7.1.2 *Let $R(\mathbf{X}) = r$, $R(\mathbf{R}) = k$ and $\mathbf{K}$ be $m \times (m - k)$ matrix generating $\mathscr{O}(\mathbf{R}')$. The model $(\mathbf{Y}, \mathbf{X}\boldsymbol{\beta}|\mathbf{R}\boldsymbol{\beta} = \mathbf{c}, \boldsymbol{\Sigma})$ is equivalent to $(\mathbf{Y}_1, \mathbf{X}_1\boldsymbol{\theta}, \boldsymbol{\Sigma})$, where $\boldsymbol{\theta}$ is $(m - k)$ vector of new parameters without any restrictions, and*

$$\mathbf{Y}_1 = \mathbf{Y} - \mathbf{X}\mathbf{R}^-\mathbf{c}, \textit{for any g-inverse of } \mathbf{R},$$

$$\mathbf{X}_1 = \mathbf{X}\mathbf{K}. \tag{7.1.8}$$

Further,

$$R(\mathbf{X}\mathbf{K}) = R(\mathbf{X}' \vdots \mathbf{R}') - R(\mathbf{R}'). \tag{7.1.9}$$

Proof: Since $\mathbf{R}\boldsymbol{\beta} = \mathbf{c}$ is consistent, we have using the result for a general solution of linear equations

$$\boldsymbol{\beta} = \mathbf{R}^-\mathbf{c} + \mathbf{K}\boldsymbol{\theta},$$

with one to one correspondence between $\boldsymbol{\theta}$, which is arbitrary, and $\boldsymbol{\beta}$, which satisfies the equation $\mathbf{R\beta} = \mathbf{c}$. Then

$$E(\mathbf{Y} - \mathbf{XR}^{-}\mathbf{c}) = \mathbf{XK\theta}, \quad D(\mathbf{Y} - \mathbf{XR}^{-}\mathbf{c}) = \mathbf{\Sigma}.$$

Consider any parametric function $\mathbf{p}'\boldsymbol{\beta}$. In terms of $\boldsymbol{\theta}$

$$\mathbf{p}'\boldsymbol{\beta} = \mathbf{p}'\mathbf{R}^{-}\mathbf{c} + \mathbf{p}'\mathbf{K}\boldsymbol{\theta},$$

and it is easy to see that $\mathbf{p}'\boldsymbol{\beta}$ is u.e. under the model $(\mathbf{Y}, \mathbf{X}\boldsymbol{\beta}|\mathbf{R}\boldsymbol{\beta} = \mathbf{c}, \mathbf{\Sigma})$, if $\mathbf{p}'\mathbf{K}\boldsymbol{\theta}$ is u.e. under the model $(\mathbf{Y}_1, \mathbf{X}_1\boldsymbol{\theta}, \mathbf{\Sigma})$. Thus the equivalence of the two models is established.

To prove (7.1.9) consider

$$\begin{pmatrix}\mathbf{K}'\\ \mathbf{R}\end{pmatrix}(\mathbf{X}' \vdots \mathbf{R}') = \begin{pmatrix}\mathbf{K}'\mathbf{X}' & \mathbf{0}\\ \mathbf{RX}' & \mathbf{RR}'\end{pmatrix}.$$

Since $(\mathbf{K} : \mathbf{R}')$ has full rank,

$$\begin{aligned} R(\mathbf{X}' : \mathbf{R}') &= R\begin{pmatrix}\mathbf{K}'\mathbf{X}' & \mathbf{0}\\ \mathbf{RX}' & \mathbf{RR}'\end{pmatrix}\\ &= R\begin{pmatrix}\mathbf{K}'\mathbf{X}' & \mathbf{0}\\ \mathbf{RX}' & \mathbf{RR}'\end{pmatrix}\begin{pmatrix}\mathbf{I} & \mathbf{0}\\ -(\mathbf{RR}')^{-}\mathbf{RX}' & \mathbf{I}\end{pmatrix}\\ &= R\begin{pmatrix}\mathbf{K}'\mathbf{X}' & \mathbf{0}\\ \mathbf{0} & \mathbf{RR}'\end{pmatrix}\\ &= R(\mathbf{K}'\mathbf{X}') + R(\mathbf{RR}') = R(\mathbf{XK}) + R(\mathbf{R}). \end{aligned}$$

Thus Lemma 7.1.2 is established.

Definition 2: Under the model $(\mathbf{Y}, \mathbf{X}\boldsymbol{\beta}|\mathbf{R}\boldsymbol{\beta} = \mathbf{c}, \mathbf{\Sigma})$, a linear function $\mathbf{L}_m'\mathbf{Y} + d$ is said to be a minimum bias estimator of $\mathbf{p}'\boldsymbol{\beta}$ if $d = \boldsymbol{\lambda}_m'\mathbf{c}$ and $\mathbf{L}_m$, $\boldsymbol{\lambda}_m$ are such that

$$\|\mathbf{X}'\mathbf{L}_m + \mathbf{R}'\boldsymbol{\lambda}_m - \mathbf{p}\| = \inf_{\mathbf{L},\boldsymbol{\lambda}} \|\mathbf{X}'\mathbf{L} + \mathbf{R}'\boldsymbol{\lambda} - \mathbf{p}\| \qquad (7.1.10)$$

for a suitable definition of norm (see also Chipman, 1964).

When there are no restrictions on $\boldsymbol{\beta}$, a linear function $\mathbf{L}_m'\mathbf{Y}$ is said to be minimum bias estimator of $\mathbf{p}'\boldsymbol{\beta}$ if

$$\|\mathbf{X}'\mathbf{L}_m - \mathbf{p}\| = \inf_{\mathbf{L}} \|\mathbf{X}'\mathbf{L} - \mathbf{p}\|. \qquad (7.1.11)$$

LEMMA 7.1.3 *Let* $\|\mathbf{x}\| = (\mathbf{x}'\mathbf{Mx})^{1/2}$ *where* $\mathbf{M}$ *is p.d. Under the model* $(\mathbf{Y}, \mathbf{X}\boldsymbol{\beta}, \mathbf{\Sigma})$ *a minimum bias estimator of* $\mathbf{p}'\boldsymbol{\beta}$ *is*

$$\mathbf{p}'[(\mathbf{X}')^{-}_{l(\mathbf{M})}]'\mathbf{Y} = \mathbf{p}'\mathbf{X}^{-}_{m(\mathbf{M}^{-1})}\mathbf{Y}, \qquad (7.1.12)$$

where $(\mathbf{X}')^-_{l(\mathbf{M})}$ *is a* **M**-*least-squares g-inverse of* $\mathbf{X}'$, *the transpose of which is same as a minimum* $\mathbf{M}^{-1}$-*norm g-inverse of* **X**.

Under the model $(\mathbf{Y}, \mathbf{X\beta}|\mathbf{R\beta} = \mathbf{c}, \mathbf{\Sigma})$ *a minimum bias estimator of* $\mathbf{p}'\mathbf{\beta}$ *is*

$$\mathbf{p}'\mathbf{M}[(\mathbf{C}_1\mathbf{X} + \mathbf{C}_2\mathbf{R})'\mathbf{Y} + (\mathbf{C}_3\mathbf{X} + \mathbf{C}_4\mathbf{R})'\mathbf{c}] \tag{7.1.13}$$

where

$$\begin{pmatrix} \mathbf{C}_1 & \mathbf{C}_2 \\ \mathbf{C}_3 & \mathbf{C}_4 \end{pmatrix} = \begin{pmatrix} \mathbf{XMX}' & \mathbf{XMR}' \\ \mathbf{RMX}' & \mathbf{RMR}' \end{pmatrix}^-$$

for any choice of g-inverse.

Proof: The results (7.1.12) and (7.1.13) follow from the definition of minimum bias estimators. Explicit expressions for $\mathbf{C}_1$, $\mathbf{C}_2$, $\mathbf{C}_3$, and $\mathbf{C}_4$ in terms of $\mathbf{XMX}'$, $\mathbf{XMR}'$, $\mathbf{RMR}'$ are given in Example 6 of the complements to Chapter 2.

Definition 3: Under the model $(\mathbf{Y}, \mathbf{X\beta}|\mathbf{R\beta} = \mathbf{c}, \mathbf{\Sigma})$, the estimator $\mathbf{L}'\mathbf{Y} + d$ is said to be BLUE (best linear unbiased estimator, best in the sense of minimum variance) of the parametric function $\mathbf{p}'\mathbf{\beta}$, if

$$E(\mathbf{L}'\mathbf{Y} + d) = \mathbf{p}'\mathbf{\beta} \text{ given } \mathbf{R\beta} = \mathbf{c},$$

and $V(\mathbf{L}'\mathbf{Y} + d)$ is a minimum in the class of linear unbiased estimators of $\mathbf{p}'\mathbf{\beta}$.

Definition 4: The estimator $\mathbf{L}'\mathbf{Y} + d$ is said to be BLMBE (best linear minimum bias estimator) of $\mathbf{p}'\mathbf{\beta}$ if $\mathbf{V}(\mathbf{L}'\mathbf{Y} + d)$ is the least in the class of linear minimum bias estimators of $\mathbf{p}'\mathbf{\beta}$.

In the next sections we shall consider problems of BLUE and BLMBE in different situations, depending on the availability of knowledge on $\mathbf{\Sigma}$, the dispersion matrix of **Y**.

7.2 MODEL: $(\mathbf{Y}, \mathbf{X\beta}, \sigma^2\mathbf{I}), \mathbf{\beta}, \sigma^2$ UNKNOWN

Let **Y** be a n-vector, **X** a $n \times m$ matrix and $\mathbf{\beta}$ an m-vector. We denote by $\mathbf{X}^-$ and $(\mathbf{X}'\mathbf{X})^-$ any g-inverses of **X** and $(\mathbf{X}'\mathbf{X})$, by $\mathbf{X}_l^-$, $\mathbf{X}_m^-$, and $\mathbf{X}^+$, the least squares, minimum norm and minimum norm least-squares g-inverses, respectively.

THEOREM 7.2.1 (Condition for unbiased estimation) *A parametric function* $\mathbf{p}'\mathbf{\beta}$ *is unbiasedly estimable* (*u.e.*) *by a linear function of* **Y** *iff*

$$\mathbf{p}'\mathbf{X}^-\mathbf{X} = \mathbf{p}' \quad \text{or } \mathbf{p}'(\mathbf{X}'\mathbf{X})^-\mathbf{X}'\mathbf{X} = \mathbf{p}'. \tag{7.2.1}$$

Proof: If (7.2.1) holds, then $\mathbf{p}' = \mathbf{L}'\mathbf{X}$ satisfying (7.1.7), by choosi $\mathbf{L}' = \mathbf{p}'\mathbf{X}^-$ or $\mathbf{L}' = \mathbf{p}'(\mathbf{X}'\mathbf{X})^-\mathbf{X}'$. Conversely, if $\mathbf{p}' = \mathbf{L}'\mathbf{X}$ for some $\mathbf{L}$, then

$$\mathbf{p}'\mathbf{X}^-\mathbf{X} = \mathbf{L}'\mathbf{X}\mathbf{X}^-\mathbf{X} (= \mathbf{L}'\mathbf{X} = \mathbf{p}')$$

$$\mathbf{p}'(\mathbf{X}'\mathbf{X})^-(\mathbf{X}'\mathbf{X}) = \mathbf{L}'\mathbf{X}(\mathbf{X}'\mathbf{X})^-(\mathbf{X}'\mathbf{X}) = \mathbf{L}'\mathbf{X} = \mathbf{p}',$$

since by (2.2.4), $\mathbf{X}(\mathbf{X}'\mathbf{X})^-(\mathbf{X}'\mathbf{X}) = \mathbf{X}$. The theorem is proved. Note that in t condition $\mathbf{p}'\mathbf{X}^-\mathbf{X} = \mathbf{p}'$ we can use *any* g-inverse. Thus the condition could written in terms of $\mathbf{X}_l^-$, $\mathbf{X}_m^-$ or $\mathbf{X}^+$.

THEOREM 7.2.2 (Why least-squares theory for BLUE?) *If* $\mathbf{p}$ *satisfi (7.2.1), then the BLUE of* $\mathbf{p}'\boldsymbol{\beta}$ *is* $\mathbf{p}'\hat{\boldsymbol{\beta}}$, *where* $\hat{\boldsymbol{\beta}}$ *is the least-squares solution* $\mathbf{Y} = \mathbf{X}\boldsymbol{\beta}$.

Let $\mathbf{L}'\mathbf{Y}$ be an unbiased estimator of $\mathbf{p}'\boldsymbol{\beta}$. Then

$$E(\mathbf{L}'\mathbf{Y}) = \mathbf{p}'\boldsymbol{\beta} \Leftrightarrow \mathbf{X}'\mathbf{L} = \mathbf{p},$$
$$V(\mathbf{L}'\mathbf{Y}) = \sigma^2\mathbf{L}'\mathbf{L}. \qquad (7.2.$$

To find the BLUE, we have to minimize $\mathbf{L}'\mathbf{L}$ subject to $\mathbf{X}'\mathbf{L} = \mathbf{p}$. Then t optimum $\mathbf{L}$ is the minimum norm solution of $\mathbf{X}'\mathbf{L} = \mathbf{p}$,

$$\mathbf{L}_0 = (\mathbf{X}')_m^-\mathbf{p},$$

in which case, the BLUE is

$$\mathbf{L}_0'\mathbf{Y} = \mathbf{p}'[(\mathbf{X}')_m^-]'\mathbf{Y} \qquad (7.2.$$
$$= \mathbf{p}'\mathbf{X}_l^-\mathbf{Y} = \mathbf{p}'\hat{\boldsymbol{\beta}},$$

by duality Theorem 3.2.4 connecting the least squares and minimum nor inverses. $\mathbf{L}_0'\mathbf{Y}$ is unique, being a minimum norm solution (Note 4 followi Theorem 3.1.1) and, hence, $\mathbf{p}'\hat{\boldsymbol{\beta}}$ is unique for any least-squares solution $\mathbf{Y} = \mathbf{X}\boldsymbol{\beta}$ (see also Sibuya, 1970).

Computation of $\hat{\boldsymbol{\beta}}$. In practice, one obtains $\hat{\boldsymbol{\beta}}$ by deriving the norm equation, that is, the equation obtained by minimizing the quadratic for

$$(\mathbf{Y} - \mathbf{X}\boldsymbol{\beta})'(\mathbf{Y} - \mathbf{X}\boldsymbol{\beta})$$

leading to

$$\mathbf{X}'\mathbf{X}\boldsymbol{\beta} = \mathbf{X}'\mathbf{Y}. \qquad (7.2.$$

The normal equation (7.2.4) is consistent and one solution is

$$\hat{\boldsymbol{\beta}} = (\mathbf{X}'\mathbf{X})^-\mathbf{X}'\mathbf{Y} \qquad (7.2.$$

for any g-inverse of $\mathbf{X}'\mathbf{X}$. Indeed $(\mathbf{X}'\mathbf{X})^-\mathbf{X}'$ is one choice for $\mathbf{X}_l^-$, a $(\mathbf{Y} - \mathbf{X}\boldsymbol{\beta})'(\mathbf{Y} - \mathbf{X}\boldsymbol{\beta})$ attains a minimum at $\hat{\boldsymbol{\beta}}$ as in (7.2.5) (see Note 4 followi Theorem 3.2.1).

The following alternative expressions for minimum value of

$$(\mathbf{Y} - \mathbf{X\beta})'(\mathbf{Y} - \mathbf{X\beta})$$

are of interest and easy to deduce.

$$\min_{\boldsymbol\beta} (\mathbf{Y} - \mathbf{X\beta})'(\mathbf{Y} - \mathbf{X\beta}) = \mathbf{Y'Y} - \hat{\boldsymbol\beta}'\mathbf{X'X}\hat{\boldsymbol\beta} = \mathbf{Y'Y} - \hat{\boldsymbol\beta}'\mathbf{X'Y}$$

$$= \mathbf{Y}'[\mathbf{I} - \mathbf{P_X}]\mathbf{Y} = \mathbf{Y}'[\mathbf{I} - \mathbf{X}(\mathbf{X'X})^-\mathbf{X}']\mathbf{Y} \quad (7.2.6)$$

where $\mathbf{P_X}$ is the projection operator onto $\mathcal{M}(\mathbf{X})$.

THEOREM 7.2.3 (Expressions for variances and covariances of BLUE estimators) *Let* $\mathbf{p}'\boldsymbol\beta$ *and* $\mathbf{q}'\boldsymbol\beta$ *be u.e. with BLUE's* $\mathbf{p}'\hat{\boldsymbol\beta}$ *and* $\mathbf{q}'\hat{\boldsymbol\beta}$. *Then*

$$V(\mathbf{p}'\hat{\boldsymbol\beta}) = \sigma^2\mathbf{p}'(\mathbf{X'X})^-\mathbf{p} = \sigma^2\mathbf{p}'\mathbf{X}_l^-(\mathbf{X}_l^-)'\mathbf{p} = \sigma^2\mathbf{p}'\mathbf{X}_l^-(\mathbf{X}')_m^-\mathbf{p} \quad (7.2.7)$$

$$\operatorname{cov}(\mathbf{p}'\hat{\boldsymbol\beta}, \mathbf{q}'\hat{\boldsymbol\beta}) = \sigma^2\mathbf{p}'(\mathbf{X'X})^-\mathbf{q} = \sigma^2\mathbf{q}'(\mathbf{X'X})^-\mathbf{p}$$

$$= \sigma^2\mathbf{p}'\mathbf{X}_l^-(\mathbf{X}_l^-)'\mathbf{q} = \sigma^2\mathbf{p}'\mathbf{X}_l^-(\mathbf{X}')_m^-\mathbf{q}. \quad (7.2.8)$$

Proof:

$$V(\mathbf{p}'\hat{\boldsymbol\beta}) = \sigma^2\mathbf{p}'(\mathbf{X'X})^-\mathbf{X'X}[(\mathbf{X'X})^-]'\mathbf{p}$$

$$= \sigma^2\mathbf{p}'[(\mathbf{X'X})^-]'\mathbf{p}, \quad \text{since } \mathbf{p}'(\mathbf{X'X})^-\mathbf{X'X} = \mathbf{p}'$$

$$= \sigma^2\mathbf{p}'(\mathbf{X'X})^-\mathbf{p}$$

The other results are similarly proved.

THEOREM 7.2.4 (Simultaneous estimation) *Let each of* $\mathbf{p}_1, \ldots, \mathbf{p}_k$ *satisfy the condition (7.2.1) and denote by* $\mathbf{P}$ *the matrix with* $\mathbf{p}_1, \ldots, \mathbf{p}_k$ *as columns. Then the BLUE of* $\mathbf{P}'\boldsymbol\beta$ *is*

$$\mathbf{P}'\hat{\boldsymbol\beta} = \mathbf{P}'(\mathbf{X'X})^-\mathbf{X'Y} \quad (7.2.9)$$

with the dispersion (variance–covariance) matrix

$$D(\mathbf{P}'\hat{\boldsymbol\beta}) = \sigma^2\mathbf{P}'(\mathbf{X'X})^-\mathbf{P}. \quad (7.2.10)$$

The estimator $\mathbf{P}'\hat{\boldsymbol\beta}$ *has the stronger property that*

$$D(\mathbf{Q}) - D(\mathbf{P}'\hat{\boldsymbol\beta}) \quad (7.2.11)$$

is an n.n.d. matrix where $D(\mathbf{Q})$ *is the dispersion matrix of any unbiased linear estimator* $\mathbf{Q}$ *of* $\mathbf{P}'\boldsymbol\beta$.

Proof: The result is obtained by considering a linear function $\mathbf{p}'\boldsymbol\beta = m_1\mathbf{p}_1'\boldsymbol\beta + \cdots + m_k\mathbf{p}_k'\boldsymbol\beta$ and expressing the condition that $\mathbf{p}'\hat{\boldsymbol\beta}$ has minimum variance as an estimator of $\mathbf{p}'\boldsymbol\beta$ for any choice of constants $m_1, \ldots, m_k$.

THEOREM 7.2.5 (Estimation of σ^2) *Let* $\hat{\boldsymbol{\beta}}$ *be a l.s. solution of* $\mathbf{Y} = \mathbf{X}\boldsymbol{\beta}$. *Then*

$$E[(\mathbf{Y} - \mathbf{X}\hat{\boldsymbol{\beta}})'(\mathbf{Y} - \mathbf{X}\hat{\boldsymbol{\beta}})] = (n - r)\sigma^2, \tag{7.2.12}$$

where $r = R(\mathbf{X})$.

Proof: Using the expression (7.2.6),

$$(\mathbf{Y} - \mathbf{X}\hat{\boldsymbol{\beta}})'(\mathbf{Y} - \mathbf{X}\hat{\boldsymbol{\beta}}) = \mathbf{Y}'(\mathbf{I} - \mathbf{P_X})\mathbf{Y}$$

$$E[\mathbf{Y}'(\mathbf{I} - \mathbf{P_X})\mathbf{Y}] = \sigma^2 \operatorname{trace}(\mathbf{I} - \mathbf{P_X}) = \sigma^2(n - r).$$

Note Let us write the normal equations (7.2.4), $\mathbf{X}'\mathbf{X}\boldsymbol{\beta} = \mathbf{X}'\mathbf{Y}$, as

$$\mathbf{S}\boldsymbol{\beta} = \mathbf{Q}. \tag{7.2.13}$$

Let us multiply both sides of (7.2.13) by a vector $\boldsymbol{\lambda}'$ to obtain the equation $\boldsymbol{\lambda}'\mathbf{S}\boldsymbol{\beta} = \boldsymbol{\lambda}'\mathbf{Q}$. If $\boldsymbol{\lambda}'\mathbf{S}\boldsymbol{\beta} \equiv \mathbf{p}'\boldsymbol{\beta}$, then the BLUE of $\mathbf{p}'\boldsymbol{\beta}$ is $\boldsymbol{\lambda}'\mathbf{Q}$ and $V(\boldsymbol{\lambda}'\mathbf{Q}) = \sigma^2\mathbf{p}'\boldsymbol{\lambda}$. In many practical situations it is possible to obtain estimates of desired parametric functions as indicated above without going through the process of solving normal equations and computing $(\mathbf{X}'\mathbf{X})^-$ to find the variance of the estimator. Further, suppose that we find $\mathbf{p}'\boldsymbol{\beta}$ is estimated by $\boldsymbol{\lambda}'\mathbf{Q}$ and $\mathbf{q}'\boldsymbol{\beta}$ by $\boldsymbol{\mu}'\mathbf{Q}$, then

$$V(\boldsymbol{\mu}'\mathbf{Q}) = \sigma^2\mathbf{q}'\boldsymbol{\mu}, \qquad \operatorname{cov}(\boldsymbol{\lambda}'\mathbf{Q}, \boldsymbol{\mu}'\mathbf{Q}) = \sigma^2\mathbf{p}'\boldsymbol{\mu} = \sigma^2\mathbf{q}'\boldsymbol{\lambda}. \tag{7.2.14}$$

We shall consider tests of hypotheses concerning the vector parameter $\boldsymbol{\beta}$. Let the null hypothesis to be tested be $\mathbf{H}\boldsymbol{\beta} = \mathbf{h}$, where $\mathbf{H}$ is a $k \times m$ matrix of rank s. We suppose that each linear function of $\boldsymbol{\beta}$ in $\mathbf{H}\boldsymbol{\beta}$ is u.e. Then the BLUE of $\mathbf{H}\boldsymbol{\beta}$ is $\mathbf{H}\hat{\boldsymbol{\beta}}$, where $\hat{\boldsymbol{\beta}} = (\mathbf{X}'\mathbf{X})^-\mathbf{X}'\mathbf{Y}$. The estimated deviation from the hypothesis is

$$\mathbf{d} = \mathbf{H}\hat{\boldsymbol{\beta}} - \mathbf{h} \tag{7.2.15}$$

with the dispersion matrix

$$\mathbf{D} = \sigma^2\mathbf{H}(\mathbf{X}'\mathbf{X})^-\mathbf{H}'. \tag{7.2.16}$$

We prove the following theorem.

THEOREM 7.2.6 *Let* $\mathbf{Y}$ *be distributed as* $N_n(\mathbf{X}\boldsymbol{\beta}, \sigma^2\mathbf{I})$ *and consider the expressions*

$$Q = \sigma^{-2}\mathbf{d}'\mathbf{D}^-\mathbf{d} \tag{7.2.17}$$

$$R^2 = \sigma^{-2}\mathbf{Y}'[\mathbf{I} - \mathbf{X}(\mathbf{X}'\mathbf{X})^-\mathbf{X}]\mathbf{Y}. \tag{7.2.18}$$

The following hold:

(i) R^2 *is a central* χ^2 *on* $(n - r)$ *d.f. where* $r = R(\mathbf{X})$ *and* Q *is a central* χ^2, *if the hypothesis* $\mathbf{H}\boldsymbol{\beta} = \mathbf{h}$ *is true and noncentral* χ^2 *if* $\mathbf{H}\boldsymbol{\beta} \neq \mathbf{h}$, *on* s *d.f. where* $s = R(\mathbf{H})$.

(ii) *Q and R^2 are independently distributed.*

(iii) *The distribution of* $F = (Q/s) \div R^2/(n-r)$ (7.2.19)
is central or noncentral F according to whether the hypothesis is true or not true.

(iv) $Q = R_1^2 - R^2$, *where*

$$R_1^2 = \min_{\mathbf{H\beta}=\mathbf{h}} (\mathbf{Y} - \mathbf{X\beta})'(\mathbf{Y} - \mathbf{X\beta})\sigma^{-2} \tag{7.2.20}$$

and R^2 as defined in (7.2.18) *is seen to be*

$$R^2 = \min_{\boldsymbol{\beta}} (\mathbf{Y} - \mathbf{X\beta})'(\mathbf{Y} - \mathbf{X\beta})\sigma^{-2}.$$

Proof of (i): Let us observe that

$$R^2 = \sigma^{-2}(\mathbf{Y} - \mathbf{X\beta})'[\mathbf{I} - \mathbf{X}(\mathbf{X}'\mathbf{X})^-\mathbf{X}'](\mathbf{Y} - \mathbf{X\beta}),$$

that is, the introduction of $\mathbf{X\beta}$ in (7.2.18) makes no difference. Now $\mathbf{Y} - \mathbf{X\beta} \sim N_n(\mathbf{0}, \sigma^2\mathbf{I})$. Hence R^2 has a central χ^2 distribution if $\mathbf{I} - \mathbf{X}(\mathbf{X}'\mathbf{X})^-\mathbf{X}'$ is idempotent, which is true as shown in Lemma 2.2.6(b).

Since $\mathbf{d} \sim N_k(\mathbf{H\beta} - \mathbf{h}, \sigma^2\mathbf{H}(\mathbf{X}'\mathbf{X})^-\mathbf{H}')$, where $R[\mathbf{H}(\mathbf{X}'\mathbf{X})^-\mathbf{H}'] = s = R(\mathbf{H})$, it follows from Theorem 9.2.1 that Q is a central χ^2 on s d.f., if $\mathbf{H\beta} - \mathbf{h} = \mathbf{0}$ and noncentral, if $\mathbf{H\beta} - \mathbf{h} \neq \mathbf{0}$.

We shall now consider the estimation of a parametric function $\mathbf{p}'\boldsymbol{\beta}$ which is not u.e. In such a case we may look for an estimator which has least variance in the class of linear minimum bias estimators (BLMBE), as given in Definition 4.

Let $\mathbf{L}'\mathbf{Y}$ be an estimator of $\mathbf{p}'\boldsymbol{\beta}$. Then

$$E(\mathbf{L}'\mathbf{Y} - \mathbf{p}'\boldsymbol{\beta}) = (\mathbf{L}'\mathbf{X} - \mathbf{p}')\boldsymbol{\beta}\ (\text{bias})$$

$$V(\mathbf{L}'\mathbf{Y}) = \sigma^2\mathbf{L}'\mathbf{L}.$$

The problem is to minimize $\mathbf{L}'\mathbf{L}$ in the class of $\mathbf{L}$ for which $\|\mathbf{X}'\mathbf{L} - \mathbf{p}\|$ is a minimum. We need to specify the norm function in $\mathscr{E}^m$.

THEOREM 7.2.7 *Let the norm in $\mathscr{E}^m$ be defined by* $\|\mathbf{x}\| = (\mathbf{x}'\mathbf{M}\mathbf{x})^{1/2}$ *where* $\mathbf{M}$ *is p.d. matrix. Then the BLMBE is*

$$\mathbf{p}'[(\mathbf{X}')^+_{\mathbf{MI}}]'\mathbf{Y} = \mathbf{p}'\mathbf{X}^+_{\mathbf{IM}^{-1}}\mathbf{Y}, \tag{7.2.21}$$

where $\mathbf{X}^+_{\mathbf{IM}^{-1}}$ *is the g-inverse of* $\mathbf{X}$, *as defined in Section* 3.3. *The variance of* (7.2.21) *is*

$$\sigma^2\mathbf{p}'\mathbf{X}^+_{\mathbf{IM}^{-1}}(\mathbf{X}^+_{\mathbf{IM}^{-1}})'\mathbf{p} = \sigma^2\mathbf{p}'\mathbf{X}^+_{\mathbf{IM}^{-1}}(\mathbf{X}')^+_{\mathbf{MI}}\mathbf{p} = \sigma^2\mathbf{p}'[(\mathbf{X}')^+_{\mathbf{MI}}]'[(\mathbf{X}')^+_{\mathbf{MI}}]\mathbf{p}. \tag{7.2.22}$$

Proof: The problem is to minimize $\|\mathbf{L}\|_n$ in the class of $\mathbf{L}$, for which $\|\mathbf{X}'\mathbf{L} - \mathbf{p}\|_m$ is a minimum. The norms in $\mathscr{E}^n$ and $\mathscr{E}^m$ are defined by the p.d.

matrices $\mathbf{I}$ and $\mathbf{M}$, respectively. The answer is provided by a minimu
$\mathbf{I}$-norm $\mathbf{M}$-least-squares solution of $\mathbf{X}'\mathbf{L} = \mathbf{p}$ (may be inconsistent),

$$\mathbf{L} = (\mathbf{X}')^+_{\mathbf{MI}}\mathbf{p}.$$

The estimator of $\mathbf{p}'\boldsymbol{\beta}$ is

$$\mathbf{L}'\mathbf{Y} = \mathbf{p}'[(\mathbf{X}')^+_{\mathbf{MI}}]'\mathbf{Y} = \mathbf{p}'\mathbf{X}^+_{\mathbf{IM}^{-1}}\mathbf{Y},$$

using the duality result (3.3.8), which proves (7.2.21). The expression (7.2.2
for variance is easily written down.

Remark 1. An explicit expression for $\mathbf{X}^+_{\mathbf{IM}^{-1}}$ is given by

$$\mathbf{M}\mathbf{X}'\mathbf{X}(\mathbf{X}'\mathbf{X}\mathbf{M}\mathbf{X}'\mathbf{X})^-\mathbf{X}', \qquad (7.2.2$$

which is unique for any choice of the g-inverse.

Remark 2. In the special case $\mathbf{M} = \mathbf{I}$, the estimator (7.2.21) reduces
$\mathbf{p}'\mathbf{X}^+\mathbf{Y}$ with variance $\sigma^2\mathbf{p}'\mathbf{X}^+(\mathbf{X}^+)'\mathbf{p}$. An explicit form for

$$\mathbf{X}^+ = \mathbf{X}'(\mathbf{X}\mathbf{X}'\mathbf{X}\mathbf{X}')^-\mathbf{X}\mathbf{X}' = \mathbf{X}'\mathbf{X}(\mathbf{X}'\mathbf{X}\mathbf{X}'\mathbf{X})^-\mathbf{X}'. \qquad (7.2.2$$

7.3 MODEL: $(\mathbf{Y}, \mathbf{X}\boldsymbol{\beta}|\mathbf{R}\boldsymbol{\beta} = \mathbf{c}, \sigma^2\mathbf{I})$, σ^2 AND $\boldsymbol{\beta}$ UNKNOWN

We shall show that the model

$$(\mathbf{Y}, \mathbf{X}\boldsymbol{\beta}|\mathbf{R}\boldsymbol{\beta} = \mathbf{c}, \sigma^2\mathbf{I}) \qquad (7.3.$$

can be reduced to a model without restrictions on parameters, in which ca
the methods of Section 7.2 apply.

A general solution of $\mathbf{R}\boldsymbol{\beta} = \mathbf{c}$ is $\boldsymbol{\beta} = \mathbf{R}^-\mathbf{c} + \mathbf{U}\boldsymbol{\eta}$, where $\mathbf{U}$ is a matrix
rank $u = m -$ rank $(\mathbf{R})$ such that $\mathbf{R}\mathbf{U} = \mathbf{0}$, $\mathbf{R}^-$ is a g-inverse of $\mathbf{R}$ and
is arbitrary. We note that one choice of $\mathbf{U}$ is $(\mathbf{R}')^\perp$ which is of order $m \times$
and rank u in which case $\boldsymbol{\eta}$ is a u-vector. Another choice of $\mathbf{U}$ is $(\mathbf{I} - \mathbf{R}^-$
in which case $\boldsymbol{\eta}$ is a m-vector. There can be a variety of choices subject
the condition $\mathbf{R}\mathbf{U} = \mathbf{0}$ and $R(\mathbf{U}) = u$. For any particular choice of
$E(\mathbf{Y} - \mathbf{X}\mathbf{R}^-\mathbf{c}) = \mathbf{X}\mathbf{U}\boldsymbol{\eta}$ so that the model $(\mathbf{Y}, \mathbf{X}\boldsymbol{\beta}|\mathbf{R}\boldsymbol{\beta} = \mathbf{c}, \sigma^2\mathbf{I})$ is the sa
as

$$(\mathbf{Y} - \mathbf{X}\mathbf{R}^-\mathbf{c}, \mathbf{X}\mathbf{U}\boldsymbol{\eta}, \sigma^2\mathbf{I}) \qquad (7.3.$$

with the observation vector $\mathbf{Y} - \mathbf{X}\mathbf{R}^-\mathbf{c}$, design matrix $\mathbf{X}\mathbf{U}$, and paramet
vector $\boldsymbol{\eta}$.

It may be noted the condition of unbiased estimability (u.e.) of $\mathbf{p}'\boldsymbol{\beta}$ und
the model (7.3.1) is the same as that of $\mathbf{p}'\mathbf{U}\boldsymbol{\eta}$ under the model (7.3.2), that
there exists a vector $\mathbf{L}$ such that

$$\mathbf{U}'\mathbf{X}'\mathbf{L} = \mathbf{U}'\mathbf{p}. \qquad (7.3.$$

Starting from the model (7.3.2), we can write down the normal equations

$$\mathbf{U}'\mathbf{X}'\mathbf{X}\mathbf{U}\boldsymbol{\eta} = \mathbf{U}'\mathbf{X}'(\mathbf{Y} - \mathbf{X}\mathbf{R}^{-}\mathbf{c})$$

a solution of which is

$$\hat{\boldsymbol{\eta}} = (\mathbf{U}'\mathbf{X}'\mathbf{X}\mathbf{U})^{-}\mathbf{U}'\mathbf{X}'(\mathbf{Y} - \mathbf{X}\mathbf{R}^{-}\mathbf{c}).$$

The BLUE of $\mathbf{p}'\boldsymbol{\beta}$, if u.e., is

$$\mathbf{p}'(\mathbf{R}^{-}\mathbf{c} + \mathbf{U}\hat{\boldsymbol{\eta}}) = \mathbf{p}'\mathbf{U}\hat{\boldsymbol{\eta}} + \mathbf{p}'\mathbf{R}^{-}\mathbf{c}$$

and variance of the estimator is [using the formula (7.2.7)],

$$\sigma^2\mathbf{p}'\mathbf{U}(\mathbf{U}'\mathbf{X}'\mathbf{X}\mathbf{U})^{-}\mathbf{U}'\mathbf{p}.$$

An unbiased estimator of σ^2 is (using the formula (7.2.12)),

$$(\mathbf{Y} - \mathbf{X}\mathbf{R}^{-}\mathbf{c})'[\mathbf{I} - \mathbf{X}\mathbf{U}(\mathbf{U}'\mathbf{X}'\mathbf{X}\mathbf{U})^{-}\mathbf{U}'\mathbf{X}'](\mathbf{Y} - \mathbf{X}\mathbf{R}^{-}\mathbf{c}) \div f \qquad (7.3.4)$$

where

$$f = n - R(\mathbf{X}\mathbf{U}) = n - R(\mathbf{X}' \,\vdots\, \mathbf{R}') + R(\mathbf{R}'). \qquad (7.3.5)$$

Test criteria of linear hypotheses can be constructed as explained in Section 7.2, starting from the model (7.3.2).

We shall describe a different computational approach to the problem without reducing the model (7.3.1) with restrictions to the model (7.3.2) without restrictions.

THEOREM 7.3.1 *Consider the model (7.3.1). The BLUE of* $\mathbf{p}'\boldsymbol{\beta}$, *if u.e., is* $\mathbf{p}'\tilde{\boldsymbol{\beta}}$, *where* $\tilde{\boldsymbol{\beta}}$ *is a solution of the equation*

$$\left.\begin{aligned} \mathbf{X}'\mathbf{X}\boldsymbol{\beta} + \mathbf{R}'\boldsymbol{\lambda} &= \mathbf{X}'\mathbf{Y} \\ \mathbf{R}\boldsymbol{\beta} &= \mathbf{c} \end{aligned}\right\}, \qquad (7.3.6)$$

which are called normal equations obtained by minimizing $(\mathbf{Y} - \mathbf{X}\boldsymbol{\beta})'(\mathbf{Y} - \mathbf{X}\boldsymbol{\beta})$ *subject to* $\mathbf{R}\boldsymbol{\beta} = \mathbf{c}$ *(least-squares theory with restrictions).*

THEOREM 7.3.2 *Let*

$$\begin{pmatrix} \mathbf{X}'\mathbf{X} & \mathbf{R}' \\ \mathbf{R} & \mathbf{0} \end{pmatrix}^{-} = \begin{pmatrix} \mathbf{C}_1 & \mathbf{C}_2 \\ \mathbf{C}_3 & \mathbf{C}_4 \end{pmatrix} \qquad (7.3.7)$$

be one choice of g-inverse. Then

$$\tilde{\boldsymbol{\beta}} = \mathbf{C}_1\mathbf{X}'\mathbf{Y} + \mathbf{C}_2\mathbf{c} \qquad (7.3.8)$$

is one solution of (7.3.6). Further

$$\left.\begin{aligned} V(\mathbf{p}'\tilde{\boldsymbol{\beta}}) &= \sigma^2\mathbf{p}'\mathbf{C}_1\mathbf{p} \\ \operatorname{cov}(\mathbf{p}'\tilde{\boldsymbol{\beta}}, \mathbf{q}'\tilde{\boldsymbol{\beta}}) &= \sigma^2\mathbf{p}'\mathbf{C}_1\mathbf{q} = \sigma^2\mathbf{q}'\mathbf{C}_1\mathbf{p}, \end{aligned}\right. \qquad (7.3.9)$$

when $\mathbf{p}'\boldsymbol{\beta}$ *and* $\mathbf{q}'\boldsymbol{\beta}$ *are both u.e., so that* $\sigma^2\mathbf{C}_1$ *behaves as a dispersion matrix of* $\tilde{\boldsymbol{\beta}}$.

Proof of Theorems 7.3.1 *and* 7.3.2: Since $\mathbf{p}'\boldsymbol{\beta}$ is u.e., there exist vectors $\mathbf{L}$ and $\mathbf{K}$ such that

$$\mathbf{X}'\mathbf{L} + \mathbf{R}'\mathbf{K} = \mathbf{p}, \tag{7.3.10}$$

in which case an unbiased estimator of $\mathbf{p}'\boldsymbol{\beta}$ is $\mathbf{L}'\mathbf{Y} + \mathbf{K}'\mathbf{c}$ and $V(\mathbf{L}'\mathbf{Y} + \mathbf{K}'\mathbf{c}) = \sigma^2\mathbf{L}'\mathbf{L}$. Then the problem is to minimize $\mathbf{L}'\mathbf{L}$ subject to (7.3.10). Introducing Lagrangian multipliers $\mathbf{v}$, the minimizing equations are

$$\begin{aligned} \mathbf{L} - \mathbf{X}\mathbf{v} &= \mathbf{0} \\ \mathbf{R}\mathbf{v} &= \mathbf{0} \\ \mathbf{X}'\mathbf{L} + \mathbf{R}'\mathbf{K} &= \mathbf{p}. \end{aligned} \tag{7.3.11}$$

Eliminating $\mathbf{L}$, we have

$$\begin{aligned} (\mathbf{X}'\mathbf{X})\mathbf{v} + \mathbf{R}'\mathbf{K} &= \mathbf{p} \\ \mathbf{R}\mathbf{v} &= \mathbf{0}. \end{aligned} \tag{7.3.12}$$

Using (7.3.7), a solution of (7.3.12) is $\hat{\mathbf{v}} = \mathbf{C}_1\mathbf{p}$, $\mathbf{K} = \mathbf{C}_3\mathbf{p}$, and $\mathbf{L} = \mathbf{X}\mathbf{C}_1\mathbf{p}$ giving the BLUE of $\mathbf{p}'\boldsymbol{\beta}$ as

$$\begin{aligned} &\mathbf{p}'\mathbf{C}_1'\mathbf{X}'\mathbf{Y} + \mathbf{p}'\mathbf{C}_3'\mathbf{c} \\ &= \mathbf{p}'\mathbf{C}_1\mathbf{X}'\mathbf{Y} + \mathbf{p}'\mathbf{C}_2\mathbf{c} \quad \text{since the transpose of (7.3.7) is also a g-inverse} \\ &= \mathbf{p}'(\mathbf{C}_1\mathbf{X}'\mathbf{Y} + \mathbf{C}_2\mathbf{c}) = \mathbf{p}'\tilde{\boldsymbol{\beta}}. \end{aligned}$$

Thus the result of Theorem 7.3.1 is proved.

The results (7.3.9) of Theorem 7.3.2 are easily established from the properties of a g-inverse.

THEOREM 7.3.3. *An unbiased estimator of* σ^2 *is*

$$\tilde{\sigma}^2 = (\mathbf{Y}'\mathbf{Y} - \tilde{\boldsymbol{\beta}}'\mathbf{X}'\mathbf{Y} - \tilde{\boldsymbol{\lambda}}'\mathbf{c})/f, \tag{7.3.13}$$

where $\tilde{\boldsymbol{\beta}}, \tilde{\boldsymbol{\lambda}}$ *is a solution of* (7.3.6) *and* f *is the d.f. as defined in* (7.3.5).

Proof: The result (7.3.13) is obtained by substituting a solution of (7.3.6) in $(\mathbf{Y} - \mathbf{X}\boldsymbol{\beta})'(\mathbf{Y} - \mathbf{X}\boldsymbol{\beta})$. The unbiased nature of (7.3.13) follows, since it is the same as the expression (7.3.4).

We shall consider some explicit forms for g-inverse defined in (7.3.7), which may or may not be useful for computational purposes in any particular problem. It may be easier to use some computational routine to find out a g-inverse of the matrix

$$\begin{pmatrix} \mathbf{X}'\mathbf{X} & \mathbf{R}' \\ \mathbf{R} & \mathbf{0} \end{pmatrix}$$

and then use the submatrices $\mathbf{C}_1, \mathbf{C}_2, \mathbf{C}_3, \mathbf{C}_4$ of such an inverse in writing down the estimates and their variances. However, the following theorem is of some interest.

THEOREM 7.3.4 *Let* $\mathbf{U} = [\mathbf{R}(\mathbf{X}'\mathbf{X})^-\mathbf{R}']^-$ *and* $\mathbf{V} = [\mathbf{R}(\mathbf{X}'\mathbf{X} + \mathbf{R}'\mathbf{R})^-\mathbf{R}']^-$ *for any choices of the g-inverses. Then one choice of explicit expressions for* $\mathbf{C}_1, \mathbf{C}_2, \mathbf{C}_3, \mathbf{C}_4$ *in (7.3.7) is as follows*

(a) *If* $\mathcal{M}(\mathbf{R}') = \mathcal{M}(\mathbf{X}')$, *then*

$$\mathbf{C}_1 = [\mathbf{I} - (\mathbf{X}'\mathbf{X})^-\mathbf{R}'\mathbf{U}\mathbf{R}](\mathbf{X}'\mathbf{X})^-, \qquad \mathbf{C}_2 = (\mathbf{X}'\mathbf{X})^-\mathbf{R}'\mathbf{U},$$
$$\mathbf{C}_3 = \mathbf{U}\mathbf{R}(\mathbf{X}'\mathbf{X})^-, \qquad \mathbf{C}_4 = -\mathbf{U}.$$

(b) *In the general case*

$$\mathbf{C}_1 = [\mathbf{I} - (\mathbf{X}'\mathbf{X} + \mathbf{R}'\mathbf{R})^-\mathbf{R}'\mathbf{V}\mathbf{R}](\mathbf{X}'\mathbf{X} + \mathbf{R}'\mathbf{R})^-, \qquad \mathbf{C}_2 = (\mathbf{X}'\mathbf{X} + \mathbf{R}'\mathbf{R})^-\mathbf{R}'\mathbf{V},$$
$$\mathbf{C}_3 = \mathbf{V}\mathbf{R}(\mathbf{X}'\mathbf{X} + \mathbf{R}'\mathbf{R})^-, \qquad \mathbf{C}_4 = -\mathbf{V} + \mathbf{I}.$$

7.4 MODEL: $(\mathbf{Y}, \mathbf{X}\boldsymbol{\beta}, \sigma^2\mathbf{S})$, $\boldsymbol{\beta}$ AND σ^2 UNKNOWN

We shall consider two cases depending on whether $\mathbf{S}$ is nonsingular or singular.

Case 1. Nonsingular $\mathbf{S}$. In this case $\mathbf{S}^{-1/2}$, a symmetrical square root of $\mathbf{S}^{-1}$, exists. If $\mathbf{Z} = \mathbf{S}^{-1/2}\mathbf{Y}$, then

$$E(\mathbf{Z}) = \mathbf{S}^{-1/2}\mathbf{X}\boldsymbol{\beta}, \qquad V(\mathbf{Z}) = \sigma^2\mathbf{I}$$

so that the model $(\mathbf{Y}, \mathbf{X}\boldsymbol{\beta}, \sigma^2\mathbf{S})$ reduces to $(\mathbf{Z}, \mathbf{S}^{-1/2}\mathbf{X}\boldsymbol{\beta}, \sigma^2\mathbf{I})$, and the methods of Section 7.2 for estimation of parameters and tests of linear hypotheses apply.

The normal equations are

$$\mathbf{X}'\mathbf{S}^{-1/2}\mathbf{S}^{-1/2}\mathbf{X}\boldsymbol{\beta} = \mathbf{X}'\mathbf{S}^{-1/2}\mathbf{Z} \quad \text{or} \quad \mathbf{X}'\mathbf{S}^{-1}\mathbf{X}\boldsymbol{\beta} = \mathbf{X}'\mathbf{S}^{-1}\mathbf{Y}, \tag{7.4.1}$$

which can be obtained by minimizing $(\mathbf{Y} - \mathbf{X}\boldsymbol{\beta})'\mathbf{S}^{-1}(\mathbf{Y} - \mathbf{X}\boldsymbol{\beta})$. A solution of (7.4.1) is

$$\hat{\boldsymbol{\beta}} = (\mathbf{X}'\mathbf{S}^{-1}\mathbf{X})^-\mathbf{X}'\mathbf{S}^{-1}\mathbf{Y},$$

with the formal dispersion matrix

$$D(\hat{\boldsymbol{\beta}}) = \sigma^2(\mathbf{X}'\mathbf{S}^{-1/2}\mathbf{S}^{-1/2}\mathbf{X})^- = \sigma^2(\mathbf{X}'\mathbf{S}^{-1}\mathbf{X})^-.$$

If $\mathbf{p}'\boldsymbol{\beta}$ and $\mathbf{q}'\boldsymbol{\beta}$ are u.e. functions, then the variances and covariances of their BLUE'S, $\mathbf{p}'\hat{\boldsymbol{\beta}}$ and $\mathbf{q}'\hat{\boldsymbol{\beta}}$, are

$$V(\mathbf{p}'\hat{\boldsymbol{\beta}}) = \sigma^2\mathbf{p}'(\mathbf{X}'\mathbf{S}^{-1}\mathbf{X})^-\mathbf{p}$$

$$\text{cov}(\mathbf{p}'\hat{\boldsymbol{\beta}}, \mathbf{q}'\hat{\boldsymbol{\beta}}) = \sigma^2\mathbf{p}'(\mathbf{X}'\mathbf{S}^{-1}\mathbf{X})^-\mathbf{q} = \sigma^2\mathbf{q}'(\mathbf{X}'\mathbf{S}^{-1}\mathbf{X})^-\mathbf{p}$$

$$V(\mathbf{q}'\hat{\boldsymbol{\beta}}) = \sigma^2\mathbf{q}'(\mathbf{X}'\mathbf{S}^{-1}\mathbf{X})^-\mathbf{q}.$$

Let $R(\mathbf{X}) = r$. Then an unbiased estimator of σ^2 is

$$\begin{aligned}\sigma^2 &= \min_{\boldsymbol{\beta}}(\mathbf{Z} - \mathbf{S}^{-1/2}\mathbf{X}\boldsymbol{\beta})'(\mathbf{Z} - \mathbf{S}^{-1/2}\mathbf{X}\boldsymbol{\beta}) \div (n - r) \\ &= \min_{\boldsymbol{\beta}}(\mathbf{Y} - \mathbf{X}\boldsymbol{\beta})'\mathbf{S}^{-1}(\mathbf{Y} - \mathbf{X}\boldsymbol{\beta}) \div (n - r) \\ &= (\mathbf{Y}'\mathbf{S}^{-1}\mathbf{Y} - \hat{\boldsymbol{\beta}}'\mathbf{X}'\mathbf{S}^{-1}\mathbf{Y}) \div (n - r).\end{aligned}$$

It may be noted that the reducted model $(\mathbf{Z}, \mathbf{S}^{-1/2}\mathbf{X}\boldsymbol{\beta}, \sigma^2\mathbf{I})$ is used only to apply the theory developed in Section 7.2, but the final expressions for the estimates of $\boldsymbol{\beta}$ and σ^2 involve only the quantities **Y**, **X**, **S**, which occur in original model.

Case 2. Singular **S**. We shall give a direct approach to the problem which is applicable whether **S** is singular or not and later show that the model in the singular case can be reduced to the model with restrictions on parameters, as discussed in Section 7.3.

Let $\mathbf{p}'\boldsymbol{\beta}$ be u.e. Then for any vector **L** such that $\mathbf{X}'\mathbf{L} = \mathbf{p}$, $\mathbf{L}'\mathbf{Y}$ is an unbiased estimator of $\mathbf{p}'\boldsymbol{\beta}$. The variance of $\mathbf{L}'\mathbf{Y}$ is $\sigma^2\mathbf{L}'\mathbf{S}\mathbf{L}$. The problem is that of minimizing $\mathbf{L}'\mathbf{S}\mathbf{L}$ subject to $\mathbf{X}'\mathbf{L} = \mathbf{p}$. The answer is provided by a minimum **S**-seminorm solution of $\mathbf{X}'\mathbf{L} = \mathbf{p}$. Such a solution is provided by

$$\mathbf{L} = [(\mathbf{X}')^-_{m(\mathbf{S})}]\mathbf{p},$$

where the inverse is as defined in section 3.1. Then the BLUE is

$$\mathbf{L}'\mathbf{Y} = \mathbf{p}'[(\mathbf{X}')^-_{m(\mathbf{S})}]'\mathbf{Y}. \tag{7.4.2}$$

It has been shown in Theorem 3.1.4 that one choice of $(\mathbf{X}')^-_{m(\mathbf{S})}$ is

$$(\mathbf{S} + \mathbf{X}\mathbf{X}')^-\mathbf{X}[\mathbf{X}'(\mathbf{S} + \mathbf{X}\mathbf{X}')^-\mathbf{X}]^-$$

in the general case and

$$\mathbf{S}^-\mathbf{X}(\mathbf{X}'\mathbf{S}^-\mathbf{X})^-$$

when $\mathcal{M}(\mathbf{X}) \subset \mathcal{M}(\mathbf{S})$, which provides explicit expressions for the BLUE given in (7.4.2).

Let us evaluate the variance of (7.4.2).

$$\begin{aligned}V(\mathbf{L}'\mathbf{Y}) &= \sigma^2\mathbf{p}'[(\mathbf{X}')^-_{m(\mathbf{S})}]'\mathbf{S}(\mathbf{X}')^-_{m(\mathbf{S})}\mathbf{p} \\ &= \sigma^2\mathbf{p}'\{[\mathbf{X}'(\mathbf{S} + \mathbf{X}\mathbf{X}')^-\mathbf{X}] - \mathbf{I}\}\mathbf{p} \text{ in general,} && (7.4.3) \\ &= \sigma^2\mathbf{p}'(\mathbf{X}'\mathbf{S}^-\mathbf{X})^-\mathbf{p} \quad \text{if} \quad \mathcal{M}(\mathbf{X}) \subset \mathcal{M}(\mathbf{S}). && (7.4.4)\end{aligned}$$

We shall show that the model $(\mathbf{Y}, \mathbf{X}\boldsymbol{\beta}, \sigma^2\mathbf{S})$, where $\mathbf{S}$ is singular, can be reduced to a model with restrictions on parameters and a nonsingular dispersion matrix for the variables.

Let $\mathbf{N}$ be an orthogonal complement of $\mathbf{S}$, that is, $\mathbf{N}$ is a matrix of maximum rank such that $\mathbf{N}'\mathbf{S} = \mathbf{0}$. Since $\mathbf{S}$ is p.s.d., there exists a matrix $\mathbf{J}$ such that $\mathbf{S} = \mathbf{J}\mathbf{J}'$, where $R(\mathbf{S}) = R(\mathbf{J}) =$ number of columns in $\mathbf{J}$. Let $\mathbf{F}'$ be a left inverse of $\mathbf{J}$, that is, $\mathbf{F}'\mathbf{J} = \mathbf{I}$, with the additional condition $\mathbf{F}'\mathbf{N} = \mathbf{0}$. We make the transformation from $\mathbf{Y}$ to $\mathbf{Y}_1$, $\mathbf{Y}_2$ with the properties

$$\mathbf{Y}_1 = \mathbf{F}'\mathbf{Y}, \qquad E(\mathbf{Y}_1) = \mathbf{F}'\mathbf{X}\boldsymbol{\beta}, \qquad D(\mathbf{Y}_1) = \sigma^2\mathbf{I}$$

$$\mathbf{Y}_2 = \mathbf{N}'\mathbf{Y}, \qquad E(\mathbf{Y}_2) = \mathbf{N}'\mathbf{X}\boldsymbol{\beta}, \qquad D(\mathbf{Y}_2) = \mathbf{0}.$$

The vector $\mathbf{Y}_2$ is nonstochastic and the equation $\mathbf{Y}_2 = \mathbf{N}'\mathbf{X}\boldsymbol{\beta}$ may then be considered as restrictions on the parameter $\boldsymbol{\beta}$. Thus the model $(\mathbf{Y}, \mathbf{X}\boldsymbol{\beta}, \mathbf{S})$ can be written:

$$(\mathbf{Y}_1, \mathbf{F}'\mathbf{X}\boldsymbol{\beta} | \mathbf{N}'\mathbf{X}\boldsymbol{\beta} = \mathbf{Y}_2, \sigma^2\mathbf{I}), \tag{7.4.5}$$

in which case the methods of Section 7.3 apply.

In particular an unbiased estimator of σ^2 is

$$\hat{\sigma}^2 = \min_{\mathbf{N}'\mathbf{X}\boldsymbol{\beta} = \mathbf{Y}_2} (\mathbf{Y}_1 - \mathbf{F}'\mathbf{X}\boldsymbol{\beta})'(\mathbf{Y}_1 - \mathbf{F}'\mathbf{X}\boldsymbol{\beta}) \div s, \tag{7.4.6}$$

where $s = R(\mathbf{S}) - R(\mathbf{F}'\mathbf{X})$.

The formula (7.4.6) as it stands involves the matrices $\mathbf{F}$ and $\mathbf{N}$, and Theorem 7.4.1 aims at expressing it in terms of original quantities $\mathbf{Y}$, $\mathbf{X}$, and $\mathbf{S}$.

THEOREM 7.4.1 *An unbiased estimator of σ^2 is*

$$\hat{\sigma}^2 = (\mathbf{Y} - \mathbf{X}\hat{\boldsymbol{\beta}})'\mathbf{S}^-(\mathbf{Y} - \mathbf{X}\hat{\boldsymbol{\beta}}) \div s \tag{7.4.7}$$

where

$$\hat{\boldsymbol{\beta}} = \mathbf{C}\mathbf{Y}, \qquad \mathbf{C} = \mathbf{X}^-_{m(\mathbf{S})} \quad \text{and} \quad s = R(\mathbf{S} \vdots \mathbf{X}) - R(\mathbf{X}) \tag{7.4.8}$$

for any choices of the g-inverses involved.

Proof: We use the formula (7.4.6) and show its equivalence with (7.4.7). First we observe that if $\mathbf{N}'\mathbf{X}\boldsymbol{\beta} = \mathbf{Y}_2$

$$\begin{aligned}(\mathbf{Y}_1 - \mathbf{F}'\mathbf{X}\boldsymbol{\beta})'(\mathbf{Y}_1 - \mathbf{F}'\mathbf{X}\boldsymbol{\beta}) &= (\mathbf{Y} - \mathbf{X}\boldsymbol{\beta})'\mathbf{F}\mathbf{F}'(\mathbf{Y} - \mathbf{X}\boldsymbol{\beta}) \\ &= (\mathbf{Y} - \mathbf{X}\boldsymbol{\beta})'\mathbf{S}^-(\mathbf{Y} - \mathbf{X}\boldsymbol{\beta})\end{aligned} \tag{7.4.9}$$

for any choice of $\mathbf{S}^-$. It only remains to show that $\hat{\boldsymbol{\beta}} = \mathbf{C}\mathbf{Y}$ satisfies the equations

$$\mathbf{X}'\mathbf{S}\mathbf{X}\boldsymbol{\beta} + \mathbf{X}'\mathbf{N}\boldsymbol{\lambda} = \mathbf{X}'\mathbf{S}^-\mathbf{Y}, \qquad \mathbf{N}'\mathbf{X}\boldsymbol{\beta} = \mathbf{N}'\mathbf{Y}, \tag{7.4.10}$$

which are obtained by minimizing $(\mathbf{Y} - \mathbf{X}\boldsymbol{\beta})'\mathbf{S}^-(\mathbf{Y} - \mathbf{X}\boldsymbol{\beta})$, subject to the condition $\mathbf{N'X\beta} = \mathbf{Y}_2$. This can be easily verified by substituting $\mathbf{CY}$ for $\boldsymbol{\beta}$ in (7.4.10) and using the property of $\mathbf{C}$, which is a minimum seminorm g-inverse, that is

$$\mathbf{SCX} = \mathbf{XCS}.$$

The formula

$$s = R(\mathbf{S}) - R(\mathbf{F'X}) = R(\mathbf{S} \vdots \mathbf{X}) - R(\mathbf{X})$$

is easily established.

Remark. Consider the model $(\mathbf{Y}, \mathbf{X\beta}, \sigma^2\mathbf{S})$, where $\boldsymbol{\beta}$ and σ^2 are unknown and $\mathbf{S}$ is known. We have considered the problem of BLUE in cases where $\mathbf{S}$ is p.d. and p.s.d. Now let $\mathbf{p'\beta}$ be any parametric function which may be u.e. or not, in which case we may seek for BLMBE. If $\mathbf{L'Y}$ is such an estimate of $\mathbf{p'\beta}$, then we find $\mathbf{L}$ to minimize the variance

$$V(\mathbf{L'Y}) = \sigma^2\mathbf{L'SL} \tag{7.4.11}$$

in the class of $\mathbf{L}$, which minimize the bias

$$\|\mathbf{X'L} - \mathbf{p}\| = [(\mathbf{X'L} - \mathbf{p})'\mathbf{M}(\mathbf{X'L} - \mathbf{p})]^{1/2}, \tag{7.4.12}$$

where $\mathbf{M}$ is a p.d. matrix. Then $\mathbf{L}$ is minimum $\mathbf{S}$-norm $\mathbf{M}$-least square solution of $\mathbf{X'L} = \mathbf{p}$ (which may be inconsistent) defined in Section 3.3. Thus the optimum $\mathbf{L} = (\mathbf{X}')^{+}_{\mathbf{MS}}\mathbf{p}$ giving the BLMBE of $\mathbf{p'\beta}$ as

$$\mathbf{p}'[(\mathbf{X}')^{+}_{\mathbf{MS}}]'\mathbf{Y} \tag{7.4.13}$$

with variance

$$\sigma^2\mathbf{p}'[(\mathbf{X}')^{+}_{\mathbf{MS}}]'(\mathbf{X}')^{+}_{\mathbf{MS}}\mathbf{p}. \tag{7.4.14}$$

7.5 ADJUSTMENT OF LEAST-SQUARES ESTIMATES FOR ADDITION OR REMOVAL OF AN OBSERVATION

The formulae for g-inverse of partitioned matrices developed in Section 3.6 can be used to make adjustments in the least-squares estimates with minimum computations, when a new observation is added or one of the available observations is omitted. We shall illustrate the method with a numerical example.

Given below are:

(i) an observation vector $\mathbf{y}$ on a vector $\mathbf{Y}$ of random variables obeying a linear model $E(\mathbf{Y}) = \mathbf{X\beta}$ and $D(\mathbf{Y}) = \sigma^2\mathbf{I}$,

(ii) the matrix $\mathbf{X}$ of the linear model,

(iii) $\mathbf{X}_l^-$, a least-squares g-inverse of $\mathbf{X}$,

(iv) $\hat{\boldsymbol{\beta}} = \mathbf{X}_l^- \mathbf{y}$, a least-squares solution for the unknown parametric vector $\boldsymbol{\beta}$,

(v) R_0^2, the residual sum of squares, and

(vi) $\hat{\sigma}^2$, an unbiased estimate of σ^2 based on $5 - 2 = 3$ d.f.

$$\mathbf{X} = \begin{pmatrix} 2 & 1 & 0 \\ 1 & 1 & 1 \\ 3 & 2 & 1 \\ 1 & 0 & -1 \\ 0 & 1 & 2 \end{pmatrix} \qquad \mathbf{y} = \begin{pmatrix} 5.6 \\ 4.2 \\ 9.6 \\ 2.4 \\ 2.4 \end{pmatrix}$$

$$\mathbf{X}_l^- = \begin{pmatrix} 5/24 & -1/12 & 1/8 & 7/24 & -3/8 \\ -1/8 & 1/4 & 1/8 & -3/8 & 5/8 \\ 0 & 0 & 0 & 0 & 0 \end{pmatrix}$$

$$\hat{\boldsymbol{\beta}} = \begin{pmatrix} 1.8167 \\ 2.1500 \\ 0.0 \end{pmatrix}$$

$$\mathrm{R}_0^2 = \mathbf{y}'\mathbf{y} - \mathbf{y}'\mathbf{X}\hat{\boldsymbol{\beta}} = 0.51 \quad \text{and} \quad \hat{\sigma}^2 = \frac{R_0^2}{5 - 2} = 0.17.$$

Suppose it is decided to rework the least-squares solution, excluding the last observation, 2.4.

$$\text{Let } \mathbf{A} = \begin{pmatrix} 2 & 1 & 0 \\ 1 & 1 & 1 \\ 3 & 2 & 1 \\ 1 & 0 & -1 \end{pmatrix} \quad \text{and} \quad \mathbf{a} = \begin{pmatrix} 0 \\ 1 \\ 2 \end{pmatrix}.$$

Notice that we are required to work out $\mathbf{A}_l^-$, a least-squares g-inverse of $\mathbf{A}$ based on $\mathbf{X}_l^-$, where $\mathbf{X} = \begin{pmatrix} \mathbf{A} \\ \cdots \\ \mathbf{a}' \end{pmatrix}$.

$$\text{Let } \mathbf{G}' = \begin{pmatrix} 5/24 & -1/12 & 1/8 & 7/24 \\ -1/8 & 1/4 & 1/8 & -3/8 \\ 0 & 0 & 0 & 0 \end{pmatrix} \quad \text{and} \quad \mathbf{b}' = \begin{pmatrix} -3/8 \\ 5/8 \\ 0 \end{pmatrix}$$

$$\text{Thus } \begin{pmatrix} \mathbf{A} \\ \cdots \\ \mathbf{a}' \end{pmatrix}_l = (\mathbf{G}' : \mathbf{b}).$$

Since transpose of a least-squares g-inverse of a matrix is a minimum norm g-inverse of the transpose, we have

$$(\mathbf{A}' : \mathbf{a})_m^- = \begin{pmatrix} \mathbf{G} \\ \cdots \\ \mathbf{b}' \end{pmatrix}.$$

Further $\mathbf{b}'\mathbf{a} = 5/8 \neq 1$. Hence applying Theorem 3.6.2, case 2 (equation 3.6.6), we have

$$(\mathbf{A}')_m^- = \mathbf{G}\left(\mathbf{I} + \frac{\mathbf{ab}'}{1 - \mathbf{b}'\mathbf{a}}\right) = \begin{pmatrix} 1/3 & -1/3 & 0 \\ -1/3 & 2/3 & 0 \\ 0 & 1/3 & 0 \\ 2/3 & -1 & 0 \end{pmatrix}.$$

$$\text{Hence } \mathbf{A}_l^- = \begin{pmatrix} 1/3 & -1/3 & 0 & 2/3 \\ -1/3 & 2/3 & 1/3 & -1 \\ 0 & 0 & 0 & 0 \end{pmatrix}.$$

The revised estimate of $\boldsymbol{\beta}$ is

$$\hat{\boldsymbol{\beta}} = \begin{pmatrix} 2.0667 \\ 1.7333 \\ 0 \end{pmatrix}.$$

$R_0^2 = 0.35$ and $\hat{\sigma}^2 = R_0^2/(4 - 2) = 0.175$.

Suppose with the same data we are required to fit a revised model $E(\mathbf{Y}) = (\mathbf{X}:\mathbf{x})\begin{pmatrix} \boldsymbol{\beta} \\ \cdots \\ \alpha \end{pmatrix}$ where $\mathbf{x}' = (0 \quad 0 \quad 1 \quad 0 \quad 0)$. Then we proceed as follows:

We shall compute $(\mathbf{X} : \mathbf{x})_l^-$ based on $\mathbf{X}_l^-$. Since $\mathbf{XX}_l^-\mathbf{x} \neq \mathbf{x}$ it follows that $\mathbf{x} \notin \mathcal{M}(\mathbf{X})$. Applying Theorem 3.6.1, case 1 (equation 3.6.2), we have

$$(\mathbf{X} : \mathbf{x})_l^- = \begin{pmatrix} \mathbf{X}_l^- - \mathbf{db}' \\ \mathbf{b}' \end{pmatrix},$$

where

$$\mathbf{d} = \mathbf{X}_l^-\mathbf{x}, \qquad \mathbf{c} = (\mathbf{I} - \mathbf{XX}_l^-)\mathbf{x} = \mathbf{x} - \mathbf{Xd} \quad \text{and} \quad \mathbf{b} = \mathbf{c}/\mathbf{c}'\mathbf{x}.$$

Thus,

$$\mathbf{d}' = (1/8 \quad 1/8 \quad 0), \qquad \mathbf{c}' = (-3/8 \quad -1/4 \quad 3/8 \quad -1/8 \quad -1/8) \quad \text{and}$$

$$\mathbf{b}' = (-1 \quad -2/3 \quad 1 \quad -1/3 \quad -1/3).$$

Hence

$$(\mathbf{X} : \mathbf{x})_l^- = \begin{pmatrix} 1/3 & 0 & 0 & 1/3 & -1/3 \\ 0 & 1/3 & 0 & -1/3 & 2/3 \\ 0 & 0 & 0 & 0 & 0 \\ -1 & -2/3 & 1 & -1/3 & -1/3 \end{pmatrix}.$$

The revised estimate of the parametric vector

$$\begin{pmatrix} \hat{\boldsymbol{\beta}} \\ \cdots \\ \hat{\alpha} \end{pmatrix} = \begin{pmatrix} 1.8667 \\ 2.2000 \\ 0.0 \\ -0.4000 \end{pmatrix}.$$

The residual sum of squares, $R_0^2 = 0.45$ and $\hat{\sigma}^2 = R_0^2/(5 - 3) = 0.225$.

COMPLEMENTS

1. (Converse of the principle of substitution.) Consider the model $(\mathbf{Y}, \mathbf{X}\boldsymbol{\beta}, \sigma^2\mathbf{I})$. If $\hat{\boldsymbol{\beta}} = \mathbf{AY}$ is such that for each estimable $\mathbf{p}'\boldsymbol{\beta}$, $\mathbf{p}'\hat{\boldsymbol{\beta}}$ is BLUE for $\mathbf{p}'\boldsymbol{\beta}$, then $\hat{\boldsymbol{\beta}}$ satisfies the normal equations (7.2.4).

2. Also if $\mathbf{Y}'(\mathbf{I} - \mathbf{XA})'(\mathbf{I} - \mathbf{XA})\mathbf{Y} \sim \sigma^2\chi^2_{n-r}$, then $\hat{\boldsymbol{\beta}} = \mathbf{AY}$ satisfies the normal equations.

3. (A least-squares accumulation theorem.) If $\mathbf{A}^*(x)$ and $\mathbf{B}^*(x)$ are least-squares mth degree polynomial approximations to $\mathbf{A}(x)$ and $\mathbf{B}(x)$ for values x_i $(i = 1, 2, \ldots, n)$, then

$$\sum_{i=1}^{n} \mathbf{A}^*(x_i)\mathbf{B}(x_i) = \sum_{i=1}^{n} \mathbf{A}(x_i)\mathbf{B}^*(x_i)$$

$$= \sum_{i=1}^{n} \mathbf{A}^*(x_i)\mathbf{B}^*(x_i) \qquad \text{(Bleick, 1940)}$$

4. For a linear model $(\mathbf{Y}, \mathbf{X}\boldsymbol{\beta}, \boldsymbol{\Sigma})$ the problem of linear estimation is said to have a complete solution, if every estimable linear function $\mathbf{p}'\boldsymbol{\beta}$ admits a BLUE (which has least variance uniformly with respect to unknown parameters). Consider the linear model

$$E\begin{pmatrix} \mathbf{Y}_1 \\ \mathbf{Y}_2 \end{pmatrix} = \begin{pmatrix} \mathbf{X}_1 \\ \mathbf{X}_2 \end{pmatrix}\boldsymbol{\beta}, \qquad D\begin{pmatrix} \mathbf{Y}_1 \\ \mathbf{Y}_2 \end{pmatrix} = \begin{pmatrix} \sigma^2\mathbf{I} & \mathbf{0} \\ \mathbf{0} & \delta^2\mathbf{I} \end{pmatrix},$$

where $\boldsymbol{\beta}$ as well as σ^2 (>0), δ^2 (>0) are unknown (a). Here the problem of linear estimation has a complete solution if $\mathscr{M}(\mathbf{X}_1')$ and $\mathscr{M}(\mathbf{X}_2')$ are virtually disjoint.

(b) If $\mathscr{M}(\mathbf{X}_1')$ and $\mathscr{M}(\mathbf{X}_2')$ are not virtually disjoint, but $\mathbf{p}'\boldsymbol{\beta}$ has a BLUE, the BLUE is given by $\mathbf{p}'(\mathbf{X}'\mathbf{X})^-\mathbf{X}'\mathbf{Y}$ where $\mathbf{X}' = (\mathbf{X}_1' \,\vdots\, \mathbf{X}_2')$ and $\mathbf{Y}' = (\mathbf{Y}_1' \,\vdots\, \mathbf{Y}_2')$.

(c) If $\mathbf{p}'\boldsymbol{\beta}$ is estimable, $\mathbf{p}'(\mathbf{X}'\mathbf{X})^{-}\mathbf{X}'\mathbf{Y}$ is the BLUE of $\mathbf{p}'\boldsymbol{\beta}$ if it is equal to $\mathbf{p}'(\mathbf{X}_1'\mathbf{X}_1)^{-}\mathbf{X}_1'$ or $\mathbf{p}'(\mathbf{X}_2'\mathbf{X}_2)^{-}\mathbf{X}_2'\mathbf{Y}_2$ or $\mathbf{p}'\boldsymbol{\beta}$ is the sum of estimable $\mathbf{p}_1'\boldsymbol{\beta}$ and $\mathbf{p}_2'\boldsymbol{\beta}$ and

$$\mathbf{p}_1'(\mathbf{X}'\mathbf{X})^{-}\mathbf{X}'\mathbf{Y} = \mathbf{p}_1'(\mathbf{X}_1'\mathbf{X}_1)^{-}\mathbf{X}_1'\mathbf{Y}_1$$

$$\mathbf{p}_2'(\mathbf{X}'\mathbf{X})^{-}\mathbf{X}'\mathbf{Y} = \mathbf{p}_2'(\mathbf{X}_2'\mathbf{X}_2)^{-}\mathbf{X}_2'\mathbf{Y}_2.$$

(d) If $\mathbf{p}$ is a nonnull vector in $\mathcal{M}(\mathbf{X}_1') \cap \mathcal{M}(\mathbf{X}_2')$, $\mathbf{p}'\boldsymbol{\beta}$ does not have a BLUE.

(e) If $\mathbf{p}' = \mathbf{b}\mathbf{X}_1$ and $\mathcal{M}(\mathbf{X}_1'\mathbf{F}) = \mathcal{M}(\mathbf{X}_1') \cap \mathcal{M}(\mathbf{X}_2')$, $\mathbf{p}'(\mathbf{X}_1'\mathbf{X}_1)^{-}\mathbf{X}_1'\mathbf{Y}_1$ is the BLUE $\mathbf{p}'\boldsymbol{\beta}$ if $\mathbf{b}'\mathbf{P}_{\mathbf{X}_1}\mathbf{F} = \mathbf{0}$.

5. (Gauss–Markoff model—multivariate case.) Let $\mathbf{Y} = (\mathbf{Y}_1 \vdots \mathbf{Y}_2 \vdots \cdots \vdots \mathbf{Y}_p)$ be $n \times p$ matrix of random variables such that

$$E(\mathbf{Y}) = \mathbf{X}_0\mathbf{B}$$

$$D(\mathbf{Y}_i) = \sigma_{ii}\mathbf{I}$$

$$\operatorname{cov}(\mathbf{Y}_i, \mathbf{Y}_j) = \sigma_{ij}\mathbf{I}$$

$(i = 1, 2, \ldots, p, j = 1, 2, \ldots, p)$, where $\mathbf{X}_0$ is an $n \times m$ matrix of known coefficients a $\mathbf{B} = (\boldsymbol{\beta}_1 \vdots \boldsymbol{\beta}_2 \vdots \cdots \vdots \boldsymbol{\beta}_p)$ is a $m \times p$ matrix of unknown parameters. Let $\mathbf{A}$ and $\mathbf{L}$ be ma rices of order $p \times n$ and $p \times m$, respectively. Then, $tr\,\mathbf{AY}$ is the BLUE of $tr\,\mathbf{LB}$ if t ith diagonal element of $\mathbf{AY}$ is the BLUE of the corresponding element of $\mathbf{LB}$ und the model $(\mathbf{Y}_i, \mathbf{X}_0\boldsymbol{\beta}_i, \sigma_{ii}\mathbf{I})$ for all i.

CHAPTER 8

Conditions for Optimality and Validity of Least-Squares Theory

8.1 INTRODUCTION

In this chapter we examine validity of procedures developed in the previous sections under specification errors in the linear model. For our discussion, it will be convenient to introduce the following definition.

'The BLUE of an estimable parametric function $\mathbf{p}'\boldsymbol{\beta}$ as computed under the model $(\mathbf{Y}, \mathbf{X}_o\boldsymbol{\beta}, \boldsymbol{\Sigma}_o)$ or briefly $(\mathbf{X}_o, \boldsymbol{\Sigma}_o)$ is said to be $(\mathbf{X}, \boldsymbol{\Sigma})$-optimal, if it is also the BLUE under the alternative model $(\mathbf{Y}, \mathbf{X}\boldsymbol{\beta}, \boldsymbol{\Sigma})$.'

8.2 SPECIFICATION ERRORS IN THE DISPERSION MATRIX

THEOREM 8.2.1 *If for every estimable parametric function* $\mathbf{p}'\boldsymbol{\beta}$, *the BLUE under the model* $(\mathbf{X}, \sigma^2\mathbf{I})$ *is* $(\mathbf{X}, \boldsymbol{\Sigma})$*-optimal, it is n.s. that any one of the following equivalent conditions holds*

(i) $\mathbf{X}'\boldsymbol{\Sigma}\mathbf{Z} = \mathbf{0}$ (8.2.1)

(ii) $\boldsymbol{\Sigma} = \mathbf{X}\boldsymbol{\Lambda}_1\mathbf{X}' + \mathbf{Z}\boldsymbol{\Lambda}_2\mathbf{Z}'$ (8.2.2)

(iii) $\boldsymbol{\Sigma} = \mathbf{X}\boldsymbol{\Theta}\mathbf{X}' + \mathbf{Z}\boldsymbol{\Gamma}\mathbf{Z}' + \sigma^2\mathbf{I}$, (8.2.3)

where $\mathbf{Z}$ *is written for* $\mathbf{X}^\perp$ *and* $\boldsymbol{\Lambda}_1, \boldsymbol{\Lambda}_2, \boldsymbol{\Theta}, \boldsymbol{\Gamma}$ *are arbitrary symmetric matrices, subject to the condition that* $\boldsymbol{\Sigma}$ *is n.n.d.*

Proof: Let $\mathbf{N}$ be one choice for $\boldsymbol{\Sigma}^\perp$. Observe that $E(\mathbf{N}'\mathbf{Y}) = \mathbf{N}'\mathbf{X}\boldsymbol{\beta}$ and $D(\mathbf{N}'\mathbf{Y}) = \mathbf{N}'\boldsymbol{\Sigma}\mathbf{N} = \mathbf{0}$ so that with probability 1, $\mathbf{N}'\mathbf{Y}$ is a vector of constants, say $\mathbf{d}$. In such a case $\mathbf{N}'\mathbf{X}\boldsymbol{\beta} = \mathbf{d}$ may be considered as a restriction on the unknown parameters. By Theorem 2.3.1(c) a n.s. condition that a linear function $\mathbf{F}'\mathbf{Y}$ has constant expectation independently of the unknown parameters but subject to $\mathbf{N}'\mathbf{X}\boldsymbol{\beta} = \mathbf{d}$ is that $\mathscr{M}(\mathbf{X}'\mathbf{F}) \subset \mathscr{M}(\mathbf{X}'\mathbf{N})$ or $\mathbf{X}'\mathbf{F} = \mathbf{X}'\mathbf{N}\mathbf{M}$ for some $\mathbf{M}$. Then $\mathbf{F}'\mathbf{Y}$ can be written

$$\begin{aligned} \mathbf{F}'\mathbf{Y} &= (\mathbf{F}'\mathbf{Y} - \mathbf{M}'\mathbf{N}'\mathbf{Y}) + \mathbf{M}'\mathbf{N}'\mathbf{Y} \\ &= \mathbf{H}'\mathbf{Y} + \mathbf{M}'\mathbf{N}'\mathbf{Y}, \end{aligned} \tag{8.2.4}$$

where $\mathbf{H}'\mathbf{Y}$ has zero expectation independently of the restrictions, that is, $\mathbf{X}'\mathbf{H} = \mathbf{0}$ or $\mathbf{H} = \mathbf{Z}\mathbf{B}$ for some $\mathbf{B}$.

Now the BLUE of an estimable parametric function under the model $(\mathbf{X}, \sigma^2\mathbf{I})$ is of the form $\boldsymbol{\lambda}'\mathbf{X}'\mathbf{Y}$ and, conversely, for any given $\boldsymbol{\lambda}$, $\boldsymbol{\lambda}'\mathbf{X}'\mathbf{Y}$ is the BLUE of its expectation. Let $\mathbf{F}'\mathbf{Y}$ be a function of the type considered in (8.2.4). Then for $\boldsymbol{\lambda}'\mathbf{X}'\mathbf{Y}$ to be a BLUE under $(\mathbf{Y}, \mathbf{X}\boldsymbol{\beta}, \boldsymbol{\Sigma})$ it is n.s. (See Rao, 1965, p. 257)

$$\begin{aligned} \text{cov}(\boldsymbol{\lambda}'\mathbf{X}'\mathbf{Y}, \mathbf{F}'\mathbf{Y}) &= \text{cov}(\boldsymbol{\lambda}'\mathbf{X}'\mathbf{Y}, \mathbf{H}'\mathbf{Y}) \\ &= \boldsymbol{\lambda}'\mathbf{X}'\boldsymbol{\Sigma}\mathbf{Z}\mathbf{B} = \mathbf{0}. \end{aligned} \tag{8.2.5}$$

In (8.2.5) $\mathbf{F}'\mathbf{Y}$ is replaced by $\mathbf{H}'\mathbf{Y}$, since the other part in $\mathbf{F}'\mathbf{Y}$, namely $\mathbf{M}'\mathbf{N}'\mathbf{Y} = \mathbf{M}'\mathbf{d}$, is a constant. If (8.2.5) is to hold for all $\boldsymbol{\lambda}$ and $\mathbf{B}$, then $\mathbf{X}'\boldsymbol{\Sigma}\mathbf{Z} = \mathbf{0}$. Thus (8.2.1) is established.

Let us now write the n.n.d. matrix $\boldsymbol{\Sigma}$ as $\boldsymbol{\Sigma} = \mathbf{C}\mathbf{C}'$. Since $\mathscr{E}^n = \mathscr{M}(\mathbf{X}) \oplus \mathscr{M}(\mathbf{Z})$, there exist matrices $\mathbf{U}_1$ and $\mathbf{U}_2$ such that $\mathbf{C} = \mathbf{X}\mathbf{U}_1 + \mathbf{Z}\mathbf{U}_2$, giving

$$\boldsymbol{\Sigma} = \mathbf{X}\boldsymbol{\Lambda}_1\mathbf{X}' + \mathbf{Z}\boldsymbol{\Lambda}_2\mathbf{Z}' + \mathbf{X}\boldsymbol{\Lambda}_3\mathbf{Z}' + \mathbf{Z}\boldsymbol{\Lambda}_3'\mathbf{X}', \tag{8.2.6}$$

where

$$\boldsymbol{\Lambda}_1 = \mathbf{U}_1\mathbf{U}_1', \qquad \boldsymbol{\Lambda}_2 = \mathbf{U}_2\mathbf{U}_2' \quad \text{and} \quad \boldsymbol{\Lambda}_3 = \mathbf{U}_1\mathbf{U}_2'.$$

Then

$$\mathbf{X}'\boldsymbol{\Sigma}\mathbf{Z} = \mathbf{X}'\mathbf{X}\boldsymbol{\Lambda}_3\mathbf{Z}'\mathbf{Z} = \mathbf{0} \Leftrightarrow \mathbf{X}(\mathbf{X}'\mathbf{X})^-\mathbf{X}'\mathbf{X}\boldsymbol{\Lambda}_3\mathbf{Z}'\mathbf{Z}(\mathbf{Z}'\mathbf{Z})^-\mathbf{Z}' = \mathbf{X}\boldsymbol{\Lambda}_3\mathbf{Z}' = \mathbf{0}.$$

Hence

$$\boldsymbol{\Sigma} = \mathbf{X}\boldsymbol{\Lambda}_1\mathbf{X}' + \mathbf{Z}\boldsymbol{\Lambda}_2\mathbf{Z}'.$$

It would be of interest to write $\boldsymbol{\Sigma}$ as the sum of $\sigma^2\mathbf{I}$ and another matrix which is easily done by using the identity

$$\mathbf{X}(\mathbf{X}'\mathbf{X})^-\mathbf{X}' + \mathbf{Z}(\mathbf{Z}'\mathbf{Z})^-\mathbf{Z}' = \mathbf{I}.$$

Multiplying this identity by an arbitrary positive scalar σ^2 and subtracting from (8.2.2), we have, by rearrangement of terms, the expression

$$\boldsymbol{\Sigma} = \mathbf{X}\boldsymbol{\Theta}\mathbf{X}' + \mathbf{Z}\boldsymbol{\Gamma}\mathbf{Z}' + \sigma^2\mathbf{I}$$

where

$$\boldsymbol{\Theta} = \boldsymbol{\Lambda}_1 - \sigma^2(\mathbf{X}'\mathbf{X})^-, \qquad \boldsymbol{\Gamma} = \boldsymbol{\Lambda}_2 - \sigma^2(\mathbf{Z}'\mathbf{Z})^-.$$

Note. The BLUE of $\mathbf{p}'\boldsymbol{\beta}$ under the model $(\mathbf{X}, \sigma^2\mathbf{I})$ will be referred to as a simple least-squares estimator (SLSE). Using Theorem 7.2.2, the SLSE of $\mathbf{p}'\boldsymbol{\beta}$ can be written as $\mathbf{p}'(\mathbf{X}'\mathbf{X})^-\mathbf{X}'\mathbf{Y}$. When $\boldsymbol{\Sigma}$ has the structure as given in Theorem 8.2.1, then the BLUE under the model $(\mathbf{X}, \boldsymbol{\Sigma})$ is same as the SLSE. This does not necessarily mean that the variances of the common estimator are the same under the two models. Now

$$V(\mathbf{p}'(\mathbf{X}'\mathbf{X})^-\mathbf{X}'\mathbf{Y}|\sigma^2\mathbf{I}) = \sigma^2\mathbf{p}'(\mathbf{X}'\mathbf{X})^-\mathbf{p}$$

$$\begin{aligned} V(\mathbf{p}'(\mathbf{X}'\mathbf{X})^-\mathbf{X}'\mathbf{Y}|\boldsymbol{\Sigma}) &= \mathbf{p}'(\mathbf{X}'\mathbf{X})^-\mathbf{X}'[\mathbf{X}\boldsymbol{\Theta}\mathbf{X}' + \mathbf{Z}\boldsymbol{\Gamma}\mathbf{Z}' + \sigma^2\mathbf{I}]\mathbf{X}(\mathbf{X}'\mathbf{X})^-\mathbf{p} \\ &= \mathbf{p}'\boldsymbol{\Theta}\mathbf{p} + \sigma^2\mathbf{p}'(\mathbf{X}'\mathbf{X})^-\mathbf{p}. \end{aligned}$$

The two expressions are equal only when $\boldsymbol{\Theta} = \mathbf{0}$, that is, when $\boldsymbol{\Sigma}$ is of the form $\mathbf{Z}\boldsymbol{\Gamma}\mathbf{Z}' + \sigma^2\mathbf{I}$.

COROLLARY 1: *If the conditions of Theorem 8.2.1 hold, then the BLUE under* $(\mathbf{X}, \boldsymbol{\Sigma})$ *is also* $(\mathbf{X}, \sigma^2\mathbf{I})$*-optimal for every estimable parametric function.*

COROLLARY 2: *If* $\boldsymbol{\Sigma}_o$ *obeys the conditions of Theorem 8.2.1, then for every BLUE under* $(\mathbf{X}, \boldsymbol{\Sigma}_o)$ *to be* $(\mathbf{X}, \boldsymbol{\Sigma})$*-optimal it is n.s. that* $\boldsymbol{\Sigma}$ *obeys the conditions of Theorem 8.2.1.*

COROLLARY 3: *Each of the following conditions is equivalent to any one of the conditions* (*i*), (*ii*), (*iii*) *stated in Theorem 8.2.1.*

(iv) $\boldsymbol{\Sigma}\mathbf{X} = \mathbf{X}\mathbf{B}$ *for some* $\mathbf{B}$,
(v) $\mathcal{M}(\mathbf{X})$ *is spanned by a subset of eigenvectors of* $\boldsymbol{\Sigma}$,
(vi) $\mathbf{P_X}\boldsymbol{\Sigma}$ *is symmetric, where* $\mathbf{P_X}$ *is the projection operator onto* $\mathcal{M}(\mathbf{X})$.

Proof: Since $\mathbf{Z} = \mathbf{X}^\perp$, $\mathbf{X}'\boldsymbol{\Sigma}\mathbf{Z} = \mathbf{0} \Leftrightarrow \mathcal{M}(\boldsymbol{\Sigma}\mathbf{X}) \subset \mathcal{M}(\mathbf{X})$. Hence (iv) is established. This form of the condition is given in Kruskal (1968).

To obtain (v), consider the proper eigenvectors of $\mathbf{A} = \mathbf{X}'\mathbf{X}\boldsymbol{\Lambda}_1\mathbf{X}'\mathbf{X}$ with respect to $\mathbf{B} = \mathbf{X}'\mathbf{X}$. Since $\mathcal{M}(\mathbf{A}) \subset \mathcal{M}(\mathbf{B})$, the conditions of Theorem 6.3.1 are trivially satisfied. Hence if $R(\mathbf{B}) = R(\mathbf{X}) = r$, by Theorem 6.3.1 there exists an $m \times r$ matrix $\mathbf{W}$ such that

$$\mathbf{W}'\mathbf{B}\mathbf{W} = \mathbf{I}_r \qquad \text{and} \qquad \mathbf{A}\mathbf{W} = \mathbf{B}\mathbf{W}\mathbf{D}, \tag{8.2.7}$$

where the matrix $\mathbf{D}$ is diagonal. Now

$$\mathbf{X}'\mathbf{X}\boldsymbol{\Lambda}_1\mathbf{X}'\mathbf{X}\mathbf{W} = \mathbf{X}'\mathbf{X}\mathbf{W}\mathbf{D} \Rightarrow \mathbf{X}\boldsymbol{\Lambda}_1\mathbf{X}'\mathbf{X}\mathbf{W} = \mathbf{X}\mathbf{W}\mathbf{D} \Rightarrow \boldsymbol{\Sigma}\mathbf{X}\mathbf{W} = \mathbf{X}\mathbf{W}\mathbf{D}$$

i.e., the columns of $\mathbf{X}\mathbf{W}$ are eigenvectors of $\boldsymbol{\Sigma}$. Also (8.2.7) $\Rightarrow R(\mathbf{X}\mathbf{W}) = r = R(\mathbf{X})$. Hence $\mathcal{M}(\mathbf{X}) = \mathcal{M}(\mathbf{X}\mathbf{W})$ and (v) is established. Condition (v) is stated by Zyskind (1967).

To establish (vi), observe that (8.2.6) $\Rightarrow \mathbf{P_X}\boldsymbol{\Sigma} = \mathbf{X}\boldsymbol{\Lambda}_1\mathbf{X}' + \mathbf{X}\boldsymbol{\Lambda}_3\mathbf{Z}'$. Hence (vi) $\Leftrightarrow \mathbf{X}\boldsymbol{\Lambda}_3\mathbf{Z}' = \mathbf{0}$. Corollary 3 is proved.

Let $\mathbf{\Sigma} = \lambda_1\mathbf{E}_1 + \lambda_2\mathbf{E}_2 + \cdots + \lambda_k\mathbf{E}_k$ be a spectral representation of $\mathbf{\Sigma}$, where $\lambda_1, \lambda_2, \ldots, \lambda_k$ are the nonnull eigenvalues of $\mathbf{\Sigma}$ (some of which could be repeated) and $\mathbf{E}_1, \mathbf{E}_2, \ldots, \mathbf{E}_k$ be the corresponding idempotents. Put $\mathbf{E}_{k+1} = \mathbf{I} - \mathbf{E}_1 - \mathbf{E}_2 - \cdots - \mathbf{E}_k$. Consider the linear manifold $\mathcal{M}(\mathbf{E}_i)$ and let $\mathcal{S}_i$ be the projection of $\mathcal{M}(\mathbf{E}_i)$ on $\mathcal{M}(\mathbf{X})$, that is, $\mathcal{S}_i = \mathcal{M}(\mathbf{P_X E}_i)$. Let us group $\mathbf{E}_i$ $(i = 1, 2, \ldots, k + 1)$ into n (as large a number as possible) sets, such that if $\mathbf{E}_i$ and $\mathbf{E}_j$ belong to two different sets, then $\mathcal{S}_i$ is orthogonal to $\mathcal{S}_j$ (that is, $\mathbf{E}_i\mathbf{P_X}\mathbf{E}_j = \mathbf{0}$). We shall now prove another equivalent condition stated by Watson (1967). It is seen that all these equivalent conditions are simple consequences of the main result $\mathbf{X'\Sigma Z} = \mathbf{0}$, presented by Rao in 1965 at the Fifth Berkeley Symposium on Prob. and Math. Stat.

COROLLARY 4: *For every estimable parametric function the BLUE under* $(\mathbf{X}, \sigma^2\mathbf{I})$ *to be* $(\mathbf{X}, \mathbf{\Sigma})$*-optimal it is n.s. that the eigenvalues associated with* $\mathbf{E}_i$ *in any given set are equal.*

Proof: We consider the representation (8.2.2) of $\mathbf{\Sigma}$ and observe

$$\mathbf{P_X\Sigma} = \mathbf{X\Lambda_1X'} = \mathbf{H}(\text{say})$$
$$\Rightarrow \mathbf{P_X}(\lambda_1\mathbf{E}_1 + \cdots + \lambda_k\mathbf{E}_k) = \mathbf{H}$$
$$\Rightarrow \lambda_i\mathbf{P_X E}_i = \mathbf{HE}_i \Rightarrow \lambda_i\mathbf{E}_j\mathbf{P_X E}_i = \mathbf{E}_j\mathbf{HE}_i.$$

Symmetry of the matrices give

$$(\mathbf{E}_j\mathbf{P_X E}_i)' = \mathbf{E}_i\mathbf{P_X E}_j, \qquad (\mathbf{E}_j\mathbf{HE}_i)' = \mathbf{E}_i\mathbf{HE}_j.$$

Hence

$$(\lambda_i - \lambda_j)\mathbf{E}_j\mathbf{P_X E}_i = \mathbf{0} \Rightarrow \lambda_i = \lambda_j \quad \text{if} \quad \mathbf{E}_j\mathbf{P_X E}_i \neq \mathbf{0}.$$

This proves the necessity part of corollary 4.

To prove the sufficiency part, consider the spectral representation $\mathbf{\Sigma} = \lambda_1\mathbf{F}_1 + \lambda_2\mathbf{F}_2 + \cdots + \lambda_u\mathbf{F}_u$, corresponding to the eigenvalues and idempotents associated with the u sets. Observe that $\mathbf{F}_i\mathbf{P_X F}_j = \mathbf{0}\forall i \neq j \Rightarrow \Sigma\mathbf{F}_i\mathbf{P_X F}_i = \mathbf{P_X}$, since $\mathbf{I} = \Sigma\mathbf{F}_i$. Also since $\mathbf{F}_i\mathbf{P_X F}_i$ is n.n.d.,

$$\mathbf{P_X Z} = \mathbf{0} \Rightarrow \mathbf{F}_i\mathbf{P_X F}_i\mathbf{Z} = \mathbf{0}\forall i \Rightarrow \mathbf{P_X F}_i\mathbf{Z} = \mathbf{0}\forall i$$
$$\Rightarrow \lambda_i\mathbf{X'F}_i\mathbf{Z} = \mathbf{0}\forall i \Rightarrow \mathbf{X'\Sigma Z} = \mathbf{0}$$

which is condition (8.2.1) of Theorem 8.2.1.

COROLLARY 5: *Let* $\mathbf{X}_r$ *be a matrix whose columns form a basis of* $\mathcal{M}(\mathbf{X})$ *and similarly let* $\mathbf{Z}_{n-r}$ *be a basis of* $\mathcal{M}(\mathbf{Z})$, *where* $r = R(\mathbf{X})$ *and* $(n - r) = R(\mathbf{Z})$. *The n.s. conditions of Theorem 8.2.1 can also be stated as follows.*

(i)′ $\mathbf{X}_r'\mathbf{\Sigma Z}_{n-r} = \mathbf{0}$
(ii)′ $\mathbf{\Sigma} = \mathbf{X}_r\mathbf{M}_1\mathbf{X}_r' + \mathbf{Z}_{n-r}\mathbf{M}_2\mathbf{Z}_{n-r}'$

(iii)′ $\boldsymbol{\Sigma} = \mathbf{X}_r\mathbf{N}_1\mathbf{X}_r' + \mathbf{Z}_{n-r}\mathbf{N}_2\mathbf{Z}_{n-r}' + \sigma^2\mathbf{I}$,

where $\mathbf{M}_1, \mathbf{M}_2, \mathbf{N}, \mathbf{N}_2$ *are arbitrary symmetric matrices and* σ^2 *is an arbitrary scalar.*

THEOREM 8.2.2 *Let* $\mathbf{Q}$ *be a specified matrix such that* $\mathscr{M}(\mathbf{Q}) \subset \mathscr{M}(\mathbf{X}')$. *If for every* $\mathbf{p} \in \mathscr{M}(\mathbf{Q})$, *the BLUE of* $\mathbf{p}'\boldsymbol{\beta}$ *under the model* $(\mathbf{X}, \sigma^2\mathbf{I})$ *is* $(\mathbf{X}, \boldsymbol{\Sigma})$-*optimal, it is n.s. that*

$$\boldsymbol{\Sigma} = \mathbf{X}\boldsymbol{\Theta}\mathbf{X}' + \mathbf{Z}\boldsymbol{\Gamma}\mathbf{Z}' + \sigma^2\mathbf{I} + \mathbf{X}\boldsymbol{\Lambda}_3\mathbf{Z}' + \mathbf{Z}\boldsymbol{\Lambda}_3'\mathbf{X}' \tag{8.2.8}$$

where $\boldsymbol{\Theta}, \boldsymbol{\Gamma}, \boldsymbol{\Lambda}_3, \sigma^2$ *are arbitrary except that* $\mathbf{Q}'\boldsymbol{\Lambda}_3\mathbf{Z}' = \mathbf{0}$ *and* $\boldsymbol{\Sigma}$ *is n.n.d. The condition* $\mathbf{Q}'\boldsymbol{\Lambda}_3\mathbf{Z}' = \mathbf{0}$ *could be replaced by* $\mathbf{Q}'\boldsymbol{\Lambda}_3 = \mathbf{0}$, *if* $\mathbf{Z}$ *in (8.2.8) is chosen so that* $\mathbf{Z}'\mathbf{Z}$ *is nonsingular.*

Proof: We refer to equation (8.2.5) and write $\boldsymbol{\lambda}'\mathbf{X}'\mathbf{Y}$ in the form $\mathbf{p}'(\mathbf{X}'\mathbf{X})^-\mathbf{X}'\mathbf{Y}$ to obtain $\mathbf{p}'(\mathbf{X}'\mathbf{X})^-\mathbf{X}'\boldsymbol{\Sigma}\mathbf{Z}\mathbf{B} = \mathbf{0}$. If this is to hold for all $\mathbf{p} \in \mathscr{M}(\mathbf{Q})$ and all $\mathbf{B}$ one must have

$$\mathbf{Q}'(\mathbf{X}'\mathbf{X})^-\mathbf{X}'\boldsymbol{\Sigma}\mathbf{Z} = \mathbf{0}. \tag{8.2.9}$$

Substituting the general expression (8.2.6) for $\boldsymbol{\Sigma}$ in (8.2.9) and observing that $\mathbf{Q}'(\mathbf{X}'\mathbf{X})^-\mathbf{X}'\mathbf{X} = \mathbf{Q}'$ we have

$$\mathbf{Q}'\boldsymbol{\Lambda}_3\mathbf{Z}'\mathbf{Z} = \mathbf{0} \Leftrightarrow \mathbf{Q}'\boldsymbol{\Lambda}_3\mathbf{Z}' = \mathbf{0}.$$

Rest of the proof follows as in the proof of Theorem 8.2.1. We note that

$$\mathbf{Q}'\boldsymbol{\Lambda}_3\mathbf{Z}'\mathbf{Z} = \mathbf{0} \Leftrightarrow \mathbf{Q}'\boldsymbol{\Lambda}_3 = \mathbf{0}$$

if $\mathbf{Z}'\mathbf{Z}$ is nonsingular.

THEOREM 8.2.3 *For every estimable parametric function the BLUE under* $(\mathbf{X}, \boldsymbol{\Sigma}_o)$ *to be* $(\mathbf{X},\boldsymbol{\Sigma})$-*optimal and vice versa, it is necessary and sufficient that any one of the following conditions hold*

(i) $\mathscr{M}(\mathbf{0} \vdots \mathbf{0} \vdots \mathbf{X})' \subset \mathscr{M}(\boldsymbol{\Sigma}\mathbf{Z} \vdots \boldsymbol{\Sigma}_0\mathbf{Z} \vdots \mathbf{X})'$

(ii) $\mathscr{M}(\boldsymbol{\Sigma}\mathbf{Z} \vdots \boldsymbol{\Sigma}_0\mathbf{Z})$ and $\mathscr{M}(\mathbf{X})$ *are virtually disjoint.*

(iii) $\mathscr{M}\begin{pmatrix}\mathbf{Z}'\boldsymbol{\Sigma}\\ \mathbf{Z}'\boldsymbol{\Sigma}_o\end{pmatrix} \subset \mathscr{M}\begin{pmatrix}\mathbf{Z}'\boldsymbol{\Sigma}\mathbf{Z}\\ \mathbf{Z}'\boldsymbol{\Sigma}_o\mathbf{Z}\end{pmatrix}$.

Proof: Let $\mathbf{L}'\mathbf{Y}$ be the BLUE for $\mathbf{p}'\boldsymbol{\beta}$ under $(\mathbf{X},\boldsymbol{\Sigma})$. Then $\mathbf{L}$ and $\boldsymbol{\lambda}$ satisfy the equation

$$\boldsymbol{\Sigma}\mathbf{L} - \mathbf{X}\boldsymbol{\lambda} = \mathbf{0}, \quad \mathbf{X}'\mathbf{L} = \mathbf{p}.$$

If $\mathbf{L}'\mathbf{Y}$ is also the BLUE for $\mathbf{p}'\boldsymbol{\beta}$ under $(\mathbf{Y}, \mathbf{X}\boldsymbol{\beta}, \boldsymbol{\Sigma}_o)$, then the following is true

$$\boldsymbol{\Sigma}_o\mathbf{L} - \mathbf{X}\boldsymbol{\mu} = \mathbf{0}, \quad \mathbf{X}'\mathbf{L} = \mathbf{p},$$

where $\boldsymbol{\mu}$ is a vector of unknowns. Hence a n.s. condition that $\mathbf{p}'\boldsymbol{\beta}$ has a common

BLUE under both models is that the linear equations

$$\mathbf{\Sigma L} = \mathbf{X\lambda}, \qquad \mathbf{\Sigma}_o\mathbf{L} = \mathbf{X\mu}, \qquad \mathbf{X'L} = \mathbf{p}$$

are soluble in $(\mathbf{L}, \boldsymbol{\lambda}, \boldsymbol{\mu})$ which is equivalent to saying that

$$\mathbf{a}_1'\mathbf{\Sigma} + \mathbf{a}_2'\mathbf{\Sigma}_o + \mathbf{a}_3'\mathbf{X}' = \mathbf{0}' \quad \mathbf{a}_1'\mathbf{X} = \mathbf{0}', \ \ \mathbf{a}_2'\mathbf{X} = \mathbf{0}'$$

$\Rightarrow \mathbf{a}_3'\mathbf{p} = \mathbf{0}'$ or $\mathbf{a}_3'\mathbf{X} = \mathbf{0}'$ if the BLUEs are to coincide for every estimable function. But $\mathbf{a}_1'\mathbf{X} = \mathbf{0}' \Rightarrow \mathbf{a}_1' = \mathbf{b}_1'\mathbf{Z}'$ and $\mathbf{a}_2'\mathbf{X} = \mathbf{0}' \Rightarrow \mathbf{a}_2' = \mathbf{b}_2'\mathbf{Z}'$. Hence,

$$\mathbf{b}_1'\mathbf{Z}'\mathbf{\Sigma} + \mathbf{b}_2'\mathbf{Z}'\mathbf{\Sigma}_0 + \mathbf{a}_3'\mathbf{X}' = \mathbf{0}' \Rightarrow \mathbf{a}_3'\mathbf{X}' = \mathbf{0}',$$

which is (ii); (i) and (iii) follow easily.

COROLLARY 1: *A sufficient condition that for each estimable parametric function the BLUE under* $(\mathbf{X}, \mathbf{\Sigma}_o)$ *is* $(\mathbf{X}, \mathbf{\Sigma})$ *optimal is that* $\mathbf{\Sigma}$ *is of the form*

$$\mathbf{\Sigma} = \mathbf{X\Theta X}' + \mathbf{\Sigma}_o\mathbf{Z\Gamma Z\Sigma}_o + \mathbf{\Sigma}_o. \tag{8.2.10}$$

COROLLARY 2: *The condition given in Corollary 1 is also necessary if* $\mathcal{M}(\mathbf{X}) + \mathcal{M}(\mathbf{\Sigma}_o\mathbf{Z}) = \mathscr{E}^n$

Proof: As in (8.2.6) we write

$$\mathbf{\Sigma} = \mathbf{X\Lambda}_1\mathbf{X}' + \mathbf{\Sigma}_o\mathbf{Z\Lambda}_2\mathbf{Z}'\mathbf{\Sigma}_o + \mathbf{X\Lambda}_3\mathbf{Z}'\mathbf{\Sigma}_0 + \mathbf{\Sigma}_o\mathbf{Z\Lambda}_3'\mathbf{X}'$$

and observe that the conditions of Theorem 8.2.3 imply $\mathbf{X\Lambda}_3\mathbf{Z}' = \mathbf{0}$. Hence

$$\mathbf{\Sigma} = \mathbf{X\Lambda}_1\mathbf{X}' + \mathbf{\Sigma}_o\mathbf{Z\Lambda}_2\mathbf{Z}'\mathbf{\Sigma}_o,$$

which is equivalent to the form given in Corollary 1.

In Theorems 8.2.1—8.2.3 we considered the conditions under which the BLUEs under $(\mathbf{X}, \mathbf{\Sigma}_o)$ and $(\mathbf{X}, \mathbf{\Sigma})$ are the same although their variances may be different. We shall now examine in some detail the conditions under which various procedures in the simple least-squares theory (i.e., assuming $\mathbf{\Sigma} = \sigma^2\mathbf{I}$), such as estimation of σ^2, tests of linear hypotheses, etc., remain valid when $\mathbf{\Sigma} \neq \sigma^2\mathbf{I}$. In other words, we determine the class of $\mathbf{\Sigma}$ for which different aspects of the simple least-squares theory remain completely valid.

THEOREM 8.2.4 *Consider alternative models* $(\mathbf{X}, \sigma^2\mathbf{I})$ *and* $(\mathbf{X}, \mathbf{\Sigma})$. *Let* $\mathbf{\Lambda}_1$, $\mathbf{\Lambda}_2$ *and* $\mathbf{\Lambda}_3$ *be defined as in (8.2.6). Then the following are true*

(a) $D(\mathbf{X'Y}|\sigma^2\mathbf{I}) = D(\mathbf{X'Y}|\mathbf{\Sigma}) \Rightarrow \mathbf{\Lambda}_1 = \sigma^2(\mathbf{X'X})^-$
(b) Let $R_o^2 = \min_{\boldsymbol{\beta}}(\mathbf{Y} - \mathbf{X\beta})'(\mathbf{Y} - \mathbf{X\beta}) = \mathbf{Y}'(\mathbf{I} - \mathbf{P_X})\mathbf{Y}$. Then $E(R_o^2|\sigma^2\mathbf{I}) = E(R_o^2|\mathbf{\Sigma}) \Rightarrow tr\mathbf{Z\Lambda}_2\mathbf{Z}' = (n-r)\sigma^2$ *where* $r = R(\mathbf{X})$ *and* n *is the number of elements in* $\mathbf{Y}$.
(c) *Let* $\mathbf{Y}$ *have a multivariate normal distribution. Then the distribution of* R_o^2/σ^2 *under* $(\mathbf{X}, \mathbf{\Sigma})$ *is* χ^2 *on* $(n-r)$ *d.f. if* $\mathbf{Z\Lambda}_2\mathbf{Z}'/\sigma^2 = (\mathbf{I} - \mathbf{P_X})$. *If*

further $\mathbf{X}'\mathbf{Y}$ *and* R_o^2 *are to be independently distributed, then* $\mathbf{X}\boldsymbol{\Lambda}_3\mathbf{Z}' = \mathbf{Z}\boldsymbol{\Lambda}_3'\mathbf{X}' = \mathbf{0}$.

Proof: The result (a) is easily established. To prove (b), we observe that $E(\mathbf{Y}'(\mathbf{I} - \mathbf{P_X})\mathbf{Y}|\boldsymbol{\Sigma}) = tr(\mathbf{I} - \mathbf{P_X})\boldsymbol{\Sigma}$

$$= tr(\mathbf{Z}\boldsymbol{\Lambda}_2\mathbf{Z}' + \mathbf{Z}\boldsymbol{\Lambda}_3'\mathbf{X}') = tr\mathbf{Z}\boldsymbol{\Lambda}_2\mathbf{Z}'.$$

Since $E(R_o^2|\mathbf{X}, \sigma^2\mathbf{I}) = (n - r)\sigma^2$, the result (b) is established. The necessary and sufficient condition for $\mathbf{Y}'(\mathbf{I} - \mathbf{P_X})\mathbf{Y}/\sigma^2$ to have a chisquare distribution is, using Lemma 9.2.1 (i)

$$\boldsymbol{\Sigma}(\mathbf{I} - \mathbf{P_X})\boldsymbol{\Sigma}(\mathbf{I} - \mathbf{P_X})\boldsymbol{\Sigma} = \sigma^2\boldsymbol{\Sigma}(\mathbf{I} - \mathbf{P_X})\boldsymbol{\Sigma}$$

$$\Leftrightarrow (\mathbf{Z}\boldsymbol{\Lambda}_2\mathbf{Z}' + \mathbf{X}\boldsymbol{\Lambda}_3\mathbf{Z}')\mathbf{Z}\boldsymbol{\Lambda}_2\mathbf{Z}'(\mathbf{Z}\boldsymbol{\Lambda}_2\mathbf{Z}' + \mathbf{Z}\boldsymbol{\Lambda}_3'\mathbf{X}')$$

$$= \sigma^2(\mathbf{Z}\boldsymbol{\Lambda}_2\mathbf{Z}' + \mathbf{X}\boldsymbol{\Lambda}_3\mathbf{Z}')(\mathbf{Z}\boldsymbol{\Lambda}_2\mathbf{Z}' + \mathbf{Z}\boldsymbol{\Lambda}_3'\mathbf{X}')$$

$$\Rightarrow \mathbf{Z}'\mathbf{Z}\boldsymbol{\Lambda}_2\mathbf{Z}'\mathbf{Z}\boldsymbol{\Lambda}_2\mathbf{Z}'\mathbf{Z}\boldsymbol{\Lambda}_2\mathbf{Z}'\mathbf{Z} = \sigma^2\mathbf{Z}'\mathbf{Z}\boldsymbol{\Lambda}_2\mathbf{Z}'\mathbf{Z}\boldsymbol{\Lambda}_2\mathbf{Z}'\mathbf{Z}$$

$$\Leftrightarrow (\mathbf{Z}\boldsymbol{\Lambda}_2\mathbf{Z}')(\mathbf{Z}\boldsymbol{\Lambda}_2\mathbf{Z}')(\mathbf{Z}\boldsymbol{\Lambda}_2\mathbf{Z}') = \sigma^2(\mathbf{Z}\boldsymbol{\Lambda}_2\mathbf{Z}')(\mathbf{Z}\boldsymbol{\Lambda}_2\mathbf{Z}')$$

$$\Leftrightarrow (\mathbf{Z}\boldsymbol{\Lambda}_2\mathbf{Z}')/\sigma^2 \text{ is idempotent.}$$

The d.f. of χ^2 is

$$tr(\mathbf{I} - \mathbf{P_X})\boldsymbol{\Sigma}/\sigma^2 = \text{tr}(\mathbf{Z}\boldsymbol{\Lambda}_2\mathbf{Z}' + \mathbf{Z}\boldsymbol{\Lambda}_3'\mathbf{X}')/\sigma^2 = tr\mathbf{Z}\boldsymbol{\Lambda}_2\mathbf{Z}'\sigma^2.$$

If $tr\mathbf{Z}\boldsymbol{\Lambda}_2\mathbf{Z}' = (n - r)\sigma^2$, then we should have $\mathbf{Z}\boldsymbol{\Lambda}_2\mathbf{Z}' = \sigma^2(\mathbf{I} - \mathbf{P_X})$. If $\mathbf{X}'\mathbf{Y}$ and $\mathbf{Y}'(\mathbf{I} - \mathbf{P_X})\mathbf{Y}$ are to be independently distributed, then $\mathbf{X}'\boldsymbol{\Sigma}(\mathbf{I} - \mathbf{P_X}) = \mathbf{0} \Leftrightarrow \mathbf{X}\boldsymbol{\Lambda}_3\mathbf{Z}' = \mathbf{Z}\boldsymbol{\Lambda}_3'\mathbf{X}' = \mathbf{0}$. Thus (c) is proved.

The following table gives the necessary and sufficient conditions on $\boldsymbol{\Sigma}$ for different procedures in simple least-squares theory [i.e. based on $(\mathbf{X}, \sigma^2\mathbf{I})$] to be optimal or valid for $(\mathbf{X}, \boldsymbol{\Sigma})$. The necessity is proved in Theorem 8.2.4. Sufficiency may be easily verified.

Property	Representation of $\boldsymbol{\Sigma}$
(i) Every SLSE is BLUE	$\mathbf{X}\boldsymbol{\Lambda}_1\mathbf{X}' + \mathbf{Z}\boldsymbol{\Lambda}_2\mathbf{Z}'$
(ii) Variance of SLSE is unchanged	$\sigma^2\mathbf{P_X} + \mathbf{Z}\boldsymbol{\Lambda}_2\mathbf{Z}' + \mathbf{X}\boldsymbol{\Lambda}_3\mathbf{Z}' + \mathbf{Z}\boldsymbol{\Lambda}_3'\mathbf{X}'$
(iii) $R_o^2/\sigma^2 \sim \chi^2(n - r)$	$\mathbf{X}\boldsymbol{\Lambda}_1\mathbf{X}' + \sigma^2(\mathbf{I} - \mathbf{P_X}) + \mathbf{X}\boldsymbol{\Lambda}_3\mathbf{Z}' + \mathbf{Z}\boldsymbol{\Lambda}_3'\mathbf{X}'$
(iv) Every SLSE and R_o^2 are independently distributed	$\mathbf{X}\boldsymbol{\Lambda}_1\mathbf{X}' + \mathbf{Z}\boldsymbol{\Lambda}_2\mathbf{Z}'$
(v) In addition to (iv), (iii) holds	$\mathbf{X}\boldsymbol{\Lambda}_1\mathbf{X}' + \sigma^2(\mathbf{I} - \mathbf{P_X})$
(vi) In addition to (v), (ii) holds	$\sigma^2\mathbf{I}$
(vii) In addition to (v), variance of particular SLSE say that of $\mathbf{p}'\boldsymbol{\beta}$ is the same	$\sigma^2\mathbf{I} + \mathbf{X}\mathbf{A}\mathbf{X}'$ where $\mathbf{p}'\mathbf{A}\mathbf{p} = 0$

From the table we observe that, while some of the procedures based on simple least-squares theory are valid for a wider class of $\mathbf{\Sigma}$, for the full battery of procedures on estimation and testing to be applicable (case vi), it is necessary that $\mathbf{\Sigma} = \sigma^2\mathbf{I}$. Case (vii) shows that, if the procedures are to be valid for particular parametric functions, $\mathbf{\Sigma}$ can belong to a wider class. A well-known example, which falls in this category, is the treatment of the two-way classification (mixed model) discussed in Rao (1965, p. 261).

8.3 SPECIFICATION ERRORS IN THE DESIGN MATRIX

In Section 8.2 we considered alternative linear models which differed in the dispersion matrices of the observations but not in their expectations. Now we consider two alternative models $(\mathbf{Y}, \mathbf{X}_o\boldsymbol{\beta}, \sigma^2\mathbf{I})$ and $(\mathbf{Y}, \mathbf{X}\boldsymbol{\beta}, \sigma^2\mathbf{I})$, which differ in the expectations and not in the dispersion of observations, and examine to what extent estimators based on one model are valid with respect to the other.

THEOREM 8.3.1 *If every BLUE under* $(\mathbf{X}_o, \sigma^2\mathbf{I})$ *is unbiased for the corresponding parametric function under* $(\mathbf{X}, \sigma^2\mathbf{I})$, *then it is n.s. that*

$$\mathbf{X} = \mathbf{X}_o + (\mathbf{I} - \mathbf{P}_{\mathbf{X}_o})\mathbf{A}, \tag{8.3.1}$$

where $\mathbf{A}$ *is arbitrary. If in addition every BLUE under* $(\mathbf{X}_o, \sigma^2\mathbf{I})$ *has also minimum variance under* $(\mathbf{X}, \sigma^2\mathbf{I})$, *then it is n.s. that the matrix* $\mathbf{A}$ *in (8.3.1) satisfies the restriction*

$$\mathcal{M}(\mathbf{A}') \cap \mathcal{M}(\mathbf{X}_o') = \{\mathbf{0}\} \tag{8.3.2}$$

that is a set containing exclusively the null vector.

Proof: We note from the standard Gauss–Markoff theory that it suffices to consider the set of linear functions $\mathbf{X}_o'\mathbf{Y}$ which are BLUE'S for their corresponding expectations under $(\mathbf{X}_o, \sigma^2\mathbf{I})$. Then

$$E(\mathbf{X}_o'\mathbf{Y}|\mathbf{X}, \sigma^2\mathbf{I}) = \mathbf{X}_o'\mathbf{X}\boldsymbol{\beta} = \mathbf{X}_o'\mathbf{X}_o\boldsymbol{\beta} = E(\mathbf{X}_o'\mathbf{Y}|\mathbf{X}_o, \sigma^2\mathbf{I})$$

$$\Leftrightarrow \mathbf{X}_o'\mathbf{X} = \mathbf{X}_o'\mathbf{X}_o \Leftrightarrow \mathbf{X}_o'(\mathbf{X} - \mathbf{X}_o) = \mathbf{0} \Leftrightarrow (\mathbf{X} - \mathbf{X}_o) = (\mathbf{I} - \mathbf{P}_{\mathbf{X}_o})\mathbf{A}$$

which establishes the n.s. condition (8.3.1).

If each linear function in $\mathbf{X}_o'\mathbf{Y}$ has minimum variance under $(\mathbf{X}, \sigma^2\mathbf{I})$, where $\mathbf{X}$ is of the form (8.3.1), then it is n.s. that

$$\text{cov}(\mathbf{X}_o'\mathbf{Y}, \mathbf{Z}'\mathbf{Y}) = \mathbf{0} \Leftrightarrow \mathbf{X}_o'\mathbf{Z} = \mathbf{0},$$

where $\mathbf{Z} = \mathbf{X}^{\perp}$, where $\mathbf{X}$ is of the form (8.3.1). The condition $\mathbf{X}_o'\mathbf{Z} = \mathbf{0} \Leftrightarrow \mathcal{M}(\mathbf{X}_o) \subset \mathcal{M}(\mathbf{X}) = \mathcal{M}(\mathbf{X}_o + (\mathbf{I} - \mathbf{P}_{\mathbf{X}_o})\mathbf{A}) \Leftrightarrow \mathcal{M}(\mathbf{X}_o') \cap \mathcal{M}(\mathbf{A}'(\mathbf{I} - \mathbf{P}_{\mathbf{X}_o})) = \{\mathbf{0}\}$. The condition (8.3.2) follows, since we can take $\mathbf{A}$ to be $(\mathbf{I} - \mathbf{P}_{\mathbf{X}_o})\mathbf{A}$.

COROLLARY 1: *If for every estimable parametric function the BLUE under* $(\mathbf{X}_o, \sigma^2\mathbf{I})$ *is* $(\mathbf{X}, \sigma^2\mathbf{I})$*-optimal it is n.s. that* $\mathbf{X}$ *is of the form*

$$\mathbf{X} = \mathbf{X}_o + (\mathbf{I} - \mathbf{P}_{\mathbf{X}_o})\mathbf{F}[\mathbf{I} + (\mathbf{I} - \mathbf{P}_{\mathbf{X}'_o})\mathbf{D}]^{\perp\prime}, \tag{8.3.3}$$

where $\mathbf{D}$ *and* $\mathbf{F}$ *are arbitrary.*

Proof: Observe that $\mathscr{M}[\{\mathbf{I} + (\mathbf{I} - \mathbf{P}_{\mathbf{X}'_o})\mathbf{D}\}^{\perp}\mathbf{F}'] \subset \mathscr{O}[\mathbf{I} + (\mathbf{I} - \mathbf{P}_{\mathbf{X}'_o})\mathbf{D}]$. If this has a vector $\mathbf{X}'_o\mathbf{u}$ in common with $\mathscr{M}(\mathbf{X}'_o)$, we have

$$\mathbf{u}'\mathbf{X}_o[\mathbf{I} + (\mathbf{I} - \mathbf{P}_{\mathbf{X}'_o})\mathbf{D}] = \mathbf{u}'\mathbf{X}_o = \mathbf{0}'.$$

Hence (8.3.3) ⇒ (8.3.1) and (8.3.2). Conversely, assume (8.3.1) and (8.3.2) are true and observe that by example 6 in complements to Chapter 5, (8.3.2) ⇒ $\mathbf{A}'[\mathbf{A}(\mathbf{I} - \mathbf{P}_{\mathbf{X}'_o})]^{\perp} = \mathbf{0} \Rightarrow \mathscr{M}(\mathbf{A}) \subset \mathscr{M}[\mathbf{A}(\mathbf{I} - \mathbf{P}_{\mathbf{X}'_o})] \Rightarrow \mathbf{A} = -\mathbf{A}(\mathbf{I} - \mathbf{P}_{\mathbf{X}'_o})\mathbf{D}$ (say) $\Rightarrow \mathbf{A}[\mathbf{I} + (\mathbf{I} - \mathbf{P}_{\mathbf{X}'_o})\mathbf{D}] = \mathbf{0}$. Therefore, (8.3.1) and (8.3.2) ⇒ (8.3.3) and the proof of Corollary 1 is concluded.

COROLLARY 2: *A sufficient condition that for every estimable function the BLUE under* $(\mathbf{X}_o, \sigma^2\mathbf{I})$ is $(\mathbf{X}, \sigma^2\mathbf{I})$*-optimal is*

$$\mathbf{X} = \mathbf{X}_o + [\mathbf{I} - \mathbf{X}_o(\mathbf{X}'_o\mathbf{X}_o)^-\mathbf{X}'_o]\mathbf{B}[\mathbf{I} - \mathbf{X}'_o\mathbf{X}_o(\mathbf{X}'_o\mathbf{X}_o)^-],$$

where $\mathbf{B}$ *is arbitrary.*

COROLLARY 3: *If* $R(\mathbf{X}_o) = m$, *the number of columns in* $\mathbf{X}_o$, *then the conditions of Theorem 8.3.1 reduce to* $\mathbf{X} = \mathbf{X}_o$.

Proof: Note that here $\mathscr{M}(\mathbf{X}'_o) = \mathscr{E}^m$. Hence $\mathscr{M}(\mathbf{A}') \cap \mathscr{M}(\mathbf{X}'_o) = \mathscr{M}(\mathbf{A}')$. $\mathscr{M}(\mathbf{A}')$ is a set consisting exclusively of the null vector which is implied in (8.3.2). Hence $\mathbf{A}' = \mathbf{0}$ and Corollary 3 is established.

THEOREM 8.3.2 *Let* $\mathbf{P}$ *be a matrix of order* $m \times k$ *such that* $\mathscr{M}(\mathbf{P}) \subset \mathscr{M}(\mathbf{X}'_o)$ *and* $R(\mathbf{P}) = k$. *A n.s. condition for* $\mathbf{L}'\mathbf{Y} = \mathbf{P}'(\mathbf{X}'_o\mathbf{X}_o)^-\mathbf{X}'_o\mathbf{Y}$ *to be BLUE for* $\mathbf{P}'\boldsymbol{\beta}$ *under* $(\mathbf{X}, \sigma^2\mathbf{I})$ *is that*

$$\mathbf{X} = \mathbf{X}_o + [\mathbf{I} - (\mathbf{L}\mathbf{L}^-)'][\mathbf{C}\mathbf{B}^- + \mathbf{D}(\mathbf{I} - \mathbf{B}\mathbf{B}^-)], \tag{8.3.4}$$

where $\mathbf{P}'\mathbf{B} = \mathbf{L}'\mathbf{L}$, $\mathbf{C} = \mathbf{L} - \mathbf{X}_o\mathbf{B}$ *and* $\mathbf{D}$ *is an arbitrary matrix of order* $n \times m$.

Proof of sufficiency: The sufficiency part follows from the fact that $\mathbf{L}'\mathbf{X} = \mathbf{L}'\mathbf{X}_o = \mathbf{P}'$ and $\mathbf{X}\mathbf{B} = \mathbf{X}_o\mathbf{B} + [\mathbf{I} - (\mathbf{L}\mathbf{L}^-)']\mathbf{C} = \mathbf{X}_o\mathbf{B} + \mathbf{C} = \mathbf{L}$, since $R(\mathbf{P}) = k \Rightarrow R(\mathbf{L}) = k \Rightarrow R(\mathbf{B}) = k \Rightarrow \mathbf{B}^-\mathbf{B} = \mathbf{I}$ and $\mathbf{L}'\mathbf{C} = \mathbf{L}'\mathbf{L} - \mathbf{L}'\mathbf{X}_o\mathbf{B} = \mathbf{L}'\mathbf{L} - \mathbf{P}'\mathbf{B} = \mathbf{0}$.

Proof of necessity: To establish the necessity part note that $E(\mathbf{L}'\mathbf{Y}|\mathbf{X}, \sigma^2\mathbf{I}) = \mathbf{P}'\boldsymbol{\beta} \Rightarrow \mathbf{L}'(\mathbf{X} - \mathbf{X}_o) = \mathbf{0} \Rightarrow \mathbf{X} = \mathbf{X}_o + [\mathbf{I} - (\mathbf{L}\mathbf{L}^-)']\mathbf{G}$. Note further, the fact that $\mathbf{L}'\mathbf{Y}$ is BLUE for $\mathbf{P}'\boldsymbol{\beta}$ under $(\mathbf{X}, \sigma^2\mathbf{I}) \Rightarrow \mathbf{L} = \mathbf{XB}$ for some $\mathbf{B}$.

The rest of the lemma follows from the general solution (X) of the equations $\mathbf{L}'\mathbf{X} = \mathbf{L}'\mathbf{X}_0$, $\mathbf{L} = \mathbf{XB}$ given in Theorem 23.3.

Of special interest is the case when the model is underspecified, that is, where instead of $(\mathbf{Y}, \mathbf{X}_o\boldsymbol{\beta}, \sigma^2\mathbf{I})$ the true model is $(\mathbf{Y}, \mathbf{X}_o\boldsymbol{\beta} + \mathbf{U}\boldsymbol{\eta}, \sigma^2\mathbf{I})$.

THEOREM 8.3.3 *For the BLUE under* $(\mathbf{X}_o, \sigma^2\mathbf{I})$ *to be* $[(\mathbf{X}_o \vdots \mathbf{U}), \sigma^2\mathbf{I}]$*-optimal for every estimable parametric function, it is n.s. that* $\mathbf{X}_o'\mathbf{U} = \mathbf{0}$.

Proof: Necessity follows from the requirement that $E(\mathbf{X}_o'\mathbf{Y}|\mathbf{X}_o, \sigma^2\mathbf{I}) = E(\mathbf{X}_o'\mathbf{Y}|(\mathbf{X}_o \vdots \mathbf{U}), \sigma^2\mathbf{I})$. Sufficiency is trivial since $\mathscr{M}(\mathbf{X}_o) \subset \mathscr{M}(\mathbf{X}_o \vdots \mathbf{U})$.

COROLLARY: *Let* $\mathbf{X}_o = (\mathbf{X}_{o1} \vdots \mathbf{X}_{o2})$*, where* $\mathbf{X}_{o1}$ *is of order* $n \times m_1$ *and* $\mathbf{X}_{o2}$ *of order* $n \times m_2$*. Then for the BLUE under* $(\mathbf{X}_o, \sigma^2\mathbf{I})$ *to be* $(\mathbf{X}_{o1}, \sigma^2\mathbf{I})$*-optimal for every estimable linear function of the first* m_1 *parameters, it is n.s. that* $\mathbf{X}_{o1}'\mathbf{X}_{o2} = \mathbf{0}$.

THEOREM 8.3.4 *Consider alternative models* $(\mathbf{X}_o, \sigma^2\mathbf{I})$ *and* $(\mathbf{X}, \sigma^2\mathbf{I})$ *where* $\mathbf{X}$ *is as in Theorem 8.3.1. Let* $R_o^2 = \min_{\boldsymbol{\beta}}(\mathbf{Y} - \mathbf{X}_o\boldsymbol{\beta})'(\mathbf{Y} - \mathbf{X}_o\boldsymbol{\beta})$*. Then*

(a) $E(R_o^2|\mathbf{X}, \sigma^2\mathbf{I}) = (n - r)\sigma^2 + \lambda\sigma^2$ *where* $\lambda = \boldsymbol{\beta}'\mathbf{A}'(\mathbf{I} - \mathbf{P}_{\mathbf{X}_o})\mathbf{A}\boldsymbol{\beta}/\sigma^2$

(b) $E(R_o^2|\mathbf{X}, \sigma^2\mathbf{I}) = (n - r)\sigma^2$ *if* $\mathbf{X} = \mathbf{X}_o$.

(c) $R_o^2/\sigma^2 \sim \chi^2((n - r), \lambda)$.

(d) *If* $\mathbf{p}'\boldsymbol{\beta}$ *is estimable under* $(\mathbf{X}_o, \sigma^2\mathbf{I})$ *and the best linear estimate is* $\mathbf{p}'\hat{\boldsymbol{\beta}}$*, then*

$$(\mathbf{p}'\hat{\boldsymbol{\beta}} - \mathbf{p}'\boldsymbol{\beta})/[\mathbf{p}'(\mathbf{X}_o'\mathbf{X}_o')^{-}\mathbf{p}R_o^2/(n - r)]^{1/2}$$

is distributed as the ratio of a standard normal deviate to an independent root mean (noncentral)χ^2 *on* $(n - r)$ *d.f.*

(e) *The F statistic of (7.2.19), is in the null case distributed as the ratio of a (central) mean chisquare and an independent mean (noncentral) chisquare on* $(n - r)$ *d.f.*

Proof:

$$\begin{aligned} E(R_o^2|\mathbf{X}, \sigma^2\mathbf{I}) &= E(tr\mathbf{Y}'(\mathbf{I} - \mathbf{P}_{\mathbf{X}_o})\mathbf{Y}|\mathbf{X}, \sigma^2\mathbf{I}] \\ &= E(tr(\mathbf{I} - \mathbf{P}_{\mathbf{X}_o})\mathbf{Y}\mathbf{Y}'|\mathbf{X}, \sigma^2\mathbf{I}] = tr(\mathbf{I} - \mathbf{P}_{\mathbf{X}_o})(\sigma^2\mathbf{I}_o + \mathbf{X}\boldsymbol{\beta}\boldsymbol{\beta}'\mathbf{X}') \\ &= (n - r)\sigma^2 + \boldsymbol{\beta}'\mathbf{X}'(\mathbf{I} - \mathbf{P}_{\mathbf{X}_o})\mathbf{X}\boldsymbol{\beta} = (n - r)\sigma^2 + \boldsymbol{\beta}'\mathbf{A}'(\mathbf{I} - \mathbf{P}_{\mathbf{X}_o})\mathbf{A}\boldsymbol{\beta}, \end{aligned}$$

which proves (a). $E(R_o^2|\mathbf{X}, \sigma^2\mathbf{I}) = (n - r)\sigma^2$ if $(\mathbf{I} - \mathbf{P}_{\mathbf{X}_o})\mathbf{A}\boldsymbol{\beta} = \mathbf{0}$, that is, $\mathbf{X} = \mathbf{X}_o$. This establishes (b). Since the dispersion matrix of $\begin{pmatrix} \mathbf{X}_o' \\ \mathbf{Z}_o' \end{pmatrix}\mathbf{Y}$ is the same under both models, the results (c), (d) and (e) follow in a straightforward way.

8.4 SPECIFICATION ERRORS IN DESIGN AND DISPERSION MATRICES

In this section we consider alternative models differing both in expectations and dispersion matrices.

THEOREM 8.4.1 *For the BLUE under* $(\mathbf{X}_o, \sigma^2\mathbf{I})$ *to be* $(\mathbf{X}, \boldsymbol{\Sigma})$ *optimal for every estimable parametric function it is n.s. that*

$$\mathbf{X} = \mathbf{X}_o + \mathbf{Z}_o\mathbf{G} \tag{8.4.1}$$

$$\boldsymbol{\Sigma} = \mathbf{X}_o\mathbf{A}\mathbf{X}_o' + \mathbf{Z}_o\mathbf{B}\mathbf{Z}_o' + \mathbf{X}_o\mathbf{A}'\mathbf{G}'\mathbf{Z}_o' + \mathbf{Z}_o\mathbf{G}\mathbf{A}\mathbf{X}_o', \tag{8.4.2}$$

where $\mathbf{A}$, $\mathbf{B}$ *and* $\mathbf{G}$ *are arbitrary except that* $\boldsymbol{\Sigma}$ *is n.n.d. and* $\mathbf{Z}_o$ *is written for* $\mathbf{X}_o^{\perp}$.

Proof of necessity: As in Theorem 8.3.1, we consider the functions $\mathbf{X}_o'\mathbf{Y}$. The condition that $E(\mathbf{X}_o'\mathbf{Y})$ is the same for $(\mathbf{X}_o, \sigma^2\mathbf{I})$ and $(\mathbf{X}, \boldsymbol{\Sigma})$ implies that $\mathbf{X} = \mathbf{X}_o + \mathbf{Z}_o\mathbf{G}_o$ for some $\mathbf{G}$. If $\mathbf{X}_o'\mathbf{Y}$ is optimal for $(\mathbf{X}, \boldsymbol{\Sigma})$, then

$$C(\mathbf{X}_o'\mathbf{Y}, \mathbf{Z}'\mathbf{Y}|\mathbf{X}, \boldsymbol{\Sigma}) = \mathbf{X}_o'\boldsymbol{\Sigma}\mathbf{Z} = \mathbf{0},$$

where $\mathbf{Z} = \mathbf{X}^{\perp}$ and $C(.\ ,\ .)$ stands for the covariance matrix. As in (8.2.6) we write

$$\boldsymbol{\Sigma} = \mathbf{X}_o\boldsymbol{\Lambda}_1\mathbf{X}_o' + \mathbf{Z}_o\boldsymbol{\Lambda}_2\mathbf{Z}_o' + \mathbf{X}_o\boldsymbol{\Lambda}_3\mathbf{Z}_o' + \mathbf{Z}_o\boldsymbol{\Lambda}_3'\mathbf{X}_o'.$$

Then

$$\begin{aligned}
\mathbf{X}_o'\boldsymbol{\Sigma}\mathbf{Z} &= \mathbf{X}_o'(\mathbf{X}_o\boldsymbol{\Lambda}_1\mathbf{X}_o' + \mathbf{X}_o\boldsymbol{\Lambda}_3\mathbf{Z}_o')\mathbf{Z} = \mathbf{0}\\
&\Rightarrow \mathbf{X}_o\boldsymbol{\Lambda}_1\mathbf{X}_o'\mathbf{X}_o + \mathbf{Z}_o\boldsymbol{\Lambda}_3'\mathbf{X}_o'\mathbf{X}_o = \mathbf{X}\mathbf{M} = \mathbf{X}_o\mathbf{M} + \mathbf{Z}_o\mathbf{G}\mathbf{M}\\
&\Rightarrow \mathbf{X}_o\boldsymbol{\Lambda}_1\mathbf{X}_o'\mathbf{X}_o = \mathbf{X}_o\mathbf{M}, \mathbf{X}_o\boldsymbol{\Lambda}_1\mathbf{X}_o' = \mathbf{X}_o\mathbf{M}(\mathbf{X}_o'\mathbf{X}_o)^{-}\mathbf{X}_o',\\
&\quad\ \mathbf{Z}_o\boldsymbol{\Lambda}_3'\mathbf{X}_o'\mathbf{X}_o = \mathbf{Z}_o\mathbf{G}\mathbf{M}, \mathbf{Z}_o\boldsymbol{\Lambda}_3'\mathbf{X}' = \mathbf{Z}_o\mathbf{G}\mathbf{M}(\mathbf{X}_o'\mathbf{X}_o)^{-}\mathbf{X}_o'.
\end{aligned}$$

Writing $\mathbf{A} = \mathbf{M}(\mathbf{X}_o'\mathbf{X}_o)^{-}$, $\mathbf{B} = \boldsymbol{\Lambda}_2$, $\boldsymbol{\Sigma}$ can be written in the form (8.4.2). Sufficiency part is easy to prove.

Note 1. In corollary 3 of Theorem 8.3.1 it was noted that $\mathbf{X} = \mathbf{X}_o$ when $R(\mathbf{X}_o)$ is equal to the number of columns in $\mathbf{X}_o$. Such a result need not be true when a general $\boldsymbol{\Sigma}$, as in (8.4.2), is considered.

Note 2. Theorem 8.4.1 asserts that when $\mathbf{X}$ and $\boldsymbol{\Sigma}$ are as in (8.4.1), (8.4.2) a BLUE of an estimable parametric function under $(\mathbf{X}_o, \sigma^2\mathbf{I})$ is also a BLUE under $(\mathbf{X}, \boldsymbol{\Sigma})$. But this does not imply that the minimum variances attained in the two models are the same. To compare the variances let us first compute the dispersion matrices of $\mathbf{X}_o'\mathbf{Y}$ under the two models

$$D(\mathbf{X}_o'\mathbf{Y}|\mathbf{X}_o, \sigma^2\mathbf{I}) = \sigma^2\mathbf{X}_o'\mathbf{X}_o$$

$$\mathbf{D}(\mathbf{X}_o'\mathbf{Y}|\mathbf{X}, \boldsymbol{\Sigma} \text{ as in } 8.4.1, 8.4.2) = \mathbf{X}_o'\mathbf{X}_o\mathbf{A}\mathbf{X}_o'\mathbf{X}_o.$$

If the variances are to be equal a n.s. condition is

$$\sigma^2 \mathbf{X}_o'\mathbf{X}_o = \mathbf{X}_o'\mathbf{X}_o\mathbf{A}\mathbf{X}_o'\mathbf{X}_o$$

or $\mathbf{A} = \sigma^2(\mathbf{X}_o'\mathbf{X}_o)^-$, while **B** and **G** in (8.4.2) could be arbitrary.

COMPLEMENTS

1. Consider the Gauss–Markoff model $(\mathbf{Y}, \mathbf{X}\boldsymbol{\beta}, \boldsymbol{\Sigma})$ and the general representation of $\boldsymbol{\Sigma}$ given in (8.2.6)

$$\boldsymbol{\Sigma} = \mathbf{X}\boldsymbol{\Lambda}_1\mathbf{X}' + \mathbf{Z}\boldsymbol{\Lambda}_2\mathbf{Z}' + \mathbf{Z}\boldsymbol{\Lambda}_3\mathbf{X}' + \mathbf{X}\boldsymbol{\Lambda}_3'\mathbf{Z}'$$

where $\mathbf{Z} = \mathbf{X}^\perp$. Assume that $\mathbf{Y} \sim N_n(\mathbf{X}\boldsymbol{\beta}, \boldsymbol{\Sigma})$. Consider further a linear hypothesis, $H_o : \mathbf{H}\boldsymbol{\beta} = \mathbf{h}$ where $\mathbf{h}$ is a known vector and $\mathbf{H}\boldsymbol{\beta}$ is estimable. When $\boldsymbol{\Sigma} = \sigma^2\mathbf{I}$, H_o is tested using the statistic

$$\frac{(R_1^2 - R_o^2)/t}{R_o^2/(n-r)},$$

where

$$R_o^2 = \min_{\boldsymbol{\beta}} (\mathbf{Y} - \mathbf{X}\boldsymbol{\beta})'(\mathbf{Y} - \mathbf{X}\boldsymbol{\beta}),$$

$$R_1^2 = \min_{\boldsymbol{\beta},\ \text{subject to } H_o} (\mathbf{Y} - \mathbf{X}\boldsymbol{\beta})'(\mathbf{Y} - \mathbf{X}\boldsymbol{\beta}),$$

$$t = R(\mathbf{H}) \quad \text{and} \quad r = R(\mathbf{X}).$$

When $\boldsymbol{\Sigma} = \sigma^2\mathbf{I}$, the statistic has a variance ratio distribution, under H_o. Given below are the conditions which $\boldsymbol{\Sigma}$ has to satisfy for the numerator and denominator of the test statistic to have the same expectation under H_o, even though the ratio as such may not obey a variance ratio distribution

(a) $$E\left\{\frac{R_1^2 - R_o^2}{t}\middle| H_o, \boldsymbol{\Sigma}\right\} = E\left\{\frac{R_o^2}{n-r}\middle| H_o, \boldsymbol{\Sigma}\right\}$$

$$\Leftrightarrow tr\, \mathbf{P}\mathbf{X}\boldsymbol{\Lambda}_1\mathbf{X}'\mathbf{P} = t(n-r)^{-1}\, tr\, \mathbf{Z}\boldsymbol{\Lambda}_2\mathbf{Z}'.$$

where $\mathbf{P}$ is the orthogonal projector onto $\mathscr{M}[\mathbf{X}(\mathbf{X}'\mathbf{X})^-\mathbf{H}']$.

(b) When $\mathscr{M}(\mathbf{H}') = \mathscr{M}(\mathbf{X}')$, (a) reduces to

$$tr\, \mathbf{X}\boldsymbol{\Lambda}_1\mathbf{X}' = r(n-r)^{-1}\, tr\, \mathbf{Z}\boldsymbol{\Lambda}_2\mathbf{Z}',$$

(c) $$E\left\{\frac{R_1^2 - R_o^2}{t}\middle| H_o, \boldsymbol{\Sigma}\right\} = E\left\{\frac{R_o^2}{n-r}\middle| H_o, \boldsymbol{\Sigma}\right\}$$

$$\forall\, \mathbf{H} \ni \mathscr{M}(\mathbf{H}') \subset \mathscr{M}(\mathbf{X}') \Leftrightarrow \mathbf{X}\boldsymbol{\Lambda}_1\mathbf{X}' = c\mathbf{P}_{\mathbf{X}},$$

where

$$c = tr\, \mathbf{Z}\boldsymbol{\Lambda}_2\mathbf{Z}'/(n-r), \qquad \mathbf{P}_{\mathbf{X}} = \mathbf{X}(\mathbf{X}'\mathbf{X})^-\mathbf{X}'.$$

[Proofs of (a), (b) as well as of the '$\Leftarrow$' part of (c) are computational. For the '$\Rightarrow$' part of (c) observe that the eigenvectors of $\mathbf{\Lambda} = \mathbf{X}\mathbf{\Lambda}_1\mathbf{X}'$, corresponding to its nonnull eigenvalues, belong to $\mathscr{M}(\mathbf{X})$. Hence, if the condition obtained in (a) is to hold for all $\mathbf{H} \ni \mathscr{M}(\mathbf{H}') \subset \mathscr{M}(\mathbf{X}')$, it is necessary that $\mathbf{\Lambda}$ has r nonnull eigenvalues each equal to c.]

2. For a square matrix $\mathbf{A}$, diag($\mathbf{A}$) denotes the diagonal matrix whose diagonal elements are the same as those in $\mathbf{A}$. The class of all diagonal p.d. matrices of order n is denoted by $\mathscr{D}$. Other notations are as in example 1 above.

(a)
$$E\left\{\frac{R_1^2 - R_o^2}{t}\middle| H_o, \mathbf{\Sigma}\right\} \underset{\mathbf{\Sigma}\in\mathscr{D}}{\equiv} E\left\{\frac{R_o^2}{n-r}\middle| H_o, \mathbf{\Sigma}\right\}$$
$$\Leftrightarrow (n-r)\,\text{diag}(\mathbf{P}) = t\,\text{diag}(\mathbf{I} - \mathbf{P_X}).$$

(b) When $\mathscr{M}(\mathbf{H}') = \mathscr{M}(\mathbf{X}')$, (a) reduces to

$$n\,\text{diag}(\mathbf{P_X}) = r\mathbf{I}.$$

(c)
$$E\left\{\frac{R_1^2 - B_o^2}{t}\middle| H_o, \mathbf{\Sigma}\right\} \underset{\mathbf{\Sigma}\in\mathscr{D}}{\equiv} E\left\{\frac{R_o^2}{n-r}\middle| H_o, \Sigma\right\}$$

$\forall\, \mathbf{H} \notin \mathscr{M}(\mathbf{H}') \subset \mathscr{M}(\mathbf{X}')$

$\Leftrightarrow r = 1$, $\mathbf{P_X}$ is a matrix with each element equal to $1/n$.

[(a), (b) and the '$\Leftarrow$' part of (c) are easy to prove. For the '$\Rightarrow$' part of (c), make use of the condition given in example 1(c), noting that $\mathbf{X}\mathbf{\Lambda}_1\mathbf{X}' = \mathbf{P_X}\mathbf{\Sigma}\mathbf{P_X}$, $\mathbf{Z}\mathbf{\Lambda}_2\mathbf{Z}' = \mathbf{P_Z}\mathbf{\Sigma}\mathbf{P_Z}$. Let $\mathbf{J}_i$ denote a diagonal matrix of order $n \times n$, whose ith diagonal element is unity, the rest are all zeros. We assume i is so chosen that $tr\,\mathbf{P_Z}\mathbf{J}_i\mathbf{P_Z} \neq 0$. Consider a sequence $\{\mathbf{\Sigma}_v\}$ such that (i) $\mathbf{\Sigma}_v \in \mathscr{D}\ \forall\ v$, (ii) $\lim_{v\to\infty} \mathbf{\Sigma}_v = \mathbf{J}_i$. Using the condition of example 1(c) we have

$$\mathbf{P_X}\mathbf{\Sigma}_v\mathbf{P_X} = c_v\mathbf{P_X}$$
$$\Rightarrow \mathbf{P_X}\mathbf{J}_i\mathbf{P_X} = c\mathbf{P_X} \quad \text{where } c = tr\,\mathbf{P_Z}\mathbf{J}_i\mathbf{P_Z}/(n-r).]$$

3. Consider the usual regression model

$$E(\mathbf{Y}|\mathbf{X} = x) = \alpha + \beta x$$

and the SLSE $\hat{\beta}$ for β based on n pairs of observations (y_i, x_i), $i = 1, 2, \ldots, n$. Let $\hat{V}(\hat{\beta})$ denote the usual unbiased estimator of the variance of $\hat{\beta}$. Suppose the true dispersion matrix of $\mathbf{Y}$ is $\mathbf{\Sigma} = \text{diag}(\sigma_1^2, \ldots, \sigma_n^2)$ instead of $\sigma^2\mathbf{I}$, as assumed under simple least-squares theory.

$$E\{\hat{V}(\hat{\beta})|\mathbf{\Sigma}\} = V(\hat{\beta}|\mathbf{\Sigma})$$
$$\text{iff } \sum \sigma_i^2(x_i - \bar{x})^2 = \sum \sigma_i^2 \sum (x_i - \bar{x})^2/n.$$

[Observe that this condition is trivially satisfied if $\sigma_i^2 = \sigma^2\ \forall\ i$.]

4. Consider the setup for example 5 in complements to Chapter 7 and the BLUE's obtained therein. The same expressions are BLUE's under a model $E(\mathbf{Y}) = \mathbf{X}\boldsymbol{\beta}$, *ceteris paribus*, iff (8.3.1) and (8.3.2) hold.

CHAPTER 9

Distribution of Quadratic Forms

9.1 INTRODUCTION

The main problem discussed in this chapter is the distribution of quadratic expressions in normal variables which may have a singular distribution. First we shall consider some well-known results on the distribution of quadratic forms of independent normal variables.

9.1.1 The Chisquare Distribution

Definition: Let $Y_1, \ldots, Y_n$ be n independent and identically distributed normal variables with zero mean and unit variance. Then the distribution of the statistic

$$T_1 = Y_1^2 + \cdots + Y_n^2 \tag{9.1.1}$$

is known as the central χ^2 distribution on n degrees of freedom (d.f.) and that of

$$T_2 = (Y_1 - \nu_1)^2 + \cdots + (Y_n - \nu_n)^2 \tag{9.1.2}$$

is known as the noncentral χ^2 distribution on n d.f. with noncentrality parameter $\delta = \Sigma\, \nu_i^2$. Indeed, it can be easily shown that the distribution of T_2 depends only on δ and not on individual values ν_i.

In the case of (9.1.1) we shall write $T_1 \sim \chi^2(n)$ and in (9.1.2), $T_2 \sim \chi^2(n, \delta)$. We shall not have any occasion to use the explicit form of the distribution of T_1 or T_2 and all the results proved depend on the definitions (9.1.1) and (9.1.2) of the central and noncentral χ^2 distributions.

9.1.2 The Multivariate Normal Distribution

We denote the univariate normal distribution by $N(\mu, \sigma^2)$ where μ is the mean and σ^2 is the variance of the distribution, which may be zero. We define multivariate normal distribution as follows (see Rao, 1965, Chapter 9).

Definition: Let $Z_1, \ldots, Z_k$ be independent $N(0, 1)$ variables and define the vector variable

$$\mathbf{Y} = \boldsymbol{\mu} + \mathbf{CZ}, \tag{9.1.3}$$

where $\mathbf{Y}$ and $\boldsymbol{\mu}$ are p-vectors, $\mathbf{C}$ is $p \times k$ matrix and $\mathbf{Z}' = (Z_1, \ldots, Z_k)$. The distribution of $\mathbf{Y}$ defined by (9.1.3) is called p-variate normal with mean vector $\boldsymbol{\mu}$ and dispersion matrix $\boldsymbol{\Sigma} = \mathbf{CC}'$ and represented by $N_p(\boldsymbol{\mu}, \boldsymbol{\Sigma})$. $R(\boldsymbol{\Sigma}) = k$ is said to be the rank of the distribution.

It may be seen that two representations

$$\mathbf{Y} = \boldsymbol{\mu} + \mathbf{CZ}, \qquad \mathbf{Y} = \boldsymbol{\mu} + \mathbf{FZ} \tag{9.1.4}$$

lead to the same distribution provided

$$\mathbf{CC}' = \mathbf{FF}'. \tag{9.1.5}$$

The characteristic function of $\mathbf{Y}$ is

$$E[\exp i\mathbf{t}'\mathbf{Y}] = \exp\{i\mathbf{t}'\boldsymbol{\mu} - \tfrac{1}{2}\mathbf{t}'\boldsymbol{\Sigma}\mathbf{t}\}. \tag{9.1.6}$$

9.1.3 Preliminary Lemmas

We prove some lemmas on the distribution of quadratic functions of independent normal variables.

LEMMA 9.1.1 *Let $Y_i \sim N(\mu_i, 1)$, $i = 1, \ldots, p$ be independent. Then the statistic*

$$\sum \lambda_i Y_i^2 + 2\sum b_i Y_i + c \tag{9.1.7}$$

has $\chi^2(k, \delta)$ distribution iff

$$\left.\begin{array}{l} \text{(i) } \lambda_i = 0 \text{ or } 1, \quad i = 1, \ldots, p \\ \text{(ii) } b_i = 0 \text{ if } \lambda_i = 0 \textit{ and } c = \sum b_i^2 \end{array}\right\} \tag{9.1.8}$$

in which case $k = \sum \lambda_i$ *and* $\delta = \sum \lambda_i(b_i + \mu_i)^2$. (9.1.9)

Proof: The lemma is easy to establish by comparing the cumulants or the characteristic functions of (9.1.7) and the statistic

$$T = (X_1 - \nu_1)^2 + \cdots + (X_k - \nu_k)^2,$$

where $X_1, \ldots, X_k$ are k independent $N(0, 1)$ variables.

LEMMA 9.1.2 *Let $\mathbf{Y} \sim N_p(\boldsymbol{\mu}, \mathbf{I})$. Then the statistic*

$$\mathbf{Y}'\mathbf{AY} + 2\mathbf{b}'\mathbf{Y} + c \tag{9.1.10}$$

has $\chi^2(k, \delta)$ distribution iff

$$\left.\begin{array}{l} \text{(i) } \mathbf{A}^2 = \mathbf{A} \\ \text{(ii) } \mathbf{b} \in \mathcal{M}(\mathbf{A}), \quad c = \mathbf{b}'\mathbf{b} \end{array}\right\} \tag{9.1.11}$$

in which case

$$k = R(\mathbf{A}) \quad \text{and} \quad \delta = (\mathbf{b} + \boldsymbol{\mu})'\mathbf{A}(\mathbf{b} + \boldsymbol{\mu}). \tag{9.1.12}$$

Proof: There exists an orthogonal matrix $\mathbf{P}$ such that $\mathbf{P}'\mathbf{A}\mathbf{P} = \boldsymbol{\Delta}$, the diagonal matrix of eigenvalues of $\mathbf{A}$. Under the transformation $\mathbf{Z} = \mathbf{P}'\mathbf{Y}$, $\mathbf{Z} \sim N_p(\mathbf{P}'\boldsymbol{\mu}, \mathbf{I})$ and the statistic (9.1.10) transforms to

$$\mathbf{Z}'\boldsymbol{\Delta}\mathbf{Z} + 2\mathbf{b}'\mathbf{P}\mathbf{Z} + c, \tag{9.1.13}$$

which is of the type (9.1.7). Hence applying (9.1.8) the expression (9.1.13) is a $\chi^2(k, \delta)$ iff

(i)′ Each diagonal element of $\boldsymbol{\Delta}$ is 0 or 1 in which case $\boldsymbol{\Delta}^2 = \boldsymbol{\Delta}$, and
(ii)′ $\mathbf{c} = \mathbf{b}'\mathbf{P}\mathbf{P}'\mathbf{b} = \mathbf{b}'\mathbf{b}$, the ith coordinate of $\mathbf{P}'\mathbf{b}$ is 0 if the ith diagonal element of $\boldsymbol{\Delta}$ is 0, that is, $\mathbf{P}'\mathbf{b} \in \mathcal{M}(\boldsymbol{\Delta})$.

The condition $\boldsymbol{\Delta}^2 = \boldsymbol{\Delta} \Leftrightarrow \mathbf{A}^2 = \mathbf{A}$ so that (i)′ $\Leftrightarrow$ (i) of (9.1.11). The condition $\mathbf{P}'\mathbf{b} \in \mathcal{M}(\boldsymbol{\Delta}) \Leftrightarrow \mathbf{b} \in \mathcal{M}(\mathbf{A})$ so that (ii)′ $\Leftrightarrow$ (ii) of (9.1.11).

Applying (9.1.9)

$$k = \sum \lambda_i = \text{trace}\,\boldsymbol{\Delta} = \text{trace}\,\mathbf{P}'\mathbf{A}\mathbf{P} = \text{trace}\,\mathbf{A} = R(\mathbf{A}), \text{ since } \mathbf{A}^2 = \mathbf{A}$$

$$\delta = (\mathbf{P}'\mathbf{b} + \mathbf{P}'\boldsymbol{\mu})'\boldsymbol{\Delta}(\mathbf{P}'\mathbf{b} + \mathbf{P}'\boldsymbol{\mu}) = (\mathbf{b} + \boldsymbol{\mu})'\mathbf{A}(\mathbf{b} + \boldsymbol{\mu}).$$

LEMMA 9.1.3 [Craig (1943) and Sakamoto (1941)] *Let* $\mathbf{A}$ *and* $\mathbf{B}$ *be* $p \times p$ *real symmetric matrices. Then*

$$\mathbf{A}\mathbf{B} = \mathbf{0} \Leftrightarrow |\mathbf{I} - t\mathbf{A}|\,|\mathbf{I} - u\mathbf{B}| = |\mathbf{I} - t\mathbf{A} - u\mathbf{B}| \tag{9.1.14}$$

for all real scalar t and u.

Proof: The '$\Rightarrow$' part is easily proved. To prove the '$\Leftarrow$' part we observe that if $t^{-1} \notin \sigma(\mathbf{A})$, $\mathbf{I} - t\mathbf{A}$ is nonsingular, hence $\mathbf{K} = (\mathbf{I} - t\mathbf{A})^{-1}$ exists. For such t

$$\begin{aligned}
|\mathbf{I} - t\mathbf{A}|\,|\mathbf{I} - u\mathbf{B}| = |\mathbf{I} - t\mathbf{A} - u\mathbf{B}| &\Rightarrow |\mathbf{I} - u\mathbf{B}| = |\mathbf{I} - u\mathbf{K}\mathbf{B}| \\
&\Rightarrow |\mathbf{I} - u^2\mathbf{B}^2| = |\mathbf{I} - u^2(\mathbf{K}\mathbf{B})^2| \\
&\Rightarrow tr\,\mathbf{B}^2 = tr(\mathbf{K}\mathbf{B})^2 = tr[(\mathbf{I} + t\mathbf{A} + t^2\mathbf{A}^2 + \cdots)\mathbf{B}]^2 \\
&\qquad = tr\,\mathbf{B}^2 + t[tr(\mathbf{A}\mathbf{B}^2 + \mathbf{B}\mathbf{A}\mathbf{B})] + t^2[tr(\mathbf{A}\mathbf{B}\mathbf{A}\mathbf{B} + \mathbf{A}^2\mathbf{B}^2 \\
&\qquad\quad + \mathbf{B}\mathbf{A}^2\mathbf{B})] + 0(t^3)
\end{aligned}$$

for all t sufficiently small

$$\begin{aligned}
&\Rightarrow tr(\mathbf{A}\mathbf{B})^2 + 2\,tr\,\mathbf{B}\mathbf{A}(\mathbf{B}\mathbf{A})' = 0 \\
&\Rightarrow tr\,\mathbf{B}\mathbf{A}(\mathbf{B}\mathbf{A})' = 0,
\end{aligned}$$

since $tr(\mathbf{AB})^2 + tr\,\mathbf{BA}(\mathbf{BA})'$ and $tr\,\mathbf{BA}(\mathbf{BA})'$ are nonnegative.

$$tr\,\mathbf{BA}(\mathbf{BA})' = 0 \Rightarrow \mathbf{BA} = \mathbf{0}, \qquad \mathbf{AB} = \mathbf{0}.$$

9.2 QUADRATIC FUNCTIONS OF CORRELATED NORMAL VARIABLES

Let $\mathbf{Y}$ be a p-vector variable distributed as $N_p(\boldsymbol{\mu}, \boldsymbol{\Sigma})$ so that $E(\mathbf{Y}) = \boldsymbol{\mu}$ and $D(\mathbf{Y}) = \boldsymbol{\Sigma}$, where $\boldsymbol{\Sigma}$ may be singular. We shall investigate the conditions under which a quadratic function

$$\mathbf{Y}'\mathbf{AY} + 2\mathbf{b}'\mathbf{Y} + c \tag{9.2.1}$$

has a $\chi^2(k, \delta)$ distribution.

Since $\boldsymbol{\Sigma}$ is at least n.n.d., there exists a matrix $\mathbf{C}$ of order $n \times r$ and of rank $r = R(\boldsymbol{\Sigma})$ such that $\boldsymbol{\Sigma} = \mathbf{CC}'$ and the matrix $\mathbf{C}'\mathbf{C}$ of order $r \times r$ is nonsingular. Then the random variable $\mathbf{Y}$ has the representation

$$\mathbf{Y} = \boldsymbol{\mu} + \mathbf{CZ}, \tag{9.2.2}$$

where $\mathbf{Z}$ is an r-vector of independent normal variables with zero mean and unit variance. The method of investigation we adopt is to substitute for $\mathbf{Y}$ the expression (9.2.2) in the statistic (9.2.1) and apply the result of Lemma 9.1.2 on independent normal variables.

THEOREM 9.2.1 *Let* $\mathbf{Y} \sim N_p(\boldsymbol{\mu}, \boldsymbol{\Sigma})$ *where* $\boldsymbol{\Sigma}$ *may be singular. Then the statistic*

$$\mathbf{Y}'\mathbf{AY} + 2\mathbf{b}'\mathbf{Y} + c \tag{9.2.3}$$

has a $\chi^2(k, \delta)$ *distribution iff*

$$\begin{aligned} &\text{(i) } \boldsymbol{\Sigma A \Sigma A \Sigma} = \boldsymbol{\Sigma A \Sigma} \text{ or equivalently } (\boldsymbol{\Sigma A})^3 = (\boldsymbol{\Sigma A})^2 \\ &\text{(ii) } \mathcal{M}[\boldsymbol{\Sigma}(\mathbf{A}\boldsymbol{\mu} + \mathbf{b})] \in \mathcal{M}(\boldsymbol{\Sigma A \Sigma}) \\ &\text{(iii) } (\mathbf{A}\boldsymbol{\mu} + \mathbf{b})'\boldsymbol{\Sigma}(\mathbf{A}\boldsymbol{\mu} + \mathbf{b}) = \boldsymbol{\mu}'\mathbf{A}\boldsymbol{\mu} + 2\mathbf{b}'\boldsymbol{\mu} + c \end{aligned} \tag{9.2.4}$$

in which case

$$\begin{aligned} k &= tr\,\mathbf{A}\boldsymbol{\Sigma} \\ \delta &= (\mathbf{b} + \mathbf{A}\boldsymbol{\mu})'\boldsymbol{\Sigma A \Sigma}(\mathbf{b} + \mathbf{A}\boldsymbol{\mu}). \end{aligned} \tag{9.2.5}$$

Proof: In terms of $\mathbf{Z} \sim N_r(0, 1)$ defined by (9.2.2), the statistic (9.2.3) becomes

$$\begin{aligned} &(\boldsymbol{\mu} + \mathbf{CZ})'\mathbf{A}(\boldsymbol{\mu} + \mathbf{CZ}) + 2\mathbf{b}'(\boldsymbol{\mu} + \mathbf{CZ}) + c \\ &\quad = \mathbf{Z}'\mathbf{C}'\mathbf{ACZ} + 2(\mathbf{A}\boldsymbol{\mu} + \mathbf{b})'\mathbf{CZ} + \boldsymbol{\mu}'\mathbf{A}\boldsymbol{\mu} + 2\mathbf{b}'\boldsymbol{\mu} + c. \end{aligned} \tag{9.2.6}$$

Applying the result (9.1.7) of Lemma 9.1.2, the statistic (9.2.6) has $\chi^2(k, \delta)$ distribution iff

$$\begin{aligned} &(\text{i})' \quad \mathbf{C'AC} \text{ is idempotent} \\ &(\text{ii})' \quad \mathbf{C'(A\mu + b)} \in \mathscr{M}(\mathbf{C'AC}) \\ &(\text{iii})' \quad (\mathbf{A\mu + b})'\mathbf{CC'}(\mathbf{A\mu + b}) = \boldsymbol{\mu}'\mathbf{A}\boldsymbol{\mu} + 2\mathbf{b}'\boldsymbol{\mu} + c. \end{aligned} \tag{9.2.7}$$

The conditions (9.2.7) involve $\mathbf{C}$ and to express them in terms of $\boldsymbol{\Sigma}$ we proceed as follows. From (i)$'$,

$$\mathbf{C'ACC'AC} = \mathbf{C'AC}. \tag{9.2.8}$$

Multiplying both sides by $\mathbf{C}$ from the left and $\mathbf{C}'$ from the right

$$\mathbf{CC'ACC'ACC'} = \mathbf{CC'ACC'} \quad \text{or} \quad \boldsymbol{\Sigma A \Sigma A \Sigma} = \boldsymbol{\Sigma A \Sigma}, \tag{9.2.9}$$

which shows that (i)$' \Rightarrow$ (i) of the theorem. Multiplying (9.2.9) from left and right by $\mathbf{C}'$ and $\mathbf{C}$ and observing that $\mathbf{C'C}$ is nonsingular, we obtain (9.2.8), which shows that (i) $\Rightarrow$ (i)$'$.

It is easy to see that

$$\boldsymbol{\Sigma A \Sigma A \Sigma} = \boldsymbol{\Sigma A \Sigma} \Rightarrow (\boldsymbol{\Sigma}\mathbf{A})^3 = (\boldsymbol{\Sigma}\mathbf{A})^2. \tag{9.2.10}$$

Also, since $\boldsymbol{\Sigma}$ is n.n.d.,

$$R(\boldsymbol{\Sigma A \Sigma}) = R(\boldsymbol{\Sigma A \Sigma}\mathbf{A}) \quad \text{or} \quad \boldsymbol{\Sigma A \Sigma} = \boldsymbol{\Sigma A \Sigma}\mathbf{AD} \quad \text{for some } \mathbf{D}. \tag{9.2.11}$$

Then multiplying both sides of $(\boldsymbol{\Sigma}\mathbf{A})^3 = (\boldsymbol{\Sigma}\mathbf{A})^2$ by $\mathbf{D}$, $\boldsymbol{\Sigma A \Sigma A \Sigma}\mathbf{AD} = \boldsymbol{\Sigma A \Sigma}\mathbf{AD}$ or $\boldsymbol{\Sigma A \Sigma A \Sigma} = \boldsymbol{\Sigma A \Sigma}$ using (9.2.11). Thus the two conditions given in (i) of the theorem are equivalent. From (ii)$'$

$$\mathbf{C'(A\mu + b)} = \mathbf{C'AC}\boldsymbol{\xi} \text{ for some } \boldsymbol{\xi} \Leftrightarrow \mathbf{CC'(A\mu + b)} = \mathbf{CC'AC}\boldsymbol{\xi},$$

that is,

$$\boldsymbol{\Sigma}(\mathbf{A\mu + b}) = \boldsymbol{\Sigma A \Sigma}\boldsymbol{\eta} \text{ for some } \boldsymbol{\eta} \quad \text{or} \quad \boldsymbol{\Sigma}(\mathbf{A\mu + b}) \in \mathscr{M}(\boldsymbol{\Sigma A \Sigma}),$$

which shows that (ii)$' \Leftrightarrow$ (ii). From (iii)$'$,

$$(\mathbf{A\mu + b})'\boldsymbol{\Sigma}(\mathbf{A\mu + b}) = \boldsymbol{\mu}'\mathbf{A}\boldsymbol{\mu} + 2\mathbf{b}'\boldsymbol{\mu} + c,$$

which is the same as (iii).

Applying (9.1.12) on the statistic (9.2.6.)

$$\begin{aligned} k &= R(\mathbf{C'AC}) = tr(\mathbf{C'AC}) = tr(\mathbf{ACC'}) = tr(\mathbf{A}\boldsymbol{\Sigma}) \\ \delta &= (\mathbf{b + A\mu})\mathbf{CC'ACC'}(\mathbf{b + A\mu}) = (\mathbf{b + A\mu})'\boldsymbol{\Sigma A \Sigma}(\mathbf{b + A\mu}), \end{aligned} \tag{9.2.12}$$

which proves (9.2.5).

The results of Theorem 9.2.1 seem to have been obtained first by Ogasawara and Takahashi (1951). The condition $(\boldsymbol{\Sigma}\mathbf{A})^3 = (\boldsymbol{\Sigma}\mathbf{A})^2$ of Theorem 9.2.1(i) and also the result of the following corollary are due to Shanbag (1969, 1970).

COROLLARY: *The condition (i) of Theorem 9.2.1 is equivalent to*

$$tr(\mathbf{\Sigma A})^2 = tr(\mathbf{\Sigma A})^3 = tr(\mathbf{\Sigma A})^4. \tag{9.2.13}$$

Proof: $(\mathbf{\Sigma A})^2 = (\mathbf{\Sigma A})^3 \Rightarrow (\mathbf{\Sigma A})^2 = (\mathbf{\Sigma A})^3 = (\mathbf{\Sigma A})^4 \Rightarrow$ (9.2.13). Conversely, if $\lambda_1, \lambda_2, \ldots, \lambda_p$ are the eigenvalues of $\mathbf{\Sigma A}$,

$$(9.2.13) \Rightarrow \sum_{i=1}^{p} \lambda_i^2(1 - \lambda_i)^2 = 0.$$

Since a nonnull eigenvalue of $\mathbf{\Sigma A}$ is also an eigenvalue of the real symmetric matrix $\mathbf{CAC}'$, it is clear that the λ_i's are real. Hence,

$$\sum_{i=1}^{p} \lambda_i^2(1 - \lambda_i)^2 = 0 \Rightarrow \lambda_i = 0 \text{ or } 1 \;\forall\; i = 1, 2, \ldots, p$$

$\Rightarrow \mathbf{CAC}'$ is idempotent

$\Rightarrow$ condition (i) of Theorem 9.2.1 as seen in the proof of this theorem.

THEOREM 9.2.2 *Let* $\mathbf{Y} \sim N_p(\mathbf{0}, \mathbf{\Sigma})$. *Then* $\mathbf{Y}'\mathbf{\Sigma}^-\mathbf{Y}$ *has* $\chi^2(k)$ *distribution for any choice of* $\mathbf{\Sigma}^-$, *where* $k = R(\mathbf{\Sigma})$.

Proof: The result follows by applying the condition (i) of (9.2.4). $\mathbf{\Sigma A \Sigma A \Sigma} = \mathbf{\Sigma A \Sigma}$. Substituting $\mathbf{\Sigma}^-$ for $\mathbf{A}$

$$\mathbf{\Sigma\Sigma^-\Sigma\Sigma^-\Sigma} = \mathbf{\Sigma\Sigma^-\Sigma}, \tag{9.2.14}$$

which is satisfied, since both sides of (9.2.14) reduce to $\mathbf{\Sigma}$.

Applying (9.2.5), $\delta = 0$ and $k = tr(\mathbf{A\Sigma}) = tr(\mathbf{\Sigma^-\Sigma}) = R(\mathbf{\Sigma^-\Sigma}) = R(\mathbf{\Sigma})$, since $\mathbf{\Sigma^-\Sigma}$ is idempotent.

THEOREM 9.2.3 *Let* $\mathbf{Y} \sim N_p(\boldsymbol{\mu}, \mathbf{\Sigma})$. *Then* $\mathbf{Y}'\mathbf{\Sigma}^-\mathbf{Y}$ *has* $\chi^2(k, \delta)$ *distribution if either* $\boldsymbol{\mu} \in \mathcal{M}(\mathbf{\Sigma})$ *whatever may be the choice of the g-inverse* $\mathbf{\Sigma}^-$, *or* $\mathbf{\Sigma}^-$ *is a symmetric reflexive g-inverse of* $\mathbf{\Sigma}$, *whatever* $\boldsymbol{\mu}$ *may be.*

In either case $k = R(\mathbf{\Sigma})$ *and* $\delta = \boldsymbol{\mu}'\mathbf{\Sigma}^-\boldsymbol{\mu}$.

Proof: The condition (i) of (9.2.4) is satisfied for any choice of $\mathbf{\Sigma}^-$; (ii) is automatically satisfied since $\mathbf{b} = \mathbf{0}$. It is easy to see that (iii) is satisfied under each of the conditions stated in the theorem. The value of δ in each case is seen to be $\boldsymbol{\mu}'\mathbf{\Sigma}^-\boldsymbol{\mu}$ applying the formula (9.2.5).

Let $\mathbf{Y} \sim N_p(\mathbf{0}, \mathbf{\Sigma})$. It may be of some interest to determine explicitly the class of all symmetric matrices $\mathbf{A}$ for which $\mathbf{Y}'\mathbf{AY}$ has a χ^2 distribution. Here one has to solve the matrix equation $\mathbf{\Sigma A \Sigma A \Sigma} = \mathbf{\Sigma A \Sigma}$ for $\mathbf{A}$.

Given a quadratic form $\mathbf{Y}'\mathbf{AY}$, the dual problem is to determine the class of all n.n.d. matrices $\mathbf{\Sigma}$ for which $\mathbf{Y}'\mathbf{AY}$ has a χ^2 distribution when $\mathbf{Y} \sim N_p(\mathbf{0}, \mathbf{\Sigma})$. Here one has to solve the matrix equation $\mathbf{\Sigma A \Sigma A \Sigma} = \mathbf{\Sigma A \Sigma}$ for $\mathbf{\Sigma}$.

For the relevant solutions to the above matrix equations, the reader is referred to Section 3.4.

9.3 SOME FURTHER THEOREMS ON THE DISTRIBUTION OF QUADRATIC FUNCTIONS

We shall first prove a general lemma which enables us to prove some theorems on the distribution of quadratic functions, similar to that of Cochran–Fisher theorem, etc.

LEMMA 9.3.1 *Let* $\mathbf{A}, \mathbf{A}_1, \mathbf{A}_2, \ldots, \mathbf{A}_k$ *be* $p \times p$ *symmetric matrices such that* $\mathbf{A} = \mathbf{A}_1 + \cdots + \mathbf{A}_k$. *Further let* $\mathbf{\Sigma}$ *be an n.n.d. matrix. Consider the following conditions:*

(a) $(\mathbf{\Sigma A}_i)^3 = (\mathbf{\Sigma A}_i)^2, \quad i = 1, \ldots, k$
(b) $\mathbf{\Sigma A}_i\mathbf{\Sigma A}_j\mathbf{\Sigma} = \mathbf{0} \quad$ for $i \neq j$
(c) $(\mathbf{\Sigma A})^3 = (\mathbf{\Sigma A})^2$
(d) $R(\mathbf{\Sigma A \Sigma}) = R(\mathbf{\Sigma A}_1\mathbf{\Sigma}) + \cdots + R(\mathbf{\Sigma A}_k\mathbf{\Sigma})$.

Then any two of the conditions (a), (b), (c) imply all the conditions, and (c) and (d) imply (a) and (b).

Proof: Let $\mathbf{\Sigma} = \mathbf{CC}'$ where $\mathbf{C}$ is $p \times r$ matrix of rank $r = R(\mathbf{\Sigma})$. Then

$$\mathbf{C}'\mathbf{AC} = \sum_{i=1}^{k} \mathbf{C}'\mathbf{A}_i\mathbf{C}.$$

Let us write $\mathbf{H}_i = \mathbf{C}'\mathbf{A}_i\mathbf{C}$ and $\mathbf{H} = \mathbf{C}'\mathbf{AC}$, then the conditions (a), (b), (c), (d) are equivalent to

(a)′ $\mathbf{H}_i$ is idempotent for $i = 1, \ldots, k$
(b)′ $\mathbf{H}_i\mathbf{H}_j = \mathbf{0} \quad$ for $i \neq j$
(c)′ $\mathbf{H}$ is idempotent
(d)′ $R(\mathbf{H}) = R(\mathbf{H}_1) + \cdots + R(\mathbf{H}_k)$.

Then the results of the theorem follow from those of Lemma 5.4.1.

A generalization of Cochran–Fisher theorem (see p. 149 of Rao, 1965) on the distribution of quadratic forms is given in Theorem 9.3.1.

THEOREM 9.3.1 *Let* $\mathbf{Y} \sim N_p(\mathbf{0}, \mathbf{\Sigma})$, *where* $\mathbf{\Sigma}$ *may be singular. Further let*

$$\mathbf{Y}'\mathbf{AY} = \sum_{i=1}^{n} \mathbf{Y}'\mathbf{A}_i\mathbf{Y}. \tag{9.3.1}$$

Then for $\mathbf{Y}'\mathbf{A}_i\mathbf{Y}$, $i = 1, \ldots, k$, *to be distributed independently as* $\chi^2(\cdot, 0)$ *the conditions*

(a) $(\boldsymbol{\Sigma}\mathbf{A})^3 = (\boldsymbol{\Sigma}\mathbf{A})^2$
(b) $R(\boldsymbol{\Sigma}\mathbf{A}\boldsymbol{\Sigma}) = R(\boldsymbol{\Sigma}\mathbf{A}_1\boldsymbol{\Sigma}) + \cdots + R(\boldsymbol{\Sigma}\mathbf{A}_k\boldsymbol{\Sigma})$

are necessary and sufficient.

Proof: As in Lemma 9.3.1, we make the transformation $\mathbf{Y} = \mathbf{CZ}$, where $\boldsymbol{\Sigma} = \mathbf{CC}'$. The condition (9.3.1) transforms to

$$\mathbf{Z}'\mathbf{C}'\mathbf{ACZ} = \mathbf{Z}'\mathbf{C}'\mathbf{A}_1\mathbf{CZ} + \cdots + \mathbf{Z}'\mathbf{C}'\mathbf{A}_k\mathbf{CZ}, \tag{9.3.2}$$

where $\mathbf{Z} \sim N_r(\mathbf{0}, \mathbf{I})$ and $r = R(\boldsymbol{\Sigma})$. Applying the well-known Cochran–Fisher theorem to equation (9.3.2), n.s. conditions for $\mathbf{Z}'\mathbf{C}'\mathbf{A}_i\mathbf{CZ}$, $i = 1, \ldots, k$, to be independently distributed as central χ^2's are

(a)′ $\mathbf{C}'\mathbf{AC}$ is idempotent
(b)′ $R(\mathbf{C}'\mathbf{AC}) = R(\mathbf{C}'\mathbf{A}_1\mathbf{C}) + \cdots + R(\mathbf{C}'\mathbf{A}_k\mathbf{C})$.

It is already shown that (a)′ ⇔ (a) of the theorem. To show the equivalence (b)′ ⇔ (b), we observe that

$$R(\mathbf{C}'\mathbf{A}_i\mathbf{C}) \geq R(\mathbf{CC}'\mathbf{A}_i\mathbf{CC}'),$$
$$R(\mathbf{CC}'\mathbf{A}_i\mathbf{CC}') \geq R(\mathbf{C}'\mathbf{CC}'\mathbf{A}_i\mathbf{CC}'\mathbf{C}) = R(\mathbf{C}'\mathbf{A}_i\mathbf{C}),$$

since $\mathbf{C}'\mathbf{C}$ is nonsingular. The theorem is established.

THEOREM 9.3.2 *Let* $\mathbf{Y} \sim N_p(\mathbf{0}, \boldsymbol{\Sigma})$ *and*

$$\mathbf{Y}'\mathbf{AY} = \sum_1^k \mathbf{Y}'\mathbf{A}_i\mathbf{Y}. \tag{9.3.3}$$

Then $\mathbf{Y}'\mathbf{A}_i\mathbf{Y}$, $i = 1, \ldots, k$, *have independent* $\chi^2(\cdot, 0)$ *distributions iff any two of the following conditions are satisfied.*

(a) $(\boldsymbol{\Sigma}\mathbf{A}_i)^3 = (\boldsymbol{\Sigma}\mathbf{A}_i)^2, \quad i = 1, \ldots, k$
(b) $\boldsymbol{\Sigma}\mathbf{A}_i\boldsymbol{\Sigma}\mathbf{A}_j\boldsymbol{\Sigma} = \mathbf{0}$ for $i \neq j$
(c) $(\boldsymbol{\Sigma}\mathbf{A})^3 = (\boldsymbol{\Sigma}\mathbf{A})^2$.

Proof: The theorem follows since, according to Lemma 9.3.1, any two of the conditions (a), (b), and (c) are equivalent to the conditions (a) and (b) of Theorem 9.3.1.

THEOREM 9.3.3 *Let* $\mathbf{Y} \sim N_p(\boldsymbol{\mu}, \boldsymbol{\Sigma})$ *and*

$$\mathbf{Y}'\mathbf{AY} = \mathbf{Y}'\mathbf{A}_1\mathbf{Y} + \cdots + \mathbf{Y}'\mathbf{A}_k\mathbf{Y}. \tag{9.3.4}$$

Consider the following conditions:

(a) $\mathbf{\Sigma A \mu} \in \mathcal{M}(\mathbf{\Sigma A \Sigma})$, $\mathbf{\mu' A \mu} = \mathbf{\mu' A \Sigma A \mu}$
(b) $\mathbf{\Sigma A}_i\mathbf{\mu} \in \mathcal{M}(\mathbf{\Sigma A}_i\mathbf{\Sigma})$, $\mathbf{\mu' A}_i\mathbf{\mu} = \mathbf{\mu' A}_i\mathbf{\Sigma A}_i\mathbf{\mu}$

and

(i) $(\mathbf{\Sigma A}_i)^3 = (\mathbf{\Sigma A}_i)^2$, $i = 1, \ldots, k$
(ii) $\mathbf{\Sigma A}_i\mathbf{\Sigma A}_j\mathbf{\Sigma} = \mathbf{0}$ for $i \neq j$
(iii) $(\mathbf{\Sigma A})^3 = (\mathbf{\Sigma A})^2$
(iv) $R(\mathbf{\Sigma A \Sigma}) = R(\mathbf{\Sigma A}_1\mathbf{\Sigma}) + \cdots + R(\mathbf{\Sigma A}_k\mathbf{\Sigma})$.

Let S be the statement that $\mathbf{Y'A}_i\mathbf{Y}$, $i = 1, \ldots, k$, *are independently distributed each as* $\chi^2(\cdot,\cdot)$. *Then the following are true.*

$$S \Rightarrow \text{(a), (b), (i), (ii), (iii), (iv).}$$

$$\text{(b), (iii), (iv)} \Rightarrow S \textit{ and } \text{(a).}$$

Proof: The results follow from those of Lemma 9.3.1 and Theorem 9.2.1.

THEOREM 9.3.4 *Let* $\mathbf{Y} \sim N_p(\mathbf{\mu}, \mathbf{\Sigma})$ *and*

$$\mathbf{Y'AY} + 2\mathbf{b'Y} + c = \sum_{i=1}^{k} (\mathbf{Y'A}_i\mathbf{Y} + 2\mathbf{b}_i'\mathbf{Y} + c_i). \tag{9.3.5}$$

Then for $(\mathbf{Y'A}_i\mathbf{Y} + 2\mathbf{b}_i'\mathbf{Y} + c_i)$, $i = 1, 2, \ldots, k$ *to be independently distributed each as* $\chi^2(.,.)$, *the results of Theorem 9.3.3 hold good if the conditions (a) and (b) are replaced by the following:*

(a) $\mathbf{\Sigma A \mu} \in \mathcal{M}(\mathbf{\Sigma A \Sigma})$, $(\mathbf{A\mu} + \mathbf{b})'\mathbf{\Sigma}(\mathbf{A\mu} + \mathbf{b}) = \mathbf{\mu'A\mu} + 2\mathbf{b'\mu} + c$
(b) $\mathbf{\Sigma A}_i\mathbf{\mu} \in \mathcal{M}(\mathbf{\Sigma A}_i\mathbf{\Sigma})$, $(\mathbf{A}_i\mathbf{\mu} + \mathbf{b})'\mathbf{\Sigma}(\mathbf{A}_i\mathbf{\mu} + \mathbf{b}) = \mathbf{\mu'A}_i\mathbf{\mu} + 2\mathbf{b}_i'\mathbf{\mu} + c_i$
$i = 1, \ldots, k$.

THEOREM 9.3.5 *Let* $\mathbf{Y} \sim N_p(\mathbf{\mu}, \mathbf{\Sigma})$, $T_1 = \mathbf{Y'A}_1\mathbf{Y} + 2\mathbf{b}_1'\mathbf{Y} + c_1 \sim \chi^2(k_1, \delta_1)$ *and* $T_2 = \mathbf{Y'A}_2\mathbf{Y} + 2\mathbf{b}_2'\mathbf{Y} + c_2 \sim \chi^2(k_2, \delta_2)$. *Then* T_1 *and* T_2 *are independently distributed iff*

$$\mathbf{\Sigma A}_1\mathbf{\Sigma A}_2\mathbf{\Sigma} = \mathbf{0}. \tag{9.3.6}$$

Proof: Since T_1 and T_2 are distributed as χ^2's, we have by applying Theorem 9.2.1.

$$\mathbf{\Sigma A}_i\mathbf{\Sigma A}_i\mathbf{\Sigma} = \mathbf{\Sigma A}_i\mathbf{\Sigma}$$

$$\mathbf{\Sigma}(\mathbf{A}_i\mathbf{\mu} + \mathbf{b}_i) \in \mathcal{M}(\mathbf{\Sigma A}_i\mathbf{\Sigma})$$

$$(\mathbf{A}_i\mathbf{\mu} + \mathbf{b}_i)'\mathbf{\Sigma}(\mathbf{A}_i\mathbf{\mu} + \mathbf{b}_i) = \mathbf{\mu'A}_i\mathbf{\mu} + 2\mathbf{b}_i'\mathbf{\mu} + c_i \quad \text{for } i = 1, 2.$$

By applying Lemma 9.3.1,

$$\left.\begin{array}{r}\Sigma A_i\Sigma A_i\Sigma = \Sigma A_i\Sigma \\ i = 1, 2. \\ \Sigma A_1\Sigma A_2\Sigma = \mathbf{0}\end{array}\right\} \Leftrightarrow \left\{\begin{array}{l}\Sigma A\Sigma A\Sigma = \Sigma A\Sigma \\ \\ R(\Sigma A\Sigma) = R(\Sigma A_1\Sigma) + R(\Sigma A_2\Sigma)\end{array}\right.$$

where $\mathbf{A} = \mathbf{A}_1 + \mathbf{A}_2$. Further,

$$\left.\begin{array}{l}\Sigma(A_i\mu + b_i) \in \mathcal{M}(\Sigma A_i\Sigma) \\ \quad (A_i\mu + b_i)'\Sigma(A_i\mu + b_i) \\ \quad = \mu' A_i\mu + 2b_i'\mu + c_i \\ i = 1, 2.\end{array}\right\} \Rightarrow \left\{\begin{array}{l}\Sigma(A\mu + b) \in \mathcal{M}(\Sigma A\Sigma) \\ \quad (A\mu + b)'\Sigma(A\mu + b) \\ \quad = \mu' A\mu + 2b'\mu + c\end{array}\right.$$

Then the conditions of the theorem are equivalent to conditions (b), (iii) and (iv) of Theorem 9.3.3, which proves the desired result.

THEOREM 9.3.6 *Let* $\mathbf{Y} \sim N_p(\boldsymbol{\mu}, \boldsymbol{\Sigma})$ *and* $T = T_1 + T_2$, *where* $T = \mathbf{Y}'\mathbf{A}\mathbf{Y} + 2\mathbf{b}'\mathbf{Y} + c$, $T_1 = (\mathbf{Y}'\mathbf{A}_1\mathbf{Y} + 2\mathbf{b}_1'\mathbf{Y} + c_1)$ *and* $T_2 = (\mathbf{Y}'\mathbf{A}_2\mathbf{Y} + 2\mathbf{b}_2'\mathbf{Y} + c_2)$. *Assume further that* T *and* T_1 *are* $\chi^2(k, \delta)$, $\chi^2(k_1, \delta_1)$ *variables, and* T_2 *is non-negative definite. Then* $T_2 \sim \chi^2(k - k_1, \delta - \delta_1)$.

Proof: Since T is a noncentral chisquare, there exists an affine linear transformation $\mathbf{X} = \mathbf{PY} + \mathbf{d}$ where P is nonsingular, such that

$$T = X_1^2 + X_2^2 + \cdots + X_k^2$$

and $X_1, X_2, \ldots, X_k$ are independent normal variables, $X_i \sim N(\eta_i, 1)$. Naturally $\delta = \sum \eta_i^2 = \boldsymbol{\eta}'\boldsymbol{\eta}$. Since $T = T_1 + T_2$ and both T_1 and T_2 are non-negative, it is clear that both T_1 and T_2 involve $X_1, X_2, \ldots, X_k$ only. If $\mathbf{X}_{(1)}' = (X_1, X_2, \ldots, X_k)$, we can write

$$T_1 = \mathbf{X}_{(1)}'\mathbf{B}\mathbf{X}_{(1)} + 2\mathbf{b}'\mathbf{X}_{(1)} + c$$

$$T_2 = \mathbf{X}_{(1)}'(\mathbf{I} - \mathbf{B})\mathbf{X}_{(1)} - 2\mathbf{b}'\mathbf{X}_{(1)} - c.$$

Since $T_1 \sim \chi^2(k_1, \delta_1)$, we have $\mathbf{B}^2 = \mathbf{B}$, $\mathbf{b} \in \mathcal{M}(\mathbf{B})$. Hence, $\mathbf{b} \in \mathcal{M}(\mathbf{I} - \mathbf{B})$. This shows T_2 cannot be non-negative, unless $\mathbf{b} = \mathbf{0}$. Hence, $c = \mathbf{b}'\mathbf{b} = 0$. Since $(\mathbf{I} - \mathbf{B})^2 = \mathbf{I} - \mathbf{B}$ by Lemma 9.1.2, $T_2 = \mathbf{X}_{(1)}'(\mathbf{I} - \mathbf{B})\mathbf{X}_{(1)} \sim \chi^2(k - k_1, \delta_2)$, where $\delta_2 = \boldsymbol{\eta}'(\mathbf{I} - \mathbf{B})\boldsymbol{\eta} = \boldsymbol{\eta}'\boldsymbol{\eta} - \boldsymbol{\eta}'\mathbf{B}\boldsymbol{\eta} = \delta - \delta_1$.

9.4 INDEPENDENCE OF QUADRATIC FORMS

In the previous sections of this chapter, we considered conditions on quadratic forms to be distributed as χ^2. We shall now consider conditions for independent distribution of quadratic forms whether they have χ^2 distributions or not.

THEOREM 9.4.1 *Let* $\mathbf{Y} \sim N_p(\boldsymbol{\mu}, \boldsymbol{\Sigma})$ *and* $Q_1 = \mathbf{Y}'\mathbf{A}\mathbf{Y}$ *and* $Q_2 = \mathbf{Y}'\mathbf{B}\mathbf{Y}$ *be two quadratic forms. Then n.s. conditions for* Q_1 *and* Q_2 *to be independently distributed are*

(a) $\boldsymbol{\Sigma}\mathbf{A}\boldsymbol{\Sigma}\mathbf{B}\boldsymbol{\Sigma} = \mathbf{0}$, $\boldsymbol{\Sigma}\mathbf{A}\boldsymbol{\Sigma}\mathbf{B}\boldsymbol{\mu} = \mathbf{0} = \boldsymbol{\Sigma}\mathbf{B}\boldsymbol{\Sigma}\mathbf{A}\boldsymbol{\mu}$ *and* $\boldsymbol{\mu}'\mathbf{A}\boldsymbol{\Sigma}\mathbf{B}\boldsymbol{\mu} = \mathbf{0}$, *when* $\mathbf{A}$ *and* $\mathbf{B}$ *are symmetric, not necessarily semidefinite, and* $\boldsymbol{\Sigma}$ *possibly singular.*
(b) $\mathbf{A}\boldsymbol{\Sigma}\mathbf{B}\boldsymbol{\Sigma} = \mathbf{0}$, $\mathbf{A}\boldsymbol{\Sigma}\mathbf{B}\boldsymbol{\mu} = \mathbf{0}$ *when* $\mathbf{A}$ *is semi-definite.*
(c) $\mathbf{A}\boldsymbol{\Sigma}\mathbf{B} = \mathbf{0}$ *when* $\mathbf{A}$ *and* $\mathbf{B}$ *are both semi-definite.*
(d) $\mathbf{A}\boldsymbol{\Sigma}\mathbf{B} = \mathbf{0}$ *when* $\boldsymbol{\Sigma}$ *is nonsingular,* $\mathbf{A}$ *and* $\mathbf{B}$ *are symmetric not necessarily semi-definite.*

THEOREM 9.4.2 *Let* $\mathbf{Y}$ *be distributed as in Theorem 9.4.1, and* $Q_1 = \mathbf{Y}'\mathbf{A}\mathbf{Y} + 2\mathbf{a}'\mathbf{Y} + \alpha$ *and* $Q_2 = \mathbf{Y}'\mathbf{B}\mathbf{Y} + 2\mathbf{b}'\mathbf{Y} + \beta$; *n.s. conditions for* Q_1 *and* Q_2 *to be independently distributed are*

(a) $\boldsymbol{\Sigma}\mathbf{A}\boldsymbol{\Sigma}\mathbf{B}\boldsymbol{\Sigma} = \mathbf{0}$, $\boldsymbol{\Sigma}\mathbf{A}\boldsymbol{\Sigma}\mathbf{b} = \mathbf{0}$, $\boldsymbol{\Sigma}\mathbf{B}\boldsymbol{\Sigma}\mathbf{a} = \mathbf{0}$ and $\mathbf{a}'\boldsymbol{\Sigma}\mathbf{b} = 0$ when $\boldsymbol{\mu} = \mathbf{0}$ and $\boldsymbol{\Sigma}$, possibly singular.
(b) $\mathbf{A}\boldsymbol{\Sigma}\mathbf{B} = \mathbf{0}$, $\mathbf{B}\boldsymbol{\Sigma}\mathbf{a} = \mathbf{0}$, $\mathbf{A}\boldsymbol{\Sigma}\mathbf{b} = \mathbf{0}$ and $\mathbf{a}'\boldsymbol{\Sigma}\mathbf{b} = 0$ when $\boldsymbol{\Sigma}$ is nonsingular and $\boldsymbol{\mu}$, not necessarily null.

The proofs of Theorems 9.4.1 and 9.4.2 follow from Lemma 9.1.3 after expressing the condition that the joint characteristic function (c.f.) of Q_1 and Q_2 is the product of the individual c.f.'s of Q_1 and Q_2.

Some of the results of Chapter 9 are new and the rest are generalizations of previous work by Cochran (1934), Craig (1943), Hogg and Craig (1958), Khatri (1963), Ogawa (1949), Sakamoto (1944), Rao (1951, 1953, 1961), Graybill and Marsaglia (1961), and others.

COMPLEMENTS

In all the examples given below, $\boldsymbol{\Sigma}$ represents a n.n.d. matrix.

1. $\boldsymbol{\Sigma}\mathbf{A}\boldsymbol{\Sigma}\mathbf{A}\boldsymbol{\Sigma} = \boldsymbol{\Sigma}\mathbf{A}\boldsymbol{\Sigma} \Rightarrow (\boldsymbol{\Sigma}\mathbf{A})^2 = \boldsymbol{\Sigma}\mathbf{A}$

iff any one of the following conditions hold.

(a) $R(\boldsymbol{\Sigma}\mathbf{A}) = tr\,\boldsymbol{\Sigma}\mathbf{A}$
(b) $R(\boldsymbol{\Sigma}\mathbf{A}) = R(\boldsymbol{\Sigma}\mathbf{A}\boldsymbol{\Sigma})$
(c) $R(\boldsymbol{\Sigma}\mathbf{A}) = R(\boldsymbol{\Sigma}\mathbf{A})^2$. (Styan, 1970).

2. If $\mathbf{A}$ is n.n.d., $\boldsymbol{\Sigma}\mathbf{A}\boldsymbol{\Sigma}\mathbf{A}\boldsymbol{\Sigma} = \boldsymbol{\Sigma}\mathbf{A}\boldsymbol{\Sigma} \Rightarrow$ (a), (b), (c) of example 1 $\Rightarrow (\boldsymbol{\Sigma}\mathbf{A})^2 = \boldsymbol{\Sigma}\mathbf{A}$.

3. If $tr\,\boldsymbol{\Sigma}\mathbf{A} = R(\mathbf{A})$, $\boldsymbol{\Sigma}\mathbf{A}\boldsymbol{\Sigma}\mathbf{A}\boldsymbol{\Sigma} = \boldsymbol{\Sigma}\mathbf{A}\boldsymbol{\Sigma} \Rightarrow (\boldsymbol{\Sigma}\mathbf{A})^2 = \boldsymbol{\Sigma}\mathbf{A} \Rightarrow \boldsymbol{\Sigma} = \mathbf{A}^-$.

4. If $tr\,\boldsymbol{\Sigma}\mathbf{A} = R(\boldsymbol{\Sigma})$, $\boldsymbol{\Sigma}\mathbf{A}\boldsymbol{\Sigma}\mathbf{A}\boldsymbol{\Sigma} = \boldsymbol{\Sigma}\mathbf{A}\boldsymbol{\Sigma} \Rightarrow (\boldsymbol{\Sigma}\mathbf{A})^2 = \boldsymbol{\Sigma}\mathbf{A} \Rightarrow \mathbf{A} = \boldsymbol{\Sigma}^-$.

5. (a) If $\mathbf{A}$ is n.n.d., $\boldsymbol{\Sigma}\mathbf{A}\boldsymbol{\Sigma}\mathbf{B}\boldsymbol{\Sigma} = \mathbf{0} \Leftrightarrow \mathbf{A}\boldsymbol{\Sigma}\mathbf{B}\boldsymbol{\Sigma} = \mathbf{0}$.
(b) If $\mathbf{A}$ and $\mathbf{B}$ are n.n.d., $\boldsymbol{\Sigma}\mathbf{A}\boldsymbol{\Sigma}\mathbf{B}\boldsymbol{\Sigma} = \mathbf{0} \Leftrightarrow \mathbf{A}\boldsymbol{\Sigma}\mathbf{B} = \mathbf{0}$.

6. Let $\mathbf{A}$ be $p \times r$ matrix and $\mathbf{Y} \sim N_p(\mathbf{0}, \mathbf{\Sigma})$, where $\mathbf{\Sigma}$ may be singular. Show that $\mathbf{Y}'\mathbf{A}(\mathbf{A}'\mathbf{\Sigma}\mathbf{A})^-\mathbf{A}'\mathbf{Y}$ has $\chi^2(s)$ distribution where $s = R(\mathbf{A}'\mathbf{\Sigma}\mathbf{A})$.

7. Let $\mathbf{Y} \sim N(\mathbf{0}, \mathbf{\Sigma})$. If $\mathbf{Y}'\mathbf{A}\mathbf{Y} \sim \chi^2(r)$, where $r = R(\mathbf{A})$ then $\mathbf{Y}'\mathbf{A}\mathbf{Y} = \mathbf{Y}'\mathbf{A}(\mathbf{A}\mathbf{\Sigma}\mathbf{A})^-\mathbf{A}\mathbf{Y}$.

8. Let $\mathbf{Y} \sim N(\mathbf{0}, \mathbf{\Sigma})$. If $\mathbf{Y}'\mathbf{A}\mathbf{Y} \sim \chi^2$, then with probability 1

$$\mathbf{Y}'\mathbf{A}\mathbf{Y} = \mathbf{Y}'\mathbf{B}'(\mathbf{B}\mathbf{\Sigma}\mathbf{B}')^-\mathbf{B}\mathbf{Y}$$

for some matrix $\mathbf{B}$. Verify that $\mathbf{B} = \mathbf{A}\mathbf{\Sigma}\mathbf{A}\mathbf{\Sigma}\mathbf{A}$ is one such choice.

9. If $\mathbf{Y} \sim N(\boldsymbol{\mu}, \mathbf{\Sigma})$ and $\mathbf{\Sigma}\mathbf{A}\boldsymbol{\mu} = \mathbf{0}$, then $\mathbf{Y}'\mathbf{A}\mathbf{Y} - \boldsymbol{\mu}'\mathbf{A}\boldsymbol{\mu} \sim \chi^2$, iff $\mathbf{\Sigma}\mathbf{A}\mathbf{\Sigma}\mathbf{A}\mathbf{\Sigma} = \mathbf{\Sigma}\mathbf{A}\mathbf{\Sigma}$. (Styan, 1970).

10. Let

$$\begin{pmatrix} \mathbf{Y}_1 \\ \mathbf{Y}_2 \end{pmatrix} \sim N\left[\begin{pmatrix} \mathbf{0} \\ \mathbf{0} \end{pmatrix}, \quad \begin{pmatrix} \mathbf{\Sigma}_{11} & \mathbf{\Sigma}_{12} \\ \mathbf{\Sigma}_{21} & \mathbf{\Sigma}_{22} \end{pmatrix}\right]$$

$\mathbf{Y}_1'\mathbf{\Sigma}_{11}^-\mathbf{Y}_1$ and $\mathbf{Y}_2'\mathbf{\Sigma}_{22}^-\mathbf{Y}_2$ are independently distributed, iff $\mathbf{\Sigma}_{12} = (\mathbf{\Sigma}_{21})' = \mathbf{0}$.

11. Let $\mathbf{A} = \mathbf{A}_1 + \cdots + \mathbf{A}_n$ and $\mathbf{Y} \sim N_p(\boldsymbol{\mu}, \mathbf{\Sigma})$. If $\mathbf{Y}'\mathbf{A}\mathbf{Y}$ and $\mathbf{Y}'\mathbf{A}_i\mathbf{Y}$, $i = 1, \ldots, n - 1$ are distributed as χ^2 and $\mathbf{A}_n$ is a n.n.d. matrix, then $\mathbf{Y}'\mathbf{A}_n\mathbf{Y}$ has a χ^2 distribution.

12. Let $\mathbf{A} = \mathbf{A}_1 + \cdots + \mathbf{A}_n$ and $\mathbf{Y} \sim N_p(\boldsymbol{\mu}, \mathbf{\Sigma})$. If $\mathbf{Y}'\mathbf{A}\mathbf{Y}$ and $\mathbf{Y}'\mathbf{A}_i\mathbf{Y}$, $i = 1, \ldots, n$ have all χ^2 distributions, then $\mathbf{Y}'\mathbf{A}_i\mathbf{Y}$, $i = 1, \ldots, n$ are all independently distributed.

13. Let $\mathbf{C} = \mathbf{A} + \mathbf{B}$ and $\mathbf{Y} \sim N_p(\boldsymbol{\mu}, \mathbf{\Sigma})$. Further, let $\mathbf{Y}'\mathbf{C}\mathbf{Y}$ have a χ^2 distribution, and $\mathbf{Y}'\mathbf{A}\mathbf{Y}$ and $\mathbf{Y}'\mathbf{B}\mathbf{Y}$ are independent. Then $\mathbf{Y}'\mathbf{A}\mathbf{Y}$ and $\mathbf{Y}'\mathbf{B}\mathbf{Y}$ have χ^2 distributions, iff $\boldsymbol{\mu}'\mathbf{A}\boldsymbol{\mu} = \boldsymbol{\mu}'\mathbf{B}\boldsymbol{\mu} = \mathbf{0}$. (The last conditions are trivial if $\mathbf{A}$ and $\mathbf{B}$ are n.n.d. or $\boldsymbol{\mu} = \mathbf{0}$ or when $\mathbf{\Sigma}$ is nonsingular).

14. Let $\mathbf{A} = \mathbf{A}_1 + \cdots + \mathbf{A}_n$ and $\mathbf{Y} \sim N_p(\boldsymbol{\mu}, \mathbf{\Sigma})$. Further let $\mathbf{Y}'\mathbf{A}\mathbf{Y}$ have a χ^2 distribution, and $\mathbf{Y}'\mathbf{A}_i\mathbf{Y}$, $i = 1, \ldots, n$, are independent. Then $\mathbf{Y}'\mathbf{A}_i\mathbf{Y}$ have χ^2 distributions iff $\boldsymbol{\mu}'\mathbf{A}_i\boldsymbol{\mu} = 0$, $i = 1, \ldots, n$.

CHAPTER 10

Miscellaneous Applications of g-Inverses

10.1 APPLICATIONS IN NETWORK THEORY

We have already considered some applications in network theory based on constrained inverse of a matrix (Section 4.11). In this section, we demonstrate the use of g-inverse in the analysis of active networks.

The indefinite admittance matrix connecting the node currents and voltages in an n-terminal network plays an important role in network analysis. Its singularity, however, poses difficult problems. Since singular matrices do not admit a regular inverse, special techniques had to be devised by network analysts to handle such matrices. It is our objective to show how the g-inverse can be brought into service to obtain all the results in an elegant way. The emphasis will be on the development of a suitable calculus, which we hope will be of use in general network theory, rather than on detailed examination of particular problems.

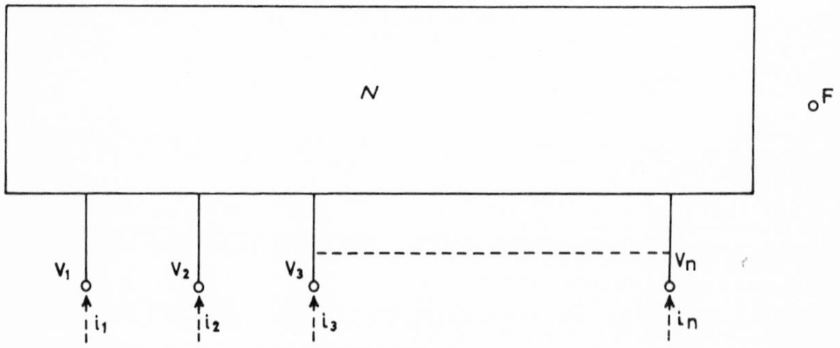

Figure 1.

10.1.1 The Indefinite Admittance Matrix and its Inverse

Consider a n-terminal network as shown in Fig. 1. Let currents $i_1, \ldots, i_n$ enter the terminals $1, \ldots, n$ from outside and let voltages $v_1, \ldots, v_n$ be measured between these terminals and an arbitrary reference terminal F.

Let $\mathbf{i} = (i_1, \ldots, i_n)'$ denote the current vector and $\mathbf{v} = (v_1, \ldots, v_n)'$ the voltage vector. Such a network is defined by a linear relationship.

$$\mathbf{i} = \mathbf{Yv}, \tag{10.1.1}$$

where the matrix $\mathbf{Y}$ is known as the indefinite admittance matrix.

By applying Kirchoff's current law and the relativity law of potentials one finds that the matrix $\mathbf{Y}$ is constrained by the relations $\mathbf{Ye} = \mathbf{0}$, $\mathbf{e}'\mathbf{Y} = \mathbf{0}'$, where $\mathbf{e} = (1, 1, \ldots, 1)'$, that is, the sum of the elements in each row and in each column of $\mathbf{Y}$ is zero. Such a matrix is said to be doubly centered. Thus $\mathbf{Y}$ is singular and the relationship between $\mathbf{i}$ and $\mathbf{v}$ induced by $\mathbf{Y}$ is not one to one, so that the inverse relationship (such as $\mathbf{v} = \mathbf{Y}^{-1}\mathbf{i}$, when $\mathbf{Y}$ is nonsingular) cannot be uniquely deduced from (10.1.1) alone. From the equation $\mathbf{i} = \mathbf{Yv}$ we find that $\mathbf{v} = \mathbf{Y}^{-}\mathbf{i}$ provides an inverse relationship for some g-inverse $\mathbf{Y}^{-}$. Can a choice of $\mathbf{Y}^{-}$ be made on some physical considerations?

By using Kirchoff's voltage law and relativity law of currents, one finds that, in a relationship such as, $\mathbf{v} = \mathbf{Zi}$, the matrix $\mathbf{Z}$ is also doubly centered, that is, $\mathbf{Ze} = \mathbf{0}$, $\mathbf{e}'\mathbf{Z} = \mathbf{0}'$. Then the problem may be posed as that of finding a g-inverse $\mathbf{Z}$ of $\mathbf{Y}$, which is also doubly centered. Theorem 10.1.1 provides the answer.

We shall consider the case where the network is fully connected, that is, $R(\mathbf{Y}) = n - 1$, so that if $\mathbf{B}$ is matrix such that $\mathbf{YB} = \mathbf{0}$, then there exists a vector $\mathbf{t}$ such that $\mathbf{B} = \mathbf{et}'$.

THEOREM 10.1.1 *The unique doubly centered inverse of* $\mathbf{Y}$ *is* $\mathbf{Y}^{+}$, *the Moore–Penrose inverse.*

Proof: Let $\mathbf{Z}$ be a g-inverse of $\mathbf{Y}$, that is,

$$\mathbf{YZY} = \mathbf{Y} \Leftrightarrow \mathbf{Y}(\mathbf{I} - \mathbf{ZY}) = \mathbf{0} \Leftrightarrow \mathbf{I} - \mathbf{ZY} = \mathbf{et}'. \tag{10.1.2}$$

If $\mathbf{Z}$ is doubly centered, then

$$\mathbf{I} - \mathbf{ZY} = \mathbf{et}' \Rightarrow \mathbf{e}'(\mathbf{I} - \mathbf{ZY}) = \mathbf{e}'\mathbf{et}' \quad \text{or} \quad \mathbf{e}' = \mathbf{e}'\mathbf{et}' = n\mathbf{t}'$$

that is, $\mathbf{e} = n\mathbf{t}$ and $\mathbf{ZY} = \mathbf{I} - n^{-1}\mathbf{ee}'$ or $\mathbf{ZY}$ is symmetrical. Similarly, $\mathbf{YZ}$ is symmetrical. Also from (10.1.2), $R(\mathbf{Z}) \geq R(\mathbf{Y})$, but $\mathbf{Ze}' = \mathbf{0}$ and, hence, $R(\mathbf{Z}) = n - 1 = R(\mathbf{Y})$. Then Lemma 2.5.1 shows that $\mathbf{ZYZ} = \mathbf{Z}$. Thus, if $\mathbf{Z}$ is doubly centered, then it satisfies all the four conditions as given in Note 4 (i) following Theorem 3.3.1, that is, $\mathbf{Z} = \mathbf{Y}^{+}$, which is unique. On the other hand if $\mathbf{Z} = \mathbf{Y}^+$, then $\mathcal{M}(\mathbf{Y}^+) = \mathcal{M}(\mathbf{Y}')$ which implies that $\mathbf{Y}^+\mathbf{e} = \mathbf{0}$. Similarly, $\mathbf{e}'\mathbf{Y}^+ = \mathbf{0}$ so that $\mathbf{Y}^+$ is doubly centered.

The inverse relationship $\mathbf{u} = \mathbf{Zi}$ to $\mathbf{i} = \mathbf{Yv}$, with the constraint that $\mathbf{Z}$ is doubly centered, is provided by $\mathbf{Z} = \mathbf{Y}^+$.

Theorem 10.1.2 relates to the computation of $\mathbf{Y}^+$ and some properties of doubly centered matrices (see Sharpe and Styan, 1965a, 1965b, 1967).

THEOREM 10.1.2 *Let* $\mathbf{Y}$ *and* $\mathbf{e}$ *be as defined above. Then the following hold:*

(a) *All cofactors of* $\mathbf{Y}$ *are equal.*
(b) *One choice of* $\mathbf{Y}^-$ *is* $(\mathbf{Y} + n^{-1}\mathbf{ee}')^{-1}$ *and*

$$\mathbf{Y}^+ = (\mathbf{Y} + n^{-1}\mathbf{ee}')^{-1} - n^{-1}\mathbf{ee}' = (\mathbf{Y} - n^{-1}\mathbf{ee}')^{-1} + n^{-1}\mathbf{ee}'.$$

(c) $\mathbf{YY}^+ = \mathbf{Y}^+\mathbf{Y} = \mathbf{I} - n^{-1}\mathbf{ee}'$.
(d) $\mathbf{Y}^+ = (\mathbf{I} - n^{-1}\mathbf{ee}')\mathbf{Y}^-(\mathbf{I} - n^{-1}\mathbf{ee}')$, *where* $\mathbf{Y}^-$ *is any g-inverse.*
(e) *If* $\mathbf{Y}_1, \mathbf{Y}_2$ *are two doubly centered matrices, each of rank* $n - 1$, *then* $(\mathbf{Y}_1\mathbf{Y}_2)^+ = \mathbf{Y}_2^+\mathbf{Y}_1^+$.

Proof of (a): Let (Y^{ij}) be the matrix of first order cofactors of $\mathbf{Y} = (Y_{ij})$. Then

$$(Y_{ij})(Y^{ij}) = \mathbf{0}.$$

But the only vector which multiplies $(\mathbf{Y}_{ij})$ to zero is $\mathbf{e}$. Thus each column of $(\mathbf{Y}^{ij})$ is a multiple of $\mathbf{e}$, and similarly each row of $(\mathbf{Y}^{ij})$ is a multiple of $\mathbf{e}'$. Then $(\mathbf{Y}^{ij}) = \lambda\mathbf{ee}'$ where λ is a scalar.

Proofs of (b), (c) and (d): The results may easily be verified from definitions by using the equations

$$(\mathbf{Y} + n^{-1}\mathbf{ee}')\mathbf{e} = \mathbf{e} \Rightarrow (\mathbf{Y} + n^{-1}\mathbf{ee}')^{-1}\mathbf{e} = \mathbf{e}$$

$$(\mathbf{Y} - n^{-1}\mathbf{ee}')\mathbf{e} = -\mathbf{e} \Rightarrow (\mathbf{Y} - n^{-1}\mathbf{ee}')^{-1}\mathbf{e} = -\mathbf{e}$$

$$(\mathbf{Y} \pm n^{-1}\mathbf{ee}')(\mathbf{I} - n^{-1}\mathbf{ee}') = \mathbf{Y} \Rightarrow (\mathbf{Y} \pm n^{-1}\mathbf{ee}')^{-1}\mathbf{Y} = (\mathbf{I} - n^{-1}\mathbf{ee}').$$

Proof of (e): We verify that all the four conditions of the Moore–Penrose inverse are satisfied by using the results of (c).

Theorem 10.1.2 is established.

We shall now prove a theorem which may be useful in situations where the constraints on $\mathbf{Y}$ are of the form

$$\mathbf{YF} = \mathbf{0}, \qquad \mathbf{F}'\mathbf{Y} = \mathbf{0}, \tag{10.1.3}$$

where $\mathbf{F}$ is a $n \times k$ matrix of rank k and $R(\mathbf{Y}) = n - k$.

THEOREM 10.1.3 *Let* $\mathbf{Y}$ *be a* $n \times n$ *matrix as defined in (10.1.3). Then* $\mathbf{Y}^+$ *is the unique inverse of* $\mathbf{Y}$ *with the property*

$$\mathbf{Y}^+\mathbf{F} = \mathbf{0}, \qquad \mathbf{F}'\mathbf{Y}^+ = \mathbf{0}. \tag{10.1.4}$$

and

$$\mathbf{Y}^{+} = (\mathbf{Y} + \mathbf{F}(\mathbf{F}'\mathbf{F})^{-1}\mathbf{F}')^{-1} - \mathbf{F}(\mathbf{F}'\mathbf{F})^{-1}\mathbf{F}'.$$

Proof: The theorem is proved along the same lines as Theorems 10.1.1 and 10.1.2.

10.1.2 Reduction of a Multipole

Suppression is an operation by which some terminals are made inaccessible. Thus the currents associated with the suppressed terminals are zero. Let $n - k$ terminals be inaccessible. Then the admittance equations for an n-pole can be written as

$$\begin{aligned}\mathbf{i}_1 &= \mathbf{Y}_{11}\mathbf{v}_1 + \mathbf{Y}_{12}\mathbf{v}_2 \\ \mathbf{i}_2 &= \mathbf{Y}_{21}\mathbf{v}_1 + \mathbf{Y}_{22}\mathbf{v}_2,\end{aligned} \tag{10.1.5}$$

where $\mathbf{i}_1$ is a k-vector, representing currents associated with k accessible terminals, and $\mathbf{i}_2$ is a $(n - k)$-vector, corresponding to the suppressed terminals. The objective is to derive the indefinite admittance matrix for the accessible k-terminal network only.

Setting $\mathbf{i}_2 = \mathbf{0}$ in (10.1.5),

$$\mathbf{i}_1 = \mathbf{Y}_{11}\mathbf{v}_1 + \mathbf{Y}_{12}\mathbf{v}_2 \tag{10.1.6}$$

$$\mathbf{0} = \mathbf{Y}_{21}\mathbf{v}_1 + \mathbf{Y}_{22}\mathbf{v}_2 \tag{10.1.7}$$

Case 1. $\mathbf{Y}_{22}$ nonsingular. In such a case from (10.1.7), $\mathbf{v}_2 = -\mathbf{Y}_{22}^{-1}\mathbf{Y}_{21}\mathbf{v}_1$ and substituting in (10.1.6)

$$\mathbf{i}_1 = (\mathbf{Y}_{11} - \mathbf{Y}_{12}\mathbf{Y}_{22}^{-1}\mathbf{Y}_{21})\mathbf{v}_1, \tag{10.1.8}$$

which is the relationship required.

Case 2. $\mathbf{Y}_{22}$ singular but $\mathcal{M}(\mathbf{Y}_{21}) = \mathcal{M}(\mathbf{Y}_{22})$ and $\mathcal{M}(\mathbf{Y}'_{12}) = \mathcal{M}(\mathbf{Y}'_{22})$. In such a case, from (10.1.7),

$$\mathbf{v}_2 = -\mathbf{Y}_{22}^{-}\mathbf{Y}_{21}\mathbf{v}_1 + (\mathbf{I} - \mathbf{Y}_{22}^{-}\mathbf{Y}_{22})\mathbf{z},$$

where $\mathbf{z}$ is arbitrary. Substituting in (10.1.6)

$$\mathbf{i}_1 = (\mathbf{Y}_{11} - \mathbf{Y}_{12}\mathbf{Y}_{22}^{-}\mathbf{Y}_{21})\mathbf{v}_1 + \mathbf{Y}_{12}(\mathbf{I} - \mathbf{Y}_{22}^{-}\mathbf{Y}_{22})\mathbf{z} \tag{10.1.9}$$

$$\Rightarrow \mathbf{i}_1 = (\mathbf{Y}_{11} - \mathbf{Y}_{12}\mathbf{Y}_{22}^{-}\mathbf{Y}_{21})\mathbf{v}_1, \tag{10.1.10}$$

as the last term in (10.1.9) vanishes due to the condition $\mathcal{M}(\mathbf{Y}'_{12}) = \mathcal{M}(\mathbf{Y}'_{22})$. The relationship (10.1.10) is unique for any choice of $\mathbf{Y}_{22}^{-}$ and is the desired one.

In other cases a relationship of the type $\mathbf{i}_1 = \mathbf{Y}\mathbf{v}_1$ may not exist.

10.1.3 Computation of Impedance Coefficients

Starting from the indefinite admittance description $\mathbf{i} = \mathbf{Y}\mathbf{v}$, where $\mathbf{Y}$ is a doubly centered $n \times n$ matrix of rank $n - 1$, we show how a g-inverse of $\mathbf{Y}$ is useful in the computation of impedance coefficients. In the literature on network theory (Sharpe and Spain, 1960, Mitra, 1969), these coefficients are expressed in terms of second order cofactors of $\mathbf{Y}$, which are difficult to compute. We obtain them as simple and elegant expressions in terms of the elements of $\mathbf{Y}^-$, any g-inverse of $\mathbf{Y}$. One choice of $\mathbf{Y}^-$ is $(\mathbf{Y} + n^{-1}\mathbf{e}\mathbf{e}')^{-1}$, noting that $(\mathbf{Y} + n^{-1}\mathbf{e}\mathbf{e}')$ is nonsingular. Thus the computation of $\mathbf{Y}^-$ itself involves only the inversion of a nonsingular matrix.

We represent the voltage difference across the terminals p, q by $v_{pq} = v_p - v_q$ when the current passing from terminal m to n is $i_{mn} = i_m = -i_n$, and the currents at other terminals are zero. The following coefficients are of interest.

Transfer impedance is the ratio $v_{pq}/i_{mn} = z_{mn}^{pq}$.

Driving point impedance is the ratio $v_{mn}/i_{mn} = z_{mn}^{mn}$.

Transfer voltage gain is the ratio $z_{mn}^{pq}/z_{mn}^{rs} = G_{rp;sq}^{mn}$.

Driving point voltage gain is the ratio $z_{mn}^{pq}/z_{mn}^{mn} = G_{mp;nq}$.

The expressions for these coefficients are given in Theorem 10.1.4.

THEOREM 10.1.4 *Let w_{ij} be the (i, j)-th element of $\mathbf{Y}^-$, a g-inverse of $\mathbf{Y}$. Then*

$$z_{mn}^{pq} = \frac{v_{pq}}{i_{mn}} = w_{pm} - w_{pn} - w_{qm} + w_{qn} \tag{10.1.11}$$

$$z_{mn}^{mn} = \frac{v_{mn}}{i_{mn}} = w_{mm} - w_{mn} - w_{nm} + w_{nn} \tag{10.1.12}$$

[Note: One choice of $\mathbf{Y}^-$ is $(\mathbf{Y} + n^{-1}\mathbf{e}\mathbf{e}')^{-1}$. Once this inverse is computed, all the impedance coefficients are obtained as 2×2 tetrad differences of suitable elements of $(\mathbf{Y} + n^{-1}\mathbf{e}\mathbf{e}')^{-1}$].

Proof: A general solution of $\mathbf{u} = \mathbf{Y}\mathbf{v}$ is

$$\mathbf{v} = \mathbf{Y}^-\mathbf{i} + c\mathbf{e}, \tag{10.1.13}$$

where c is an arbitrary scalar and $\mathbf{e} = (1, 1, \ldots, 1)'$. From (10.1.13)

$$v_{pq} = v_p - v_q = \sum_r (w_{pr} - w_{qr})i_r,$$

which is independent of the arbitrary scalar c. Setting $i_{mn} = i_m = -i_n$ and the rest of i_r to zero.

$$v_{pq} = (w_{pm} - w_{pn} - w_{qm} + w_{qn})i_{mn},$$

which establishes (10.1.11). The formula (10.1.12) follows by putting $p = m$, $q = n$ in (10.1.11).

10.1.4 Admittance—Impedance Conversion

We shall now consider the relationship between the indefinite admittance matrix $\mathbf{Y}$ connecting $\mathbf{i}$ and $\mathbf{v}$ (the node currents and voltages) and the impedance matrix $\mathbf{Z}$ connecting $\mathbf{u}$ and $\mathbf{j}$ (the mesh voltages and currents) in a network as in Figure 2. The relationships may be written

$$\mathbf{i} = \mathbf{Y}\mathbf{v} \quad \mathbf{u} = \mathbf{Z}\mathbf{j}, \tag{10.1.14}$$

where $\mathbf{Y}$ is assumed to be doubly centered.

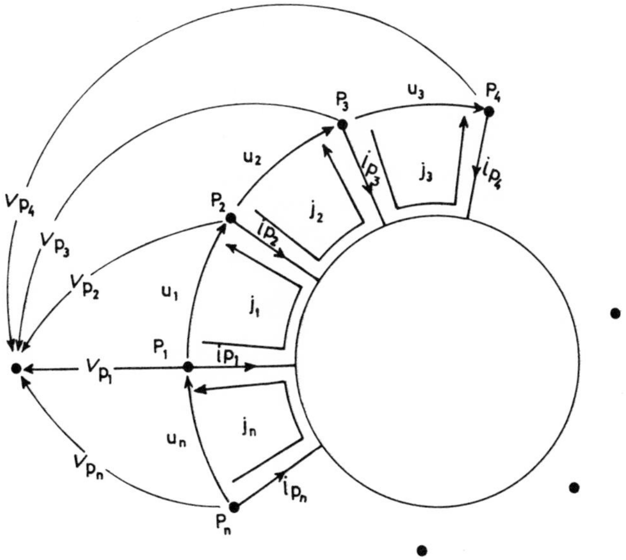

Figure 2.

From Figure 2 we have two further relationships

$$\mathbf{u} = \mathbf{T}\mathbf{v}, \qquad \mathbf{i} = \mathbf{T}'\mathbf{j} \tag{10.1.15}$$

where

$$\mathbf{T} = \begin{pmatrix} \mathbf{e}'_{p_1} - \mathbf{e}'_{p_2} \\ \mathbf{e}'_{p_2} - \mathbf{e}'_{p_3} \\ \cdot \ \cdot \cdot \cdot \ \cdot \\ \mathbf{e}'_{p_n} - \mathbf{e}'_{p_1} \end{pmatrix}.$$

In (10.1.15), $\mathbf{e}'_k$ is a row vector of order n with zero everywhere except for the k-th element, which is unity. Thus the s-th row of $\mathbf{T}$ has $+1$ in its p_s-th column and -1 in its p_{s+1}-th column. Furthermore the p_t-th column of $\mathbf{T}$ has $+1$ in its t-th row and -1 in its $(t-1)$th row.

THEOREM 10.1.5 *Let* $\mathbf{Y}$ *be doubly centered and of rank* $n-1$. *If the equations* $\mathbf{i} = \mathbf{Yv}$, $\mathbf{u} = \mathbf{Zj}$ *and* $\mathbf{u} = \mathbf{Tv}$, $\mathbf{i} = \mathbf{T'j}$ *are consistent in* $\mathbf{i}, \mathbf{v}$ *and* $\mathbf{u}$ *for any given* $\mathbf{j}$, *then* $\mathbf{Z}$ *is doubly centered and*

$$\mathbf{Z} = \mathbf{TY^-T'}, \tag{10.1.16}$$

which is unique for any choice of the g-inverse $\mathbf{Y}^-$.

Proof: Eliminating $\mathbf{i}$ and $\mathbf{u}$ from the given equations we have

$$\mathbf{Yv} = \mathbf{T'j}, \qquad \mathbf{Tv} = \mathbf{Zj}. \tag{10.1.17}$$

If the equations (10.1.17) are consistent for any $\mathbf{j}$, it follows that the columns of the matrix $\begin{pmatrix}\mathbf{T}'\\ \mathbf{Z}\end{pmatrix}$ depend on those of $\begin{pmatrix}\mathbf{Y}\\ \mathbf{T}\end{pmatrix}$. But by definition, the columns and rows of $\mathbf{T}$ depend on those of $\mathbf{Y}$ and hence by applying the following Lemma 10.1.1, $\mathbf{Z} = \mathbf{TY^-T'}$ and is unique.

LEMMA 10.1.1 *Consider the partitioned matrix*

$$\mathbf{V} = \begin{pmatrix}\mathbf{A} & \mathbf{B}\\ \mathbf{C} & \mathbf{D}\end{pmatrix},$$

where $R(\mathbf{V}) = R(\mathbf{A})$. *Then* $\mathbf{D} = \mathbf{CA^-B}$ *for any choice of g-inverse of* $\mathbf{A}$.

Proof: Since $R(\mathbf{V}) = R(\mathbf{A})$, $\mathbf{B} = \mathbf{AE}$ and $\mathbf{C} = \mathbf{FA}$ for some matrices $\mathbf{E}$ and $\mathbf{F}$.

Let $\boldsymbol{\lambda}' = -\boldsymbol{\mu}'\mathbf{CA}^-$ in which case $\boldsymbol{\lambda}'\mathbf{A} + \boldsymbol{\mu}'\mathbf{C} = \mathbf{0}'$. But $\boldsymbol{\lambda}'\mathbf{A} + \boldsymbol{\mu}'\mathbf{C} = \mathbf{0}' \Rightarrow \boldsymbol{\lambda}'\mathbf{B} + \boldsymbol{\mu}'\mathbf{D} = \mathbf{0}'$, that is, $\boldsymbol{\mu}'(\mathbf{CA^-B} - \mathbf{D}) = \mathbf{0}'$ for all $\boldsymbol{\mu}$, which implies that $\mathbf{D} = \mathbf{CA^-B}$, for any choice of $\mathbf{A}^-$.

10.1.5 Parallel Sum of Matrices (1)

In this section and the next we consider applications of g-inverse in network synthesis. The discussion is based on two extensions of the star product $y_{1*}y_2$ of two (scalar) admittances y_1 and y_2 defined by (Seshu and Reed, 1961),

$$y_1 * y_2 = \frac{y_1 y_2}{y_1 + y_2} = y_1(y_1 + y_2)^{-1}y_2 = \left(\frac{1}{y_1} + \frac{1}{y_2}\right)^{-1}, \tag{10.1.18}$$

providing the admittance of a series combination of y_1 and y_2.

One type of extension of the formula (10.1.18) to two admittance matrices $\mathbf{Y}_1, \mathbf{Y}_2$ of n-terminal networks in series is

$$(\mathbf{Y}_1^+ + \mathbf{Y}_2^+)^+, \tag{10.1.19}$$

where $+$ indicates the unique Moore–Penrose inverse. The expression (10.1.19) may be called the parallel sum of matrices.

In theory we could consider any two matrices of the same order and define parallel sum and difference and develop a calculus of handling them.

Definition: Let $\mathbf{A}$ and $\mathbf{B}$ be two matrices of the same order $m \times n$. We define their parallel sum $\mathbf{A} \mp \mathbf{B}$ and difference $\mathbf{A} \overline{\overline{-}} \mathbf{B}$ as follows

$$\mathbf{A} \mp \mathbf{B} = (\mathbf{A}^+ + \mathbf{B}^+)^+, \mathbf{A} \overline{\overline{-}} \mathbf{B} = (\mathbf{A}^+ - \mathbf{B}^+)^+. \tag{10.1.20}$$

The following results follow from definitions

(i) $\mathbf{A} \mp \mathbf{B} = \mathbf{B} \mp \mathbf{A}$
(ii) $(\mathbf{A} \mp \mathbf{B}) \mp \mathbf{C} = \mathbf{A} \mp (\mathbf{B} \mp \mathbf{C}) = \mathbf{A} \mp \mathbf{B} \mp \mathbf{C}$
(iii) $(\mathbf{A} \mp \mathbf{B} \mp \mathbf{C} \mp \cdots) = (\mathbf{A}^+ + \mathbf{B}^+ + \mathbf{C}^+ + \cdots)^+$
(iv) $(\mathbf{A} \mp \mathbf{B} \overline{\overline{-}} \mathbf{C} \ldots) = (\mathbf{A}^+ + \mathbf{B}^+ - \mathbf{C}^+ \ldots)^+$
(v) $(\mathbf{A} \mp \mathbf{B}) \overline{\overline{-}} \mathbf{B} = \mathbf{A}$
(vi) $(\mathbf{A}^+ \mp \mathbf{B}^+) = (\mathbf{A} + \mathbf{B})^+$
(vii) $(\mathbf{A}^+ \overline{\overline{-}} \mathbf{B}^+) = (\mathbf{A} - \mathbf{B})^+$
(viii) $\mathbf{A} \mp \mathbf{B} = \mathbf{C} \Leftrightarrow \mathbf{A} = \mathbf{C} \overline{\overline{-}} \mathbf{B}$.

Using the results (i)—(vii) above, we can solve equations involving parallel sum and difference operations. For instance, consider the equation

$$\mathbf{A} = \mathbf{B} + [\mathbf{C} \mp \{\mathbf{D} + (\mathbf{E} \mp \mathbf{F})\}]. \tag{10.1.21}$$

Carrying out the operations indicated we find

$$\mathbf{E} = \{(\mathbf{A} - \mathbf{B}) \overline{\overline{-}} \mathbf{C} - \mathbf{D}\} \overline{\overline{-}} \mathbf{F} \tag{10.1.22}$$

and so on.

We now prove some theorems concerning parallel sum and difference of indefinite admittance matrices.

THEOREM 10.1.6 *Let $\mathbf{Y}_1$ and $\mathbf{Y}_2$ be two doubly centered matrices such that $R(\mathbf{Y}_1) = R(\mathbf{Y}_2) = R(\mathbf{Y}_1 + \mathbf{Y}_2) = n - 1$. Then*

$$(\mathbf{Y}_1^+ + \mathbf{Y}_2^+)^+ = \mathbf{Y}_1(\mathbf{Y}_1 + \mathbf{Y}_2)^+\mathbf{Y}_2 = \mathbf{Y}_1(\mathbf{Y}_1 + \mathbf{Y}_2)^-\mathbf{Y}_2 \tag{10.1.23}$$

for any g-inverse $\mathbf{Y}^-$ of $\mathbf{Y}$.

Proof: Using the result (b) of Theorem 10.1.2.

$$\begin{aligned}\mathbf{Y}_1^+ + \mathbf{Y}_2^+ &= (\mathbf{Y}_1 + n^{-1}\mathbf{ee}')^{-1} + (\mathbf{Y}_2 - n^{-1}\mathbf{ee}')^{-1} \\ &= (\mathbf{Y}_1 + n^{-1}\mathbf{ee}')^{-1}(\mathbf{Y}_1 + \mathbf{Y}_2)(\mathbf{Y}_2 - n^{-1}\mathbf{ee}')^{-1}\end{aligned} \tag{10.1.24}$$

$$(\mathbf{Y}_1^+ + \mathbf{Y}_2^+)^+ = (\mathbf{Y}_2 - n^{-1}\mathbf{ee}')(\mathbf{Y}_1 + \mathbf{Y}_2)^+(\mathbf{Y}_1 + n^{-1}\mathbf{ee}'), \tag{10.1.25}$$

where it may be verified, in view of $R(\mathbf{Y}_1 + \mathbf{Y}_2) = n - 1$, that (10.1.25) is the Moore–Penrose inverse of (10.1.24). But $(\mathbf{Y}_1 + \mathbf{Y}_2)^+$ is also doubly centered. Therefore (10.1.25) reduces to $\mathbf{Y}_2(\mathbf{Y}_1 + \mathbf{Y}_2)^+\mathbf{Y}_1$.

If $(\mathbf{Y}_1 + \mathbf{Y}_2)^-$ is any g-inverse of $\mathbf{Y}_1 + \mathbf{Y}_2$, then

$$\begin{aligned}(\mathbf{Y}_1 + \mathbf{Y}_2)^- &= (\mathbf{Y}_1 + \mathbf{Y}_2)^+ \\ &\quad + \mathbf{U} - (\mathbf{Y}_1 + \mathbf{Y}_2)^+(\mathbf{Y}_1 + \mathbf{Y}_2)\mathbf{U}(\mathbf{Y}_1 + \mathbf{Y}_2)(\mathbf{Y}_1 + \mathbf{Y}_2)^+\end{aligned}$$

for some $\mathbf{U}$. Multiplying by $\mathbf{Y}_1$ and $\mathbf{Y}_2$ and observing that

$$(\mathbf{Y}_1 + \mathbf{Y}_2)^+(\mathbf{Y}_1 + \mathbf{Y}_2) = \mathbf{I} - n^{-1}\mathbf{ee}' = (\mathbf{Y}_1 + \mathbf{Y}_2)(\mathbf{Y}_1 + \mathbf{Y}_2)^+,$$

$$\mathbf{Ye} = \mathbf{0},\ \mathbf{e}'\mathbf{Y} = \mathbf{0},$$

we have

$$\mathbf{Y}_1(\mathbf{Y}_1 + \mathbf{Y}_2)^-\mathbf{Y}_2 = \mathbf{Y}_1(\mathbf{Y}_1 + \mathbf{Y}_2)^+\mathbf{Y}_2. \tag{10.1.26}$$

and the theorem is proved.

THEOREM 10.1.7 *Let $\mathbf{Y}_1$ and $\mathbf{Y}_2$ be doubly centered matrices such that $R(\mathbf{Y}_1) = R(\mathbf{Y}_2) = R(\mathbf{Y}_1 - \mathbf{Y}_2) = n - 1$. Then*

$$(\mathbf{Y}_1^+ - \mathbf{Y}_2^+)^+ = \mathbf{Y}_2(\mathbf{Y}_2 - \mathbf{Y}_1)^+\mathbf{Y}_1 = \mathbf{Y}_2(\mathbf{Y}_2 - \mathbf{Y}_1)^-\mathbf{Y}_1. \tag{10.1.27}$$

Proof: The result (10.1.27) is established on the same lines as in Theorem 10.1.6.

10.1.6 Parallel Sum of Matrices (2)

In this section we study a second type of extension of the star product (10.1.18) considered by Anderson and Duffin (1969). They define parallel sum of matrices, for a pair of matrices of the same order and show that the parallel sum has many interesting properties, if the matrices concerned are both n.n.d. We show here that these properties are in fact true for a much wider class of matrices. They hold for certain pairs of square matrices not necessarily n.n.d. and may even hold for a pair of rectangular matrices.

Definition: For a pair of matrices $\mathbf{A}$ and $\mathbf{B}$ of the same order, the parallel sum of $\mathbf{A}$ and $\mathbf{B}$, denoted by the symbol $\mathbf{A} \mp \mathbf{B}$, is defined by

$$\mathbf{A} \mp \mathbf{B} = \mathbf{A}(\mathbf{A} + \mathbf{B})^-\mathbf{B}. \tag{10.1.28}$$

It is clear that for this definition to be meaningful, the expression $\mathbf{A}(\mathbf{A} + \mathbf{B})^-\mathbf{B}$ must be independent of the choice of the g-inverse $(\mathbf{A} + \mathbf{B})^-$. Accordingly, a pair of matrices $\mathbf{A}$ and $\mathbf{B}$ will be said to be *parallel summable* (or briefly p.s.) if $\mathbf{A}(\mathbf{A} + \mathbf{B})^-\mathbf{B}$ is invariant under the choice of the g-inverse

$(\mathbf{A}+\mathbf{B})^-$, that is, if

$$\mathscr{M}(\mathbf{A}) \subset \mathscr{M}(\mathbf{A}+\mathbf{B}), \qquad \mathscr{M}(\mathbf{A}^*) \subset \mathscr{M}(\mathbf{A}^*+\mathbf{B}^*), \tag{10.1.29}$$

or equivalently

$$\mathscr{M}(\mathbf{B}) \subset \mathscr{M}(\mathbf{A}+\mathbf{B}), \qquad \mathscr{M}(\mathbf{B}^*) \subset \mathscr{M}(\mathbf{A}^*+\mathbf{B}^*). \tag{10.1.30}$$

Note that for a pair of n.n.d. matrices the conditions (10.1.29) and (10.1.30) are automatically satisfied and hence such matrices are necessarily parallel summable. We say the parallel sum of $\mathbf{A}$ and $\mathbf{B}$ is defined if $\mathbf{A}$ and $\mathbf{B}$ are parallel summable.

For parallel summable matrices the definition given here agrees with the definition given by Anderson and Duffin (1969). These authors use $(\mathbf{A}+\mathbf{B})^+$ in place of $(\mathbf{A}+\mathbf{B})^-$ in (10.1.28).

THEOREM 10.1.8 *For a pair of p.s. matrices* $\mathbf{A}$, $\mathbf{B}$ *of order* $m \times n$ *each*

(a) $\mathbf{A} \mp \mathbf{B} = \mathbf{B} \mp \mathbf{A}$ *(commutativity).*
(b) $\mathbf{A}^*, \mathbf{B}^*$ *are also p.s. and* $\mathbf{A}^* \mp \mathbf{B}^* = (\mathbf{A} \mp \mathbf{B})^*$.
(c) *If* $\mathbf{C}$ *is nonsingular, of order* $m \times m$, *then* $\mathbf{CA}$ *and* $\mathbf{CB}$ *are p.s. and* $\mathbf{CA} \mp \mathbf{CB} = \mathbf{C}(\mathbf{A} \mp \mathbf{B})$.
(d) $\mathbf{A}^- + \mathbf{B}^-$ is one choice of $(\mathbf{A} \mp \mathbf{B})^-$ *and conversely every g-inverse of* $(\mathbf{A} \mp \mathbf{B})^-$ *is expressible as* $\mathbf{A}^- + \mathbf{B}^-$ *for suitable choices of these g-inverses.*
(e) $\mathscr{M}(\mathbf{A} \mp \mathbf{B}) = \mathscr{M}(\mathbf{A}) \cap \mathscr{M}(\mathbf{B})$.
(f) *If* $\mathbf{P}_*$ *is the orthogonal projector onto* $\mathscr{M}(\mathbf{A}^*) \cap \mathscr{M}(\mathbf{B}^*)$ *and* $\mathbf{P}$ *is the orthogonal projector onto* $\mathscr{M}(\mathbf{A}) \cap \mathscr{M}(\mathbf{B})$, *then*

$$(\mathbf{A} \mp \mathbf{B})^+ = \mathbf{P}_*(\mathbf{A}^- + \mathbf{B}^-)\mathbf{P}.$$

(g) $(\mathbf{A} \mp \mathbf{B}) \mp \mathbf{C} = \mathbf{A} \mp (\mathbf{B} \mp \mathbf{C})$ *if all the parallel sum operations involved are defined.*
(h) *If* $\mathbf{P}$ *and* $\mathbf{Q}$ *are the orthogonal projectors onto* $\mathscr{M}(\mathbf{A})$ *and* $\mathscr{M}(\mathbf{B})$, *respectively, then the orthogonal projector onto* $\mathscr{M}(\mathbf{A}) \cap \mathscr{M}(\mathbf{B})$ *is* $2(\mathbf{P} \mp \mathbf{Q})$.

Proof of (a): Observe that in view of (10.1.29) and (10.1.30)

$$\begin{aligned}\mathbf{A}(\mathbf{A}+\mathbf{B})^-\mathbf{B} &= \mathbf{A}(\mathbf{A}+\mathbf{B})^-(\mathbf{A}+\mathbf{B}) - \mathbf{A}(\mathbf{A}+\mathbf{B})^-\mathbf{A} = \mathbf{A} - \mathbf{A}(\mathbf{A}+\mathbf{B})^-\mathbf{A}\\ &= (\mathbf{A}+\mathbf{B})(\mathbf{A}+\mathbf{B})^-\mathbf{A} - \mathbf{A}(\mathbf{A}+\mathbf{B})^-\mathbf{A} = \mathbf{B}(\mathbf{A}+\mathbf{B})^-\mathbf{A}.\end{aligned}$$

Proofs of (b) and (c): These are straightforward.

Proof of (d): $(\mathbf{A} \mp \mathbf{B})(\mathbf{A}^- + \mathbf{B}^-)(\mathbf{A} \mp \mathbf{B})$

$$= \mathbf{A}(\mathbf{A}+\mathbf{B})^-\mathbf{B}\mathbf{B}^-\mathbf{B}(\mathbf{A}+\mathbf{B})^-\mathbf{A} + \mathbf{B}(\mathbf{A}+\mathbf{B})^-\mathbf{A}\mathbf{A}^-\mathbf{A}(\mathbf{A}+\mathbf{B})^-\mathbf{B} \tag{10.1.31}$$

$$= \mathbf{A}(\mathbf{A}+\mathbf{B})^-\mathbf{B}(\mathbf{A}+\mathbf{B})^-(\mathbf{A}+\mathbf{B}) = \mathbf{A}(\mathbf{A}+\mathbf{B})^-\mathbf{B} = \mathbf{A} \mp \mathbf{B}.$$

To prove the converse, note that the most general form of a g-inverse of $\mathbf{C} = \mathbf{A} \mp \mathbf{B}$ is

$$\begin{aligned}\mathbf{G} &= \mathbf{A}^- + \mathbf{B}^- + (\mathbf{I} - \mathbf{C}^-\mathbf{C})\mathbf{V} + \mathbf{W}(\mathbf{I} - \mathbf{C}\mathbf{C}^-) \\ &= [\mathbf{A}^- + (\mathbf{I} - \mathbf{A}^-\mathbf{A})\mathbf{V}_1 + \mathbf{W}_1(\mathbf{I} - \mathbf{A}\mathbf{A}^-)] \\ &\quad + [\mathbf{B}^- + (\mathbf{I} - \mathbf{B}^-\mathbf{B})\mathbf{V}_2 + \mathbf{W}_2(\mathbf{I} - \mathbf{B}\mathbf{B}^-)]\end{aligned}$$

for suitable choices of $\mathbf{V}_1$, $\mathbf{V}_2$, $\mathbf{W}_1$, and $\mathbf{W}_2$, since $\mathscr{O}(\mathbf{C}) = \mathscr{O}(\mathbf{A}) + \mathscr{O}(\mathbf{B})$ and $\mathscr{O}(\mathbf{C}^*) = \mathscr{O}(\mathbf{A}^*) + \mathscr{O}(\mathbf{B}^*)$, in view of result (e).

Proof of (e): Clearly, $\mathscr{M}(\mathbf{A} \mp \mathbf{B}) \subset \mathscr{M}(\mathbf{A}) \cap \mathscr{M}(\mathbf{B})$. Conversely, let

$$\mathbf{x} \in \mathscr{M}(\mathbf{A}) \cap \mathscr{M}(\mathbf{B}),$$

then

$$\begin{aligned}(\mathbf{A} \mp \mathbf{B})(\mathbf{A}^- + \mathbf{B}^-)\mathbf{x} &= \mathbf{A}(\mathbf{A} + \mathbf{B})^-\mathbf{B}\mathbf{B}^-\mathbf{x} + \mathbf{B}(\mathbf{A} + \mathbf{B})^-\mathbf{A}\mathbf{A}^-\mathbf{x} \\ &= \mathbf{A}(\mathbf{A} + \mathbf{B})^-\mathbf{x} + \mathbf{B}(\mathbf{A} + \mathbf{B})^-\mathbf{x} = (\mathbf{A} + \mathbf{B})(\mathbf{A} + \mathbf{B})^-\mathbf{x} = \mathbf{x},\end{aligned}$$

since $\mathbf{x} \in \mathscr{M}(\mathbf{A}) \cap \mathscr{M}(\mathbf{B}) \subset \mathscr{M}(\mathbf{A} + \mathbf{B})$. This shows $\mathbf{x} \in \mathscr{M}(\mathbf{A} \mp \mathbf{B})$. Hence, $\mathscr{M}(\mathbf{A}) \cap \mathscr{M}(\mathbf{B}) \subset \mathscr{M}(\mathbf{A} \mp \mathbf{B})$ and (e) is established.

Proof of (f): (f) follows from the explicit forms of the minimum norm least-squares g-inverse, given in Section 3.3.

Proof of (g):

$$\begin{aligned}\mathscr{M}[(\mathbf{A} \mp \mathbf{B}) \mp \mathbf{C}] &= \mathscr{M}[\mathbf{A} \mp (\mathbf{B} \mp \mathbf{C})] \\ &= \mathscr{M}(\mathbf{A}) \cap \mathscr{M}(\mathbf{B}) \cap \mathscr{M}(\mathbf{C}).\end{aligned}$$

Also observe that $\mathbf{A}^- + \mathbf{B}^- + \mathbf{C}^-$ is a g-inverse of both $(\mathbf{A} \mp \mathbf{B}) \mp \mathbf{C}$ and $\mathbf{A} \mp (\mathbf{B} \mp \mathbf{C})$. Let $\mathbf{P}_*$ be the orthogonal projector onto $\mathscr{M}(\mathbf{A}^*) \cap \mathscr{M}(\mathbf{B}^*) \cap \mathscr{M}(\mathbf{C}^*)$ and $\mathbf{P}$ be orthogonal projector onto $\mathscr{M}(\mathbf{A}) \cap \mathscr{M}(\mathbf{B}) \cap \mathscr{M}(\mathbf{C})$. Observe as in the proof of (f) that

$$\begin{aligned}\mathbf{P}_*(\mathbf{A}^- + \mathbf{B}^- + \mathbf{C}^-)\mathbf{P} &= [(\mathbf{A} \mp \mathbf{B}) \mp \mathbf{C}]^+ \\ &= [\mathbf{A} \mp (\mathbf{B} \mp \mathbf{C})]^+ \\ \Rightarrow [\mathbf{P}_*(\mathbf{A}^- + \mathbf{B}^- + \mathbf{C}^-)\mathbf{P}]^+ &= (\mathbf{A} \mp \mathbf{B}) \mp \mathbf{C} = \mathbf{A} \mp (\mathbf{B} \mp \mathbf{C}).\end{aligned}$$

Proof of (h): Since $\mathscr{M}(\mathbf{P}) = \mathscr{M}(\mathbf{A})$, $\mathscr{M}(\mathbf{Q}) = \mathscr{M}(\mathbf{B})$, clearly $\mathscr{M}(\mathbf{P} \mp \mathbf{Q}) = \mathscr{M}(\mathbf{P}) \cap \mathscr{M}(\mathbf{Q}) = \mathscr{M}(\mathbf{A}) \cap \mathscr{M}(\mathbf{B})$. Also $\mathbf{P} = \mathbf{P}^*$, $\mathbf{Q} = \mathbf{Q}^* \Rightarrow \mathbf{P} \mp \mathbf{Q} = \mathbf{P}^* \mp \mathbf{Q}^* = (\mathbf{P} \mp \mathbf{Q})^*$ by (b). Hence to prove (h) it suffices to show that $2(\mathbf{P} \mp \mathbf{Q})$

is idempotent for which note

$$\begin{aligned}2(\mathbf{P} \mp \mathbf{Q})^2 &= \mathbf{P}(\mathbf{P}+\mathbf{Q})^-\mathbf{Q}^2(\mathbf{P}+\mathbf{Q})^-\mathbf{P} + \mathbf{Q}(\mathbf{P}+\mathbf{Q})^-\mathbf{P}^2(\mathbf{P}+\mathbf{Q})^-\mathbf{Q}\\ &= \mathbf{P}(\mathbf{P}+\mathbf{Q})^-\mathbf{Q}(\mathbf{P}+\mathbf{Q})^-(\mathbf{P}+\mathbf{Q}) = \mathbf{P}(\mathbf{P}+\mathbf{Q})^-\mathbf{Q}\\ &= \mathbf{P} \mp \mathbf{Q}.\end{aligned}$$

COROLLARY: *Let* **A** *and* **B** *be p.s. matrices. Then,* $\mathbf{A} \mp \mathbf{B}$ *is hermitian or n.n.d. according as* **A** *and* **B** *are hermitian or n.n.d.*

THEOREM 10.1.9 *If* **A**, **B** *are p.s.,* $(\mathbf{A} + \mathbf{B})$ *and* **C** *are p.s. and* $(\mathbf{A}+\mathbf{B}) \mp \mathbf{C} = \mathbf{0}$, *then*

$$(\mathbf{A}+\mathbf{C}) \mp \mathbf{B} = \mathbf{A} \mp (\mathbf{B}+\mathbf{C}) = \mathbf{A} \mp \mathbf{B}.$$

Proof: Since $\mathscr{M}(\mathbf{A}^* + \mathbf{B}^*) \subset \mathscr{M}(\mathbf{A}^* + \mathbf{B}^* + \mathbf{C}^*)$

$$(\mathbf{A}+\mathbf{B})(\mathbf{A}+\mathbf{B}+\mathbf{C})^-(\mathbf{A}+\mathbf{B}+\mathbf{C}) = (\mathbf{A}+\mathbf{B})$$

$$\Rightarrow (\mathbf{A}+\mathbf{B})(\mathbf{A}+\mathbf{B}+\mathbf{C})^-(\mathbf{A}+\mathbf{B}) + (\mathbf{A}+\mathbf{B})(\mathbf{A}+\mathbf{B}+\mathbf{C})^-\mathbf{C} = (\mathbf{A}+\mathbf{B})$$

$$\Rightarrow (\mathbf{A}+\mathbf{B}+\mathbf{C})^- = (\mathbf{A}+\mathbf{B})^- \quad \text{since } (\mathbf{A}+\mathbf{B}) \mp \mathbf{C} = \mathbf{0}.$$

Therefore, $\mathbf{A} \mp \mathbf{B}$ is equal to

$$\begin{aligned}\mathbf{A}(\mathbf{A}+\mathbf{B}+\mathbf{C})^-\mathbf{B} &= \mathbf{A}(\mathbf{A}+\mathbf{B}+\mathbf{C})^-(\mathbf{B}+\mathbf{C}) - \mathbf{A}(\mathbf{A}+\mathbf{B}+\mathbf{C})^-\mathbf{C}\\ &= \mathbf{A} \mp (\mathbf{B}+\mathbf{C}).\end{aligned}$$

since $\mathscr{M}(\mathbf{A}^*) \subset \mathscr{M}(\mathbf{A}^* + \mathbf{B}^*)$ and $(\mathbf{A}+\mathbf{B}) \mp \mathbf{C} = \mathbf{0} \Rightarrow \mathbf{A}(\mathbf{A}+\mathbf{B}+\mathbf{C})^-\mathbf{C} = \mathbf{0}$. Similarly $\mathbf{A} \mp \mathbf{B} = (\mathbf{A}+\mathbf{C}) \mp \mathbf{B}$.

THEOREM 10.1.10 *Let* **A**, **B** *be p.s. matrices of order* $n \times n$ *and* **x** *be an eigenvector of both* **A** *and* **B** *corresponding to nonnull eigenvalues* a *and* b, *respectively. If* $a + b \neq 0$, **x** *is also an eigenvector of* $\mathbf{A} \mp \mathbf{B}$, *corresponding to its nonnull eigenvalue* $ab/(a+b)$.

Proof: Observe that **x** is an eigenvector of $(\mathbf{A} + \mathbf{B})$ corresponding to its nonnull eigenvalue $a + b$. Let $(\mathbf{A}+\mathbf{B})^-$ be a g-inverse with the eigenvalue property, such as the Scoggs Odell pseudoinverse. Then

$$\begin{aligned}(\mathbf{A} \mp \mathbf{B})\mathbf{x} &= \mathbf{A}(\mathbf{A}+\mathbf{B})^-\mathbf{B}\mathbf{x} = b\mathbf{A}(\mathbf{A}+\mathbf{B})^-\mathbf{x}\\ &= [b/(a+b)]\mathbf{A}\mathbf{x} = [ab/(a+b)]\mathbf{x}.\end{aligned}$$

THEOREM 10.1.11 *Let* **A**, **B**, **C** *be matrices with the following properties*

(i) **C** *commutes with both* **A** *and* **B**,
(ii) $\mathbf{A} + \mathbf{B}$ *and* **C** *are each of index 1, and*
(iii) **A** *and* **B** *are p.s. matrices.*

Then, **AC** *and* **BC** *are also p.s. and*

$$\mathbf{AC} \mp \mathbf{BC} = (\mathbf{A} \mp \mathbf{B})\mathbf{C}.$$

Proof: The first part is easy. For the second part note that $(\mathbf{A} + \mathbf{B})_{\rho\chi}^{-}$ and $\mathbf{C}_{\rho\chi}^{-}$ exist and are polynomials in $\mathbf{A} + \mathbf{B}$ and $\mathbf{C}$, respectively. Hence, in view of (i) one choice of $(\mathbf{AC} + \mathbf{BC})^{-}$ is $(\mathbf{A} + \mathbf{B})_{\rho\chi}^{-}\mathbf{C}_{\rho\chi}^{-}$ and

$$\begin{aligned}\mathbf{AC} \mp \mathbf{BC} &= \mathbf{AC}(\mathbf{A} + \mathbf{B})_{\rho\chi}^{-}\mathbf{C}_{\rho\chi}^{-}\mathbf{BC} = \mathbf{A}(\mathbf{A} + \mathbf{B})_{\rho\chi}^{-}\mathbf{BCC}_{\rho\chi}^{-}\mathbf{C} \\ &= (\mathbf{A} \mp \mathbf{B})\mathbf{C}.\end{aligned}$$

THEOREM 10.1.12 *Definitions* 1 *and* 2 *of the parallel sum lead to identical results, that is,*

$$\mathbf{A}(\mathbf{A} + \mathbf{B})^{-}\mathbf{B} = (\mathbf{A}^{+} + \mathbf{B}^{+})^{+}$$

$$\textit{iff}\ \mathcal{M}(\mathbf{A}) = \mathcal{M}(\mathbf{B}) = \mathcal{M}(\mathbf{A} + \mathbf{B}),\ \mathcal{M}(\mathbf{A}^*) = \mathcal{M}(\mathbf{B}^*) = \mathcal{M}(\mathbf{A}^* + \mathbf{B}^*)$$

Proof: The result is easy to establish.

10.2 APPLICATIONS TO MATHEMATICAL PROGRAMMING PROBLEMS

10.2.1 Interval Linear Programming (IP)

An interval linear programming (IP) problem can be stated as follows: (Ben-Israel and Charnes, 1968):

To maximize

$$\sum_{j=1}^{n} c_j x_j = \mathbf{c}'\mathbf{x} \tag{10.2.1}$$

subject to

$$b_i \leq \sum_{j=1}^{n} a_{ij}x_j \leq d_i \ \forall\, i = 1, 2, \ldots, m. \tag{10.2.2}$$

Definition: IP is said to be *feasible* if the set S of feasible solutions given by

$$S = \{\mathbf{x} : \mathbf{x} \in \mathcal{R}^n, \qquad \mathbf{b} \leq \mathbf{Ax} \leq \mathbf{d}\}$$

is nonempty, where the matrix form $\mathbf{b} \leq \mathbf{Ax} \leq \mathbf{d}$ is used for (10.2.2). Further, a feasible IP is *bounded*, if

$$\max\{\mathbf{c}'\mathbf{x}, \mathbf{x} \in S\} < \infty$$

LEMMA 10.2.1 *A feasible IP is bounded if*

$$\mathbf{c} \in \mathcal{M}(\mathbf{A}'), \tag{10.2.3}$$

or equivalently

$$\mathbf{c}'\mathbf{A}^{-}\mathbf{A} = \mathbf{c}', \tag{10.2.4}$$

where $\mathbf{A}^{-}$ *is any g-inverse of* $\mathbf{A}$.

Proof: Lemma 10.2.1 follows from the fact that if $\mathbf{x}$ is a feasible solution of the IP, $\mathbf{x} + \mathbf{w}$ is also a feasible solution for arbitrary $\mathbf{w} \in \mathcal{O}(\mathbf{A}')$.

We shall treat separately the two cases, $R(\mathbf{A}) = m$, $R(\mathbf{A}) < m$. Observe that if $R(\mathbf{A}) = m$, the matrix $\mathbf{AA}'$ is nonsingular. Further, every g-inverse of $\mathbf{A}$ is necessarily a right inverse of $\mathbf{A}$ that is, $\mathbf{AA}^{-} = \mathbf{I}_m$.

THEOREM 10.2.1 *If the IP is feasible and bounded and* $R(\mathbf{A}) = m$, *the class of optimal solutions of IP is given by*

$$\mathbf{x} = \mathbf{A}^{-}\mathbf{e} + [\mathbf{I} - \mathbf{A}^{-}\mathbf{A}]\mathbf{z}, \tag{10.2.5}$$

where $\mathbf{z}$ *is an arbitrary vector in* $\mathcal{R}^n$ *and the vector* $\mathbf{e} \in \mathcal{R}^m$ *is determined as follows*

$$\begin{aligned} e_i &= b_i \text{ if } (\mathbf{c}'\mathbf{A}^{-})_i < 0 \\ &= d_i \text{ if } (\mathbf{c}'\mathbf{A}^{-})_i > 0 \\ &= \text{arbitrary in } [b_i, d_i] \text{ if } (\mathbf{c}'\mathbf{A}^{-})_i = 0, \end{aligned}$$

where $(\mathbf{c}'\mathbf{A}^{-})_i$ *denotes the ith co-ordinate of the row vector* $\mathbf{c}'\mathbf{A}^{-}$.

Proof: Since $\mathcal{M}[\mathbf{A}^{-} \vdots (\mathbf{I} - \mathbf{A}^{-}\mathbf{A})] = \mathcal{R}^n$, any arbitrary vector $\mathbf{x} \in \mathcal{R}^n$ can always be expressed as $\mathbf{x} = \mathbf{A}^{-}\mathbf{e} + (\mathbf{I} - \mathbf{A}^{-}\mathbf{A})\mathbf{z}$. Restrictions imposed in Theorem 10.2.1 on $\mathbf{e}$ arise because of the fact that under the assumed conditions

$$\mathbf{Ax} = \mathbf{A}[\mathbf{A}^{-}\mathbf{e} + (\mathbf{I} - \mathbf{A}^{-}\mathbf{A})\mathbf{z}] = \mathbf{AA}^{-}\mathbf{e} = \mathbf{e},$$

and

$$\mathbf{c}'\mathbf{x} = \mathbf{c}'[\mathbf{A}^{-}\mathbf{e} + (\mathbf{I} - \mathbf{A}^{-}\mathbf{A})\mathbf{z}] = \mathbf{c}'\mathbf{A}^{-}\mathbf{e}.$$

This completes the proof of Theorem 10.2.1.

We now consider the case, $R(\mathbf{A}) < m$. Let us first study the case where $R(\mathbf{A}) = m - 1$ and assume without any loss of generality that the first $m - 1$ rows of $\mathbf{A}$ are linearly independent. Consider the subproblem (SUBIP): To maximize $\mathbf{c}'\mathbf{x}$, subject to $b_i \leq \Sigma a_{ij}x_j \leq d_i$ $(i = 1, 2, \ldots, m - 1)$. If the IP is bounded and feasible so is the SUBIP. Further the constraint matrix $\hat{\mathbf{A}}$ of the SUBIP is of full row rank. Consider the class of optimal solutions of the SUBIP as determined by Theorem 10.2.1. If $b_m \leq \Sigma_j a_{mj}x_j \leq d_m$ for some $\mathbf{x}$ in this class, then clearly $\mathbf{x}$ is also an optimal solution of the IP. If not, let $\hat{\mathbf{x}}$ be an optimal solution which is closest to satisfying this constraint and suppose $\Sigma_j a_{mj}\hat{x}_j - b_m = \delta < 0$. We shall find an optimum way of modifying

$\hat{\mathbf{x}}$ so as to satisfy all the constraints. Let us write

$$\hat{\mathbf{x}} = (\hat{\mathbf{A}})^- \mathbf{e} + [\mathbf{I} - (\hat{\mathbf{A}})^- \hat{\mathbf{A}}]\mathbf{z} \tag{10.2.6}$$

$$I_{11} = \{i : (\mathbf{c}'(\hat{\mathbf{A}})^-)_i < 0,\ (\mathbf{a}'_m(\hat{\mathbf{A}})^-)_i < 0\}$$

$$I_{12} = \{i : (\mathbf{c}'(\hat{\mathbf{A}})^-)_i < 0,\ (\mathbf{a}'_m(\hat{\mathbf{A}})^-)_i > 0\}$$

$$I_{21} = \{i : (\mathbf{c}'(\hat{\mathbf{A}})^-)_i > 0,\ (\mathbf{a}'_m(\hat{\mathbf{A}})^-)_i < 0\}$$

$$I_{22} = \{i : (\mathbf{c}'(\hat{\mathbf{A}})^-)_i > 0,\ (\mathbf{a}'_m(\hat{\mathbf{A}})^-)_i > 0\}.$$

For an integer $i \in I_{11} \cup I_{12}$, $e_i = b_i$. Hence the only permissible modification here is an increase in e_i by a maximum amount of $\delta_i = d_i - b_i$. However, for $i \in I_{11}$ an increase in e_i adversely affects $\Sigma_j a_{mj} x_j$. Accordingly and by a similar reasoning, we observe that in order to seek an optimum way of modifying $\hat{\mathbf{x}}$ (so as to satisfy the constraint on $\Sigma a_{mj} x_j$) one need only consider the set of indices $I_{12} \cup I_{21}$. Let q be the cardinality of this set. For each $i \in I_{12} \cup I_{21}$ we compute $v_i = (\mathbf{c}'(\hat{\mathbf{A}})^-)_i / (\mathbf{a}'_m(\hat{\mathbf{A}})^-)_i$. Let $i_1, i_2, \ldots, i_q$ be a rearrangement of the indices $i \in I_{12} \cup I_{21}$ so that

$$v_{i_1} \geq v_{i_2} \geq \cdots \geq v_{i_q}.$$

The IP is infeasible if

$$\sum_{k=1}^{q} \delta_{i_k} |(\mathbf{a}'_m(\hat{\mathbf{A}})^-)_{i_k}| < |\delta|. \tag{10.2.7}$$

If (10.2.7) is not true, let p be the largest integer satisfying the inequality

$$S_p = \sum_{k=1}^{p} \delta_{i_k} |(\mathbf{a}'_m(\hat{\mathbf{A}})^-)_{i_k}| < |\delta|,$$

and ε be such that

$$S_p + \varepsilon\, |(\mathbf{a}'_m(\hat{\mathbf{A}})^-)_{i_{p+1}}| = |\delta|.$$

By an argument similar to that of the Neyman-Pearson lemma, an optimum solution for the IP is obtained by modifying $\hat{\mathbf{x}}$ in (10.2.6) as follows:

For $k \leq p$ modify e_{i_k} by the maximum permissible amount δ_{i_k}, that is, if $i_k \in I_{12}$, replace e_{i_k} by d_{i_k}, if $i_k \in I_{21}$ replace e_{i_k} by b_{i_k}. If $i_{p+1} \in I_{12}$ increase $e_{i_{p+1}}$ by an amount ε. If $i_{p+1} \in I_{21}$ decrease $e_{i_{p+1}}$ by an amount ε. Other co-ordinates in $\mathbf{e}$ are left undisturbed.

The case where $\Sigma a_{mj} \hat{x}_j - d_m = \delta > 0$ can be treated in a similar manner.

When $R(\mathbf{A}) < (m - 1)$, the procedure just outlined has to be iterated a number of times. For further details on this (SUBOPT) method refer to Ben-Israel, Charnes, and Roberts (1968) and other relevant references listed therein.

10.2.2 Modular Design Problem and Convex Programming

A modular design problem can be formulated as follows.
To minimize

$$\sum_{i=1}^{n} y_i \sum_{j=1}^{m} z_j \tag{10.2.8}$$

subject to

$$y_i z_j \geq r_{ij} > 0, \quad (i = 1, 2, \ldots, m), \quad (j = 1, 2, \ldots, n) \tag{10.2.9}$$

and

$$y_i, z_j > 0, (i = 1, 2, \ldots, m), (j = 1, 2, \ldots, n). \tag{10.2.10}$$

Charnes and Kirby (1965) make an interesting application of generalized inverse to reduce this problem to one of minimizing a separable convex functional subject to linear constraints on non-negative variables. Advantages gained in this process are substantial, for it opens up the possibility of using existing computation routines for linear programming for solving the modular design problem, making suitable piecewise linear approximations to the separable convex functional. We shall indicate how g-inverse can be applied in this problem and refer the reader for further computational details to the original paper by Charnes and Kirby and to references listed therein.

We make the substitution $y_i = e^{u_i}$, $z_j = e^{v_j}$ and $r_{ij} = e^{c_{ij}}$. In terms of these new variables and constants, the objective function (10.2.8) becomes

$$\sum_{i,j} e^{u_i + v_j}, \tag{10.2.11}$$

the constraints (10.2.9) are equivalent to linear constraints

$$u_i + v_j \geq c_{ij}, \tag{10.2.12}$$

while the constraints (10.2.10) become redundant since they are always true irrespective of the values assumed by the u_i's and v_j's.

The constraints (10.2.12) could be expressed in matrix form as

$$\mathbf{Ax} \geq \mathbf{c}, \tag{10.2.13}$$

where $\mathbf{A}$ is a matrix of order $mn \times (m + n)$, $\mathbf{x}$ and $\mathbf{c}$ are vectors of order $(m + n) \times 1$ and $mn \times 1$, respectively. $\mathbf{x}' = (u_1, u_2, \ldots, u_m, v_1, u_2, \ldots, u_m)$. The co-ordinates of $\mathbf{c}$ as well as the rows of $\mathbf{A}$ are numbered (ij), $(i = 1, 2, \ldots, m)$, $(j = 1, 2, \ldots, n)$ and listed in the lexicographic order. The (ij)-th co-ordinate of $\mathbf{c}$ is c_{ij} and the (ij)-th row of $\mathbf{A}$ has $+1$ in the i-th and $(m + j)$-th columns and 0 elsewhere. Let $\mathbf{w}$ be a $mn \times 1$ vector with w_{ij} as its (ij)-th co-ordinate, and $\mathbf{A}^-$ a g-inverse. In terms of $\mathbf{w}$ and $\mathbf{A}^-$, the problem could be equivalently reformulated as follows:

To minimize

$$\sum_{i,j} e^{c_{ij}+w_{ij}} = \sum_{i,j} r_{ij} e^{w_{ij}} \tag{10.2.14}$$

subject to

$$\mathbf{AA}^{-}(\mathbf{c}+\mathbf{w}) = \mathbf{c}+\mathbf{w} \Leftrightarrow (\mathbf{I}-\mathbf{AA}^{-})\mathbf{w} = -(\mathbf{I}-\mathbf{AA}^{-})\mathbf{c} \tag{10.2.15}$$

and $\mathbf{w} \geq \mathbf{0}$.

We note that once an optimal $\mathbf{w}$ is found for this problem, $\mathbf{x}$ could be determined by the equation $\mathbf{x} = \mathbf{A}^{-}(\mathbf{c}+\mathbf{w})$, leading ultimately to optimal values of y_i and z_j for the original problem. Actually, this approach works for any choice of $\mathbf{A}^{-}$. One choice of $\mathbf{A}^{-}$, a $(m+n) \times mn$ matrix is as follows. This first row is a null row. For $i = 2(1)m$, the ith row has -1 in (1, 1)-th column and $+1$ in $(i, 1)$-th column. For $j = 1, 2, \ldots, n$ the $m+j$th row has $+1$ in $(1, j)$-th column. Rest of the elements are zero. Such a choice of the g-inverse $\mathbf{A}^{-}$, leads to the following reformulation of the problem

To minimize

$$\sum_{i,j} r_{ij} e^{w_{ij}}$$

subject to

$$w_{11} - w_{i1} - w_{1j} + w_{ij} = -(c_{11} - c_{i1} - c_{1j} + c_{ij}) \ \forall\, i = 2(1)m, \quad j = 2(1)n$$

and $w_{ij} \geq 0 \ \forall\, i, j$.

An Initial Feasible Solution. From the formulation of the problem given in (10.2.11) and (10.2.12), it is to be expected that for the optimal solution, $u_i + v_j \approx c_{ij}$. This implies that in the vicinity of the optimal solution $e^{u_i+v_j} \approx e^{c_{ij}}(1 + u_i + v_j - c_{ij})$. An initial solution can therefore be obtained by minimizing

$$\sum_{i=1}^{m} a_i u_i + \sum_{j=1}^{n} b_j v_j \text{ subject to } u_i + v_j \geq c_{ij} \ \forall\, i, j$$

where

$$a_i = \sum_{j=1}^{n} e^{c_{ij}} = \sum_{j=1}^{n} r_{ij}$$

and

$$b_j = \sum_{i=1}^{m} e^{c_{ij}} = \sum_{i=1}^{m} r_{ij}.$$

This is the dual of the distribution problem.

To minimize

$$\sum_{i,j} c_{ij} x_{ij}$$

subject to

$$\sum_i x_{ij} = b_j, \qquad \sum_j x_{ij} = a_i, \quad \text{and } x_{ij} \geq 0, \forall\, i, j.$$

Since efficient algorithms are available for the distribution problem, initial feasible solutions u_i^o, v_j^o for u_i, v_j could be obtained in this manner.

10.3 VARIANCE COMPONENTS

Henderson's Methods

In many biological and industrial situations one encounters a variance components model, a fairly general form of which is

$$\mathbf{Y} = \mathbf{X}_1\boldsymbol{\beta}_1 + \mathbf{X}_2\mathbf{b}_2 + \mathbf{X}_3\mathbf{b}_3 + \mathbf{e},$$

where $\mathbf{Y}$ is a $n \times 1$ vector of responses (random variables), $\mathbf{X}_1, \mathbf{X}_2, \mathbf{X}_3$ are known matrices of order $n \times m_1, n \times m_2, n \times m_3$, respectively, $\boldsymbol{\beta}$, is a $m_1 \times 1$ vector of unknown parameters, $\mathbf{b}_2, \mathbf{b}_3$ and $\mathbf{e}$ are $m_2 \times 1$, $m_3 \times 1$ and $n \times 1$ vectors of random variables, with

$$E(\mathbf{b}_2) = \mathbf{0}, \qquad E(\mathbf{b}_3) = \mathbf{0}, \qquad E(\mathbf{e}) = \mathbf{0}$$

and dispersion matrix

$$D\begin{pmatrix} \mathbf{b}_2 \\ \mathbf{b}_3 \\ \mathbf{e} \end{pmatrix} = \begin{pmatrix} \sigma_1^2\mathbf{I} & \mathbf{0} & \mathbf{0} \\ \mathbf{0} & \sigma_2^2\mathbf{I} & \mathbf{0} \\ \mathbf{0} & \mathbf{0} & \sigma_3^2\mathbf{I} \end{pmatrix}.$$

The problem is to estimate σ_1^2, σ_2^2, σ_3^2 on the basis of available observation on $\mathbf{Y}$. Three methods of estimating variance components, obtained more or less on heuristic considerations, are described in Henderson (1953). All of Henderson's three methods involve (i) calculating mean squares of some kind (ii) obtaining their expectations and (iii) solving linear equations in the unknown variance components derived from equating calculated mean squares to their expected values.

In this section, we shall describe just one of the three methods of estimation, mainly to indicate a fruitful area of application of the g-inverse. A reader interested in a fuller treatment of the subject along these lines is referred to Searle (1968) who gives a reformulation of Henderson's methods in matrix language; our exposition will follow Rohde (1968). We note, however, that the existing methods of estimation of variance components are not very satisfactory.

Let us assume that $\mathbf{b}_2$ and $\mathbf{b}_3$ are fixed parameters and consider the following partition of the total sum of squares $\mathbf{Y}'\mathbf{Y}$, which is fairly standard in analysis of variance:

$$\mathbf{Y}'\mathbf{Y} = S_1 + S_2 + S_3 + S_4$$

where

S_1 = sum of squares due to $\boldsymbol{\beta}_1$, ignoring $\mathbf{b}_2$ and $\mathbf{b}_3$,

S_2 = sum of squares due to $\mathbf{b}_2$ eliminating $\boldsymbol{\beta}_1$ but ignoring $\mathbf{b}_3$,

S_3 = sum of squares due to $\mathbf{b}_3$ eliminating $\boldsymbol{\beta}_1$ and $\mathbf{b}_2$, and

S_4 = residual sum of squares.

Explicit algebraic expressions for these sums of squares are obtained using a partitioned form of a generalized inverse of the matrix

$$\begin{pmatrix} \mathbf{X}_1'\mathbf{X}_1 & \mathbf{X}_1'\mathbf{X}_2 & \mathbf{X}_1'\mathbf{X}_3 \\ \mathbf{X}_2'\mathbf{X}_1 & \mathbf{X}_2'\mathbf{X}_2 & \mathbf{X}_2'\mathbf{X}_3 \\ \mathbf{X}_3'\mathbf{X}_1 & \mathbf{X}_3'\mathbf{X}_2 & \mathbf{X}_3'\mathbf{X}_3 \end{pmatrix}$$

of normal equation.

We have thus

$$S_1 = \mathbf{Y}'(\mathbf{I} - \mathbf{D}_1)\mathbf{Y}, \qquad S_3 = \mathbf{Y}'(\mathbf{D}_{12} - \mathbf{D}_{123})\mathbf{Y},$$
$$S_2 = \mathbf{Y}'(\mathbf{D}_1 - \mathbf{D}_{12})\mathbf{Y}, \quad S_4 = \mathbf{Y}'\mathbf{D}_{123}\mathbf{Y},$$

where

$$\mathbf{D}_1 = \mathbf{I} - \mathbf{X}_1(\mathbf{X}_1'\mathbf{X}_1)^-\mathbf{X}_1'$$
$$\mathbf{D}_{12} = \mathbf{D}_1 - \mathbf{D}_1\mathbf{X}_2(\mathbf{X}_2'\mathbf{D}_1\mathbf{X}_2)^-\mathbf{X}_2'\mathbf{D}_1$$
$$\mathbf{D}_{123} = \mathbf{D}_{12} - \mathbf{D}_{12}\mathbf{X}_3(\mathbf{X}_3'\mathbf{D}_{12}\mathbf{X}_3)^-\mathbf{X}_3'\mathbf{D}_{12}.$$

If the vector of random variables $\mathbf{Y}$ have expectation $E(\mathbf{Y}) = \boldsymbol{\mu}$ and dispersion matrix $D(\mathbf{Y}) = \boldsymbol{\Sigma}$ it is well known that

$$E(\mathbf{Y}'\mathbf{A}\mathbf{Y}) = E(tr\, \mathbf{Y}'\mathbf{A}\mathbf{Y}) = E(tr\, \mathbf{A}\mathbf{Y}\mathbf{Y}')$$
$$= tr\, \mathbf{A}(\boldsymbol{\Sigma} + \boldsymbol{\mu}\boldsymbol{\mu}') = \boldsymbol{\mu}'\mathbf{A}\boldsymbol{\mu} + tr\, \mathbf{A}\boldsymbol{\Sigma}.$$

Hence if $\mathbf{s}' = (S_2, S_3, S_4)$ and $\boldsymbol{\theta}' = (\sigma_1^2, \sigma_2^2, \sigma_3^2)$ it is seen that since here $E(\mathbf{Y}) = \mathbf{X}_1\boldsymbol{\beta}_1$ and $\mathbf{X}_1'\mathbf{D}_1 = \mathbf{0} \Rightarrow \mathbf{X}_1'\mathbf{D}_{12} = \mathbf{0} \Rightarrow \mathbf{X}_1'\mathbf{D}_{123} = \mathbf{0}$ we have $E(\mathbf{s}) = \mathbf{E}\boldsymbol{\theta}$

where

$$\mathbf{E} = \begin{pmatrix} e_1 & e_2 - e_3 & r_2 \\ 0 & e_3 & r_3 \\ 0 & 0 & n - r_1 - r_2 - r_3 \end{pmatrix}$$

with

$$e_1 = tr\, \mathbf{X}_2'\mathbf{D}_1\mathbf{X}_2, \qquad e_2 = tr\, \mathbf{X}_3'\mathbf{D}_1\mathbf{X}_3$$
$$e_3 = tr\, \mathbf{X}_3'\mathbf{D}_{12}\mathbf{X}_3, \qquad r_1 = R(\mathbf{X}_1), r_1 + r_2 = R(\mathbf{X}_1 : \mathbf{X}_2)$$
$$r_1 + r_2 + r_3 = R(\mathbf{X}_1 : \mathbf{X}_2 : \mathbf{X}_3).$$

Since $\mathbf{X}_2'\mathbf{D}_1\mathbf{X}_2$ is a n.n.d. matrix $e_1 = 0 \Rightarrow \mathbf{X}_2'\mathbf{D}_1\mathbf{X}_2 = \mathbf{0} \Rightarrow \mathbf{X}_2'\mathbf{D}_1 = \mathbf{0}$ $\Rightarrow \mathcal{M}(\mathbf{X}_2) \subset \mathcal{M}(\mathbf{X}_1)$. Similarly, $e_3 = 0 \Rightarrow \mathcal{M}(\mathbf{X}_3) \subset \mathcal{M}(\mathbf{X}_1 : \mathbf{X}_2)$. Hence if r_2 and r_3 are integers different from 0, the matrix $\mathbf{E}$ is nonsingular. Thus when these conditions hold $\mathbf{E}^{-1}\mathbf{s}$ provides an unbiased estimate for $\boldsymbol{\theta}$ and hence of the individual variance components.

MINQUE Theory

Recently a new method has been put forward by Rao (1970, 1971) for the estimation of variance and covariance components applicable to very general situations. We consider the linear model $(\mathbf{Y}, \mathbf{X}\boldsymbol{\beta}, \boldsymbol{\Sigma})$ described in chapters 7 and 8, such that $E(\mathbf{Y}) = \mathbf{X}\boldsymbol{\beta}$ where $\mathbf{X}$ is a known matrix and $\boldsymbol{\beta}$ an unknown vector parameter and $D(\mathbf{Y}) = \boldsymbol{\Sigma}$ has the structure

$$\boldsymbol{\Sigma} = \sigma_1^2\mathbf{V}_1 + \cdots + \sigma_k^2\mathbf{V}_k \tag{10.3.1}$$

where $\sigma_1^2, \ldots, \sigma_k^2$ are unknown variance components and $\mathbf{V}_1, \ldots, \mathbf{V}_k$ are given matrices.

Definition: A quadratic function $\mathbf{Y}'\mathbf{A}\mathbf{Y}$ is said to be MINQUE (Minimum Norm Quadratic Unbiased Estimator) of $p_1\sigma_1^2 + \cdots + p_k\sigma_k^2$ (a linear function of σ_i^2) iff the matrix $\mathbf{A}$ is such that

$$\left.\begin{array}{c}\text{tr}\,\mathbf{AUAU} \text{ is a minimum} \\ \text{subject to } \mathbf{AX} = \mathbf{0},\ \text{tr}\,\mathbf{AV}_i = p_i,\ i = 1, \ldots, k\end{array}\right\}$$

where $\mathbf{U}$ is a suitably chosen p.d. matrix.

From statistical considerations $\mathbf{U}$ is chosen as $\mathbf{V}_1 + \cdots + \mathbf{V}_k$ when nothing is known about σ_i^2 and as $\alpha_1^2\mathbf{V}_1 + \cdots + \alpha_k^2\mathbf{V}_k$ when approximate ratios of σ_i's are given as α_i's. Theorems 10.3.1 and 10.3.2 provide the solution to the problem (see Rao (1970, 1971) for some other theorems).

THEOREM 10.3.1 *Min tr* $\mathbf{AUAU}$ *subject to* $\mathbf{AX} = \mathbf{0}$ *and tr* $\mathbf{AV}_i = p_i$, $i = 1, \ldots, k$ *is attained at*

$$\mathbf{A}_* = \mathbf{B}(\Sigma\lambda_i \mathbf{V}_i)\mathbf{B} \tag{10.3.2}$$

where

$$\mathbf{B} = \mathbf{U}^{-1} - \mathbf{U}^{-1}\mathbf{X}(\mathbf{X}'\mathbf{U}^{-1}\mathbf{X})^{-}\mathbf{X}'\mathbf{U}^{-1} \tag{10.3.3}$$

and λ_i *satisfy the equations*

$$p_j = \Sigma\lambda_i \operatorname{tr} \mathbf{BV}_i\mathbf{BV}_j, \quad j = 1, \ldots, k. \tag{10.3.4}$$

Proof: Let $\mathbf{A} = \mathbf{A}_* + \mathbf{D}$, where $\mathbf{DX} = \mathbf{0}$ and $\operatorname{tr} \mathbf{DV}_i = 0$, $i = 1, \ldots, k$, be an alternative choice of $\mathbf{A}$. Then

$$\begin{aligned} \operatorname{tr} \mathbf{AUAU} &= \operatorname{tr} (\mathbf{A}_* + \mathbf{D})\mathbf{U}(\mathbf{A}_* + \mathbf{D})\mathbf{U} \\ &= \operatorname{tr} \mathbf{A}_*\mathbf{UA}_*\mathbf{U} + \operatorname{tr} \mathbf{DUDU} \end{aligned}$$

since

$$\begin{aligned} \operatorname{tr} \mathbf{A}_*\mathbf{U}\,\mathbf{DU} &= \Sigma\lambda_i \operatorname{tr} \mathbf{BV}_i\mathbf{BUDU} \\ &= \Sigma\lambda_i \operatorname{tr} \mathbf{V}_i\mathbf{BUDUB} = \Sigma\lambda_i \operatorname{tr} \mathbf{V}_i\mathbf{D} = 0 \end{aligned}$$

and the result is proved, noting that with $\mathbf{B}$ as in (10.3.3) $\mathbf{BUDUB} = \mathbf{D}$ since $\mathbf{DX} = \mathbf{0}$, $\mathbf{X}'\mathbf{D} = \mathbf{0}$.

THEOREM 10.3.2 *Let* $\mathbf{Q}$ *be the k-vector with the i-th component equal to* $\mathbf{Y}'\mathbf{BV}_i\mathbf{BY}$ *and* $\mathbf{S}$ *be the matrix with (i, j)th element equal to tr* $\mathbf{BV}_i\mathbf{BV}_j$. *Then the MINQUE of* $\Sigma p_i\sigma_i^2$ *is* $\mathbf{p}'\mathbf{S}^{-}\mathbf{Q}$, *where* $\mathbf{p}$ *is the vector with* p_i *as the ith component.*

Proof: Using $\mathbf{A}_*$ as determined in (10.3.2), the MINQUE of $\Sigma p_i\sigma_i^2$ is

$$\Sigma\lambda_i \mathbf{Y}'\mathbf{BV}_i\mathbf{BY} = \boldsymbol{\lambda}'\mathbf{Q} \tag{10.3.5}$$

where $\boldsymbol{\lambda}$ satisfies the equation (10.3.4), which in matrix form is $\mathbf{S}\boldsymbol{\lambda} = \mathbf{p}$. Then (10.3.5) is the same as $\mathbf{p}'\mathbf{S}^{-}\mathbf{Q}$, and the result is proved. The result of Theorem 10.3.2 can also be expressed as stated in the corollary.

COROLLARY: *Let* $\boldsymbol{\sigma}' = (\sigma_1^2, \ldots, \sigma_k^2)$. *Then the MINQUE of* $\mathbf{p}'\boldsymbol{\sigma}$ *is* $\mathbf{p}'\hat{\boldsymbol{\sigma}}$ *where* $\hat{\boldsymbol{\sigma}}$ *satisfies the equation* $\mathbf{S}\boldsymbol{\sigma} = \mathbf{Q}$.

Mitra (1971b) provides an alternative derivation of the result of the corollary by drawing an analogy with the problem of BLUE in a linear model.

Standard Error of Estimates. We shall now indicate how the standard errors of these estimates could be obtained under normality assumption. Let $\mathbf{Y} \sim N_n(\boldsymbol{\mu}, \boldsymbol{\Sigma})$. The joint moment generating function of quadratic forms

$\mathbf{Y}'\mathbf{AY}$ and $\mathbf{Y}'\mathbf{BY}$ is given by

$$\begin{aligned}\phi(t, u) &= E\{\exp \mathbf{Y}'\mathbf{MY}\} \\ &= |\mathbf{I} - 2\mathbf{M\Sigma}|^{-1/2} \exp\left[\boldsymbol{\mu}'\left\{\mathbf{I} + \sum_{n=1}^{\infty} (2\mathbf{M\Sigma})^n\right\}\mathbf{M}\boldsymbol{\mu}\right],\end{aligned}$$

where $\mathbf{M} = t\mathbf{A} + u\mathbf{B}$.

Taking logarithms, we have therefore the following expression for the joint cumulant generating function

$$K(t, u) = \boldsymbol{\mu}'\left\{\mathbf{I} + \sum_{n=1}^{\infty} (2\mathbf{M\Sigma})^n\right\}\mathbf{M}\boldsymbol{\mu} + \frac{1}{2}\sum_{n=1}^{\infty} tr(2\mathbf{M\Sigma})^n/n,$$

from which the coefficients of $t^2/2$, $u^2/2$ and tu can be easily determined. We have thus

$$\begin{aligned}V(\mathbf{Y}'\mathbf{AY}) &= 4\boldsymbol{\mu}'\mathbf{A\Sigma A}\boldsymbol{\mu} + 2\,tr(\mathbf{A\Sigma})^2 \\ V(\mathbf{Y}'\mathbf{BY}) &= 4\boldsymbol{\mu}'\mathbf{B\Sigma B}\boldsymbol{\mu} + 2\,tr(\mathbf{B\Sigma})^2 \\ \mathrm{Cov}(\mathbf{Y}'\mathbf{AY}, \mathbf{Y}'\mathbf{BY}) &= 4\boldsymbol{\mu}'\mathbf{A\Sigma B}\boldsymbol{\mu} + 2\,tr(\mathbf{A\Sigma B\Sigma}).\end{aligned}$$

Since the estimates of variance components are quadratic forms in $\mathbf{Y}$, these formulae are useful in computing variances and covariances of these estimates. The requisite formulae are given in Rohde and Tallis (1969).

10.4 MAXIMUM LIKELIHOOD ESTIMATION WHEN THE INFORMATION MATRIX IS SINGULAR

Let $p(\mathbf{x}, \boldsymbol{\theta})$ be the probability density, where $\mathbf{x}$ stands for observed data and $\boldsymbol{\theta}$ for vector of n unknown parameters $\theta_1, \ldots, \theta_n$. The function

$$L(\boldsymbol{\theta}, \mathbf{x}) = \log p(\mathbf{x}, \boldsymbol{\theta}), \tag{10.4.1}$$

as a function of $\boldsymbol{\theta}$ for given $\mathbf{x}$, is known as log likelihood of parameters. Let $f_i(\mathbf{x}, \boldsymbol{\theta}) = \partial L/\partial\theta_i$, $i = 1, \ldots, n$ be the first derivatives, and denote the vector $[f_1(\mathbf{x}, \boldsymbol{\theta}), \ldots, f_n(\mathbf{x}, \boldsymbol{\theta})]'$ by $\mathbf{f}$. The information matrix on $\boldsymbol{\theta}$ is defined by

$$H(\boldsymbol{\theta}) = E(\mathbf{ff}'). \tag{10.4.2}$$

Maximum likelihood (m.l.) estimate of $\boldsymbol{\theta}$ is usually obtained from the equation $\mathbf{f} = \mathbf{0}$, and the asymptotic theory of estimation is well known when the matrix $\mathbf{H}$ is not singular in the neighborhood of the true value.

If $L(\boldsymbol{\theta}, \mathbf{x})$ depends essentially on $s < n$ independent functions $\phi_1, \ldots, \phi_s$ of $\boldsymbol{\theta}$, then $\mathbf{H}$ becomes singular and not all the parameters $\theta_1, \ldots, \theta_n$ are estimable. Only $\phi_1, \ldots, \phi_s$ and their functions are estimable. In such a case we can define the log likelihood $L(\boldsymbol{\theta}, \mathbf{x})$ in terms of fewer parameters as $L(\boldsymbol{\phi}, \mathbf{x})$, where $\boldsymbol{\phi} = (\phi_1, \ldots, \phi_s)'$ such that $\mathbf{J}$ the information matrix on $\boldsymbol{\phi}$ is nonsingular in the neighborhood of the true value. Then the usual theory would apply. Of course, there is some arbitrariness in the choice of $\boldsymbol{\phi}$, but this does not cause any trouble. However, the calculus of g-inverses enables us to deal with the likelihood as a function of the original parameter $\boldsymbol{\theta}$ and obtain their 'm.l. estimates and the associated asymptotic variance-covariance matrix'. When all the parameters are not estimable, the individual estimates and the variance-covariance matrix so obtained are not meaningful, but they are useful in computing m.l. estimates and standard errors of estimable parametric functions (i.e., those expressible in terms of $\boldsymbol{\phi}$ only).

Method of Scoring with a Singular Information Matrix

The m.l. estimates are obtained by solving the equations

$$f_i(\mathbf{x}, \boldsymbol{\theta}) = 0, \qquad i = 1, \ldots, n. \tag{10.4.3}$$

The equations (10.4.3) are usually complicated, in which case one obtains solutions by successive approximations using a technique such as Fishers' method of scoring [see Rao (1965) pp. 302–309]. Let $\boldsymbol{\theta}_0$ be an approximate solution and $\delta\boldsymbol{\theta}$ the correction. Then neglecting higher order terms in $\delta\boldsymbol{\theta}$

$$\mathbf{f}(\mathbf{x}, \boldsymbol{\theta}_0) = \mathbf{H}(\boldsymbol{\theta}_0)\,\delta\boldsymbol{\theta}. \tag{10.4.4}$$

Since $\mathbf{H}$ is singular, there is no unique solution to (10.4.4) and therefore, the question of choosing a suitable solution arises. A natural choice is a solution with a minimum norm

$$\delta\boldsymbol{\theta} = \mathbf{H}_m^- \mathbf{f}, \tag{10.4.5}$$

where in $\mathbf{H}$ and $\mathbf{f}$, the arguments $\mathbf{x}, \boldsymbol{\theta}_0$ are omitted for convenience of notation. We may terminate the iterative procedure when the correction needed is negligible. Let $\hat{\boldsymbol{\theta}}$ be an approximate solution thus obtained and $\mathbf{H}^-$ any g-inverse of $\mathbf{H}$ computed at $\hat{\boldsymbol{\theta}}$. As observed earlier, $\hat{\boldsymbol{\theta}}$ and $\mathbf{H}^-$ are not meaningful when $\mathbf{H}$ is singular.

When a parametric function $\psi(\boldsymbol{\theta})$ is estimable, $\boldsymbol{\xi}(\boldsymbol{\theta})$, the vector of derivatives of $\psi(\boldsymbol{\theta})$ with respect to $\theta_1, \ldots, \theta_n$, belongs to $\mathcal{M}(\mathbf{H})$. For such a function $\psi(\boldsymbol{\theta})$, $\psi(\hat{\boldsymbol{\theta}})$ is the unique m.l. estimate for any choice $\hat{\boldsymbol{\theta}}$ of m.l. estimate of $\boldsymbol{\theta}$ and the asymptotic variance of $\psi(\hat{\boldsymbol{\theta}})$ is

$$\boldsymbol{\xi}(\hat{\boldsymbol{\theta}})'\mathbf{H}^-\boldsymbol{\xi}(\hat{\boldsymbol{\theta}}), \tag{10.4.6}$$

which is unique for any choice of the g-inverse of $\mathbf{H}$. Observe that for obtaining the solution (10.4.5), we need a minimum norm g-inverse of $\mathbf{H}$, but in the formula (10.4.6) any g-inverse of $\mathbf{H}$ may be used.

Chernoff (1953) defined an inverse of a singular information matrix, which is not a g-inverse in our sense. For instance, when no individual parameter is estimable, Chernoff's inverse does not exist (all the entries become infinite), while $\mathbf{H}^-$ exists and can be used as in formula (10.4.6) to find standard errors of estimable parametric functions.

10.5 DISCRIMINANT FUNCTION IN MULTIVARIATE ANALYSIS

Singular Multivariate Normal Distribution

In the book on *Linear Statistical Inference and its applications* by Rao (1965), a density-free approach was developed to study the distribution and inference problems associated with a multivariate normal distribution. Such an approach was more general, as it included the study of the normal distribution with a singular covariance matrix as well, which does not admit density in the usual sense. The elegance of the density-free approach was further demonstrated by Mitra (1970). However, in some problems it is useful to have an explicit expression for density as in the construction of a discriminant function. The density function of a multivariate normal distribution, as is usually written, involves the inverse of the variance-covariance (dispersion) matrix, which necessitated the assumption that the dispersion matrix is nonsingular. In this section we demonstrate how the g-inverse is useful in defining the density function and in extending some of the results developed for the nonsingular case to the singular distribution.

Let $\mathbf{Y}$ be a $p \times 1$ vector random variable. In Rao (1965), $\mathbf{Y}$ is defined to have a p-variate normal distribution if $\mathbf{m}'\mathbf{Y}$ has a univariate normal distribution for every vector $\mathbf{m} \in \mathscr{R}^p$. In such a case it is shown that the distribution is characterized by the parameters

$$\boldsymbol{\mu} = E(\mathbf{Y}) \text{ and } \boldsymbol{\Sigma} = E[(\mathbf{Y} - \boldsymbol{\mu})(\mathbf{Y} - \boldsymbol{\mu})'], \tag{10.5.1}$$

called the mean vector and the dispersion matrix of $\mathbf{Y}$, respectively; the symbol $N_p(\boldsymbol{\mu}, \boldsymbol{\Sigma})$ is used to denote the p-variate distribution. The distribution is said to be singular if $R(\boldsymbol{\Sigma}) = \rho < p$, in which case ρ is called the rank of the distribution and we may use the symbol $N_p(\boldsymbol{\mu}, \boldsymbol{\Sigma}[\rho])$ to specify the rank in addition to the basic parameters.

Let $\mathbf{N}$ be a $p \times (p - \rho)$ matrix of rank $p - \rho$, such that $\mathbf{N}'\boldsymbol{\Sigma} = \mathbf{0}$, $\mathbf{N}'\mathbf{N} = \mathbf{I}$ and $\mathbf{A}$ be a $p \times \rho$ matrix of rank ρ such that $\mathbf{N}'\mathbf{A} = \mathbf{0}$ and $\mathbf{A}'\mathbf{A} = \mathbf{I}$. By construction $(\mathbf{N}:\mathbf{A})$ is an orthogonal matrix. Making the transformation $\mathbf{Z}_1 = \mathbf{N}'\mathbf{Y}$, $\mathbf{Z}_2 = \mathbf{A}'\mathbf{Y}$, we have

$$\begin{aligned} E(\mathbf{Z}_1) &= \mathbf{N}'\boldsymbol{\mu} \quad E(\mathbf{Z}_2) = \mathbf{A}'\boldsymbol{\mu} \\ D(\mathbf{Z}_1) &= \mathbf{N}'\boldsymbol{\Sigma}\mathbf{N} = \mathbf{0},\ D(\mathbf{Z}_2) = \mathbf{A}'\boldsymbol{\Sigma}\mathbf{A} \text{ (nonsingular)} \end{aligned} \tag{10.5.2}$$

so that

$$\mathbf{Z}_1 = \mathbf{N}'\mathbf{Y} = \mathbf{N}'\boldsymbol{\mu} = \boldsymbol{\zeta} \text{ (constant)} \tag{10.5.3}$$

with probability 1 and $\mathbf{Z}_2$ has the ρ-variate normal density

$$(2\pi)^{-\rho/2}|\mathbf{A}'\boldsymbol{\Sigma}\mathbf{A}|^{-1/2} \exp\{-\tfrac{1}{2}(\mathbf{Z}_2 - \mathbf{A}'\boldsymbol{\mu})'(\mathbf{A}'\boldsymbol{\Sigma}\mathbf{A})^{-1}(\mathbf{Z}_2 - \mathbf{A}'\boldsymbol{\mu})\}. \tag{10.5.4}$$

We observe that $|\mathbf{A}'\boldsymbol{\Sigma}\mathbf{A}| = \lambda_1 \cdots \lambda_\rho$, where $\lambda_1, \ldots, \lambda_\rho$ are the nonzero eigenroots of $\boldsymbol{\Sigma}$ and

$$(\mathbf{Z}_2 - \mathbf{A}'\boldsymbol{\mu})'(\mathbf{A}'\boldsymbol{\Sigma}\mathbf{A})^{-1}(\mathbf{Z}_2 - \mathbf{A}'\boldsymbol{\mu}) = (\mathbf{Y} - \boldsymbol{\mu})'\boldsymbol{\Sigma}^{-}(\mathbf{Y} - \boldsymbol{\mu}), \tag{10.5.5}$$

where $\boldsymbol{\Sigma}^{-}$ is any g-inverse of $\boldsymbol{\Sigma}$. Thus the density of $\mathbf{Y}$ on the hyperplane $\mathbf{N}'(\mathbf{Y} - \boldsymbol{\mu}) = \mathbf{0}$ or $\mathbf{N}'\mathbf{Y} = \boldsymbol{\zeta}$ is defined by

$$\frac{(2\pi)^{-\rho/2}}{\sqrt{\lambda_1 \cdots \lambda_\rho}} \exp\{-\tfrac{1}{2}(\mathbf{Y} - \boldsymbol{\mu})'\boldsymbol{\Sigma}^{-}(\mathbf{Y} - \boldsymbol{\mu})\}, \tag{10.5.6}$$

which is an explicit function of the vector $\mathbf{Y}$ and its associated parameters $\boldsymbol{\mu}$ and $\boldsymbol{\Sigma}$. The expression (10.5.6) was considered by Khatri (1968) in deriving some distributions in the case of a singular normal distribution.

Discriminant Function. The density function derived in (10.5.6) can be used in determining the discriminant function (ratio of likelihoods) for assigning an individual as a member of one of two populations to which it may belong.

Let $\mathbf{Y}$ be a $p \times 1$ vector of observations, which has the distribution $N_p(\boldsymbol{\mu}_1, \boldsymbol{\Sigma}_1[\rho_1])$ in the first population and $N_p(\boldsymbol{\mu}_2, \boldsymbol{\Sigma}_2[\rho_2])$ in the second population. We shall construct the discriminant function applicable to different situations.

Case 1. $\boldsymbol{\Sigma}_1 = \boldsymbol{\Sigma}_2 = \boldsymbol{\Sigma}$, $R(\boldsymbol{\Sigma}) = \rho < p \quad \mathbf{N}'\boldsymbol{\mu}_1 \neq \mathbf{N}'\boldsymbol{\mu}_2$.

The distribution of $\mathbf{Y}$ consists of two parts as shown in (10.5.3) and (10.5.4), of which (10.5.3) is the almost sure part. If $\mathbf{N}'\boldsymbol{\mu}_1 \neq \mathbf{N}'\boldsymbol{\mu}_2$,

$$\mathbf{N}'\mathbf{Y} = \mathbf{N}'\boldsymbol{\mu}_1 = \boldsymbol{\zeta}_1 \text{ with probability 1} \tag{10.5.7}$$

if $\mathbf{Y}$ comes from the first population, and

$$\mathbf{N}'\mathbf{Y} = \mathbf{N}'\boldsymbol{\mu}_2 = \boldsymbol{\zeta}_2 \text{ with probability 1} \tag{10.5.8}$$

if $\mathbf{Y}$ comes from the second population. Then the discriminant function is $\mathbf{N}'\mathbf{Y}$, and, in fact, it provides perfect discrimination. No use need be made of the other part (10.5.6) of the distribution of $\mathbf{Y}$.

Case 2. $\boldsymbol{\Sigma}_1 = \boldsymbol{\Sigma}_2 = \boldsymbol{\Sigma}$, $R(\boldsymbol{\Sigma}) = \rho < p$, $\mathbf{N}'\boldsymbol{\mu}_1 = \mathbf{N}'\boldsymbol{\mu}_2$.

In this case $\mathbf{N}'\mathbf{Y}$ does not provide any discrimination and we have to consider the density (10.5.6). The log densities for the two populations are (apart from a constant)

$$-\tfrac{1}{2}\log(\lambda_1 \cdots \lambda_\rho) - \tfrac{1}{2}(\mathbf{Y} - \boldsymbol{\mu}_1)'\boldsymbol{\Sigma}^-(\mathbf{Y} - \boldsymbol{\mu}_1) \tag{10.5.9}$$

$$-\tfrac{1}{2}\log(\lambda_1 \cdots \lambda_\rho) - \tfrac{1}{2}(\mathbf{Y} - \boldsymbol{\mu}_2)'\boldsymbol{\Sigma}^-(\mathbf{Y} - \boldsymbol{\mu}_2). \tag{10.5.10}$$

Taking the difference and retaining only the portion depending on $\mathbf{Y}$ we obtain the discriminant function

$$\boldsymbol{\delta}'\boldsymbol{\Sigma}^-\mathbf{Y} \quad \text{where } \boldsymbol{\delta} = \boldsymbol{\mu}_1 - \boldsymbol{\mu}_2, \tag{10.5.11}$$

which is of the same form as in the nonsingular case ($\boldsymbol{\delta}'\boldsymbol{\Sigma}^{-1}\mathbf{Y}$). Now

$$V(\boldsymbol{\delta}\boldsymbol{\Sigma}^-\mathbf{Y}) = \boldsymbol{\delta}'\boldsymbol{\Sigma}^-\boldsymbol{\delta}, \tag{10.5.12}$$

which is the analogue of Mahalanobis distance $D^2(= \boldsymbol{\delta}'\boldsymbol{\Sigma}^{-1}\boldsymbol{\delta})$ in the singular case.

Case 3. $\boldsymbol{\Sigma}_1 \neq \boldsymbol{\Sigma}_2$, $\mathcal{M}(\boldsymbol{\Sigma}_1) \neq \mathcal{M}(\boldsymbol{\Sigma}_2)$

The discrimination is perfect as in case 1. Let $\mathbf{N}$ be a matrix of maximum rank such that $\mathbf{N}'\boldsymbol{\Sigma}_1\mathbf{N} = \mathbf{0} = \mathbf{N}'\boldsymbol{\Sigma}_2\mathbf{N}$, $\mathbf{A}$ be a matrix of maximum rank such that $(\mathbf{N}:\mathbf{A})'\boldsymbol{\Sigma}_1 = \mathbf{0}$ and $\mathbf{B}$ be a matrix of maximum rank such that $(\mathbf{N}:\mathbf{B})'\boldsymbol{\Sigma}_2 = \mathbf{0}$. Finally let $\mathbf{C}$ be such that $(\mathbf{N}:\mathbf{A}:\mathbf{B}:\mathbf{C})$ is a $p \times p$ matrix of rank p. Consider the transformation

$$\mathbf{Z}_1 = \mathbf{N}'\mathbf{Y} \quad \mathbf{Z}_2 = \mathbf{A}'\mathbf{Y} \quad \mathbf{Z}_3 = \mathbf{B}'\mathbf{Y} \quad \mathbf{Z}_4 = \mathbf{C}'\mathbf{Y}. \tag{10.5.13}$$

The distributions of these variables in the two populations are as follows

Population	$\mathbf{Z}_1$	$\mathbf{Z}_2$	$\mathbf{Z}_3$	$\mathbf{Z}_4$
1	$\zeta_{11} = \mathbf{N}'\boldsymbol{\mu}_1$ with prob. 1	$\zeta_{12} = \mathbf{A}'\boldsymbol{\mu}_1$ with prob. 1	$N(\mathbf{B}'\boldsymbol{\mu}_1, \mathbf{B}'\boldsymbol{\Sigma}_1\mathbf{B})$	$N(\mathbf{C}'\boldsymbol{\mu}_1, \mathbf{C}'\boldsymbol{\Sigma}_1\mathbf{C})$
2	$\zeta_{21} = \mathbf{N}'\boldsymbol{\mu}_2$ with prob. 1	$N(\mathbf{A}'\boldsymbol{\mu}_2, \mathbf{A}'\boldsymbol{\Sigma}_2\mathbf{A})$	$\zeta_{22} = \mathbf{B}'\boldsymbol{\mu}_2$ with prob. 1	$N(\mathbf{C}'\boldsymbol{\mu}_2, \mathbf{C}'\boldsymbol{\Sigma}_2\mathbf{C})$

It is seen that the variables $\mathbf{Z}_1$, $\mathbf{Z}_2$ and $\mathbf{Z}_3$ provide perfect discrimination unless $\zeta_{11} = \zeta_{21}$, $\mathbf{A} = \mathbf{0}$ and $\mathbf{B} = \mathbf{0}$, which can happen only when $\mathcal{M}(\boldsymbol{\Sigma}_1) = \mathcal{M}(\boldsymbol{\Sigma}_2)$.

Case 4. $\boldsymbol{\Sigma}_1 \neq \boldsymbol{\Sigma}_2 \quad \mathcal{M}(\boldsymbol{\Sigma}_1) = \mathcal{M}(\boldsymbol{\Sigma}_2)$.

Let $\mathbf{N}$ be as defined in case 3 and consider $\mathbf{N}'\mathbf{Y}$, which is a constant for both the populations. If $\mathbf{N}'\boldsymbol{\mu}_1 \neq \mathbf{N}'\boldsymbol{\mu}_2$, then we have perfect discrimination. If $\mathbf{N}'\boldsymbol{\mu}_1 = \mathbf{N}'\boldsymbol{\mu}_2$, then we have to consider the densities

$$(\lambda_1 \cdots \lambda_\rho)^{-1/2} \exp\{-\tfrac{1}{2}(\mathbf{Y} - \boldsymbol{\mu}_1)'\boldsymbol{\Sigma}_1^-(\mathbf{Y} - \boldsymbol{\mu}_1)\} \tag{10.5.14}$$

$$(\lambda_1' \cdots \lambda_\rho')^{-1/2} \exp\{-\tfrac{1}{2}(\mathbf{Y} - \boldsymbol{\mu}_2)'\boldsymbol{\Sigma}_2^-(\mathbf{Y} - \boldsymbol{\mu}_2)\}, \tag{10.5.15}$$

where $\lambda_1, \ldots, \lambda_\rho$, are the nonzero eigenvalues of $\boldsymbol{\Sigma}_1$, $\lambda_1', \ldots, \lambda_\rho'$ are those of $\boldsymbol{\Sigma}_2$ and $\boldsymbol{\Sigma}_1^-$, $\boldsymbol{\Sigma}_2^-$ are any g-inverses of $\boldsymbol{\Sigma}_1, \boldsymbol{\Sigma}_2$. Taking logarithm of the ratio of densities and retaining only the terms depending on $\mathbf{Y}$, we have the quadratic discriminant function

$$(\mathbf{Y} - \boldsymbol{\mu}_1)'\boldsymbol{\Sigma}_1^-(\mathbf{Y} - \boldsymbol{\mu}_1) - (\mathbf{Y} - \boldsymbol{\mu}_2)'\boldsymbol{\Sigma}_2^-(\mathbf{Y} - \boldsymbol{\mu}_2) \tag{10.5.16}$$

analogous to the expression in the nonsingular case.

For other uses of g-inverse in studying the properties of the multivariate normal distribution the reader is referred to Chapter 8 of Rao (1965, specially pp. 437–445).

CHAPTER 11

Computational Methods

In this chapter we provide a number of formulae for g-inverses of various types, not all of which may be suitable for numerical computations. Some of them are useful in theoretical investigations involving singular or rectangular matrices.

11.1 GENERAL FORMULAE

Let **A** be a $m \times n$ matrix and **M, N** be p.d. matrices of order m and n, respectively. Then the following formulae hold, where $(\ )^-$ indicates any choice of g-inverse

(a_1) $\mathbf{A}^*(\mathbf{A}\mathbf{A}^*)^- = \mathbf{A}_m^-$, ($a_2$) $\mathbf{N}^{-1}\mathbf{A}^*(\mathbf{A}\mathbf{N}^{-1}\mathbf{A}^*)^- = \mathbf{A}_m^-(\mathbf{N})$

(b_1) $(\mathbf{A}^*\mathbf{A})^-\mathbf{A}^* = \mathbf{A}_l^-$, ($b_2$) $(\mathbf{A}^*\mathbf{M}\mathbf{A})^-\mathbf{A}^*\mathbf{M} = \mathbf{A}_{l(\mathbf{M})}^-$

(c_1) $\mathbf{A}_{\mathbf{MN}}^+ = (\mathbf{A}^*\mathbf{M}\mathbf{A})_{m(\mathbf{N})}^-\mathbf{A}^*\mathbf{M} = \mathbf{N}^{-1}\mathbf{A}^*(\mathbf{A}\mathbf{N}^{-1}\mathbf{A}^*)_{l(\mathbf{M})}^-$

(c_2) $\mathbf{A}^+ = (\mathbf{A}^*\mathbf{A})_m^-\mathbf{A}^* = \mathbf{A}^*(\mathbf{A}\mathbf{A}^*)_l^- = \mathbf{A}^*\mathbf{A}(\mathbf{A}^*\mathbf{A}\mathbf{A}^*\mathbf{A})^-\mathbf{A}^*$

$$= \mathbf{A}^*(\mathbf{A}\mathbf{A}^*\mathbf{A}\mathbf{A}^*)^-\mathbf{A}\mathbf{A}^*$$

(d) $\mathbf{A}_{MN}^+ = \mathbf{N}^{-1/2}(\mathbf{M}^{1/2}\mathbf{A}\mathbf{N}^{-1/2})^+\mathbf{M}^{1/2}$,

where $\mathbf{X}^{1/2}$ denotes symmetric square root.

(e_1) $\mathbf{A}^+ = \mathbf{A}_m^-$ if $R(\mathbf{A}) = m$, (e_2) $\mathbf{A}^+ = \mathbf{A}_l^-$ if $R(\mathbf{A}) = n$.

11.2 COMPUTATION OF g-INVERSE WHEN INDEPENDENT ROWS OR COLUMNS ARE IDENTIFIABLE

In some problems of estimation from simple linear models by least squares, the rank of the normal equations is known, and it is also possible to identify independent rows and columns. In such cases the computation of g-inverse is easy, as it involves only the inversion of suitably chosen nonsingular matrices.

This method of computation is thus available in all analyses of variance problems, such as two way classification, incomplete block designs, etc.

(a) In general let $\mathbf{A}$ be a $m \times n$ matrix of rank r and let it be possible to partition $\mathbf{A}$ in the form

$$\mathbf{A} = (\mathbf{B}_1 \vdots \mathbf{B}_2)$$

by a permutation of columns, if necessary, such that $\mathbf{B}_1$ is a $m \times r$ matrix of rank r and $\mathbf{B}_2$ is a $m \times (n - r)$ matrix. Then one choice of g-inverse of $\mathbf{A}$ is

$$\mathbf{A}^- = \begin{pmatrix} \mathbf{P} \\ \mathbf{0} \end{pmatrix} \quad \text{where } \mathbf{P} = (\mathbf{B}_1^*\mathbf{B}_1)^{-1}\mathbf{B}_1^*, \tag{11.2.1}$$

This inverse is also reflexive and provides a basic solution to $\mathbf{Ax} = \mathbf{y}$.

(b) Let, if possible, $\mathbf{A}$ be partitioned in the form,

$$\mathbf{A} = \begin{pmatrix} \mathbf{C}_1 \\ \cdots \\ \mathbf{C}_2 \end{pmatrix}$$

by a permutation of rows if necessary, where $\mathbf{C}_1$ is a $r \times n$ matrix of rank r and $\mathbf{C}_2$ is a $(m - r) \times n$ matrix. Then one choice of g-inverse of $\mathbf{A}$ is

$$\mathbf{A}^- = (\mathbf{Q} \vdots \mathbf{0}), \text{ where } \mathbf{Q} = \mathbf{C}_1^*(\mathbf{C}_1\mathbf{C}_1^*)^{-1}, \tag{11.2.2}$$

which is also reflexive.

(c) Let, if possible, $\mathbf{A}$ be partitioned in the form

$$\mathbf{A} = \begin{pmatrix} \mathbf{A}_1 & \mathbf{A}_2 \\ \mathbf{A}_3 & \mathbf{A}_4 \end{pmatrix}$$

by a permutation of rows and columns, if necessary, such that $\mathbf{A}_1$ is a $r \times r$ matrix of rank r, $\mathbf{A}_2$, $\mathbf{A}_3$, $\mathbf{A}_4$ are matrices of suitable orders such that the order of $\mathbf{A}$ is $m \times n$. In such a case $\mathbf{A}_4 = \mathbf{A}_3\mathbf{A}_1^{-1}\mathbf{A}_2$. Then one choice of g-inverse of $\mathbf{A}$ is

$$\mathbf{A}^- = \begin{pmatrix} \mathbf{A}_1^{-1} & \mathbf{0} \\ \mathbf{0} & \mathbf{0} \end{pmatrix} \tag{11.2.3}$$

and

$$\mathbf{A}^+ = \begin{pmatrix} \mathbf{A}_1^*\mathbf{P}\mathbf{A}_1^* & \mathbf{A}_1^*\mathbf{P}\mathbf{A}_3^* \\ \mathbf{A}_2^*\mathbf{P}\mathbf{A}_1^* & \mathbf{A}_2^*\mathbf{P}\mathbf{A}_3^* \end{pmatrix}, \tag{11.2.4}$$

where $\mathbf{P} = (\mathbf{A}_1\mathbf{A}_1^* + \mathbf{A}_2\mathbf{A}_2^*)^{-1}\mathbf{A}_1(\mathbf{A}_1^*\mathbf{A}_1 + \mathbf{A}_3^*\mathbf{A}_3)^{-1}$. The expression (11.2.4) for $\mathbf{A}^+$ is due to Penrose (1956).

A g-inverse of $\mathbf{A}$ of maximum rank is

$$\mathbf{A}^- = \begin{pmatrix} \mathbf{A}_1^{-1} & -\mathbf{A}_1^{-1}\mathbf{A}_2 \\ \mathbf{0} & \mathbf{I} \end{pmatrix}. \tag{11.2.5}$$

11.3 g-INVERSES BASED ON FACTORIZATION OF MATRICES

Now we consider some formulae based on factorizations of matrices. Let $\mathbf{\Delta}$ be a diagonal matrix, not necessarily square, with nonzero elements only in the main diagonal. Define $\mathbf{\Delta}^-$ as the matrix obtained from $\mathbf{\Delta}$ by taking the transpose and replacing the nonzero elements by their reciprocals. It is seen that $\mathbf{\Delta}^-$ is also $\mathbf{\Delta}^+$.

(a) Rank Factorization. Let $\mathbf{A}$ be a $m \times n$ matrix of rank r. Consider the rank factorization $\mathbf{A} = \mathbf{BC}^*$, where $\mathbf{B}$ is $m \times r$ and $\mathbf{C}$ is $n \times r$ matrices each of rank r. Then

$$\mathbf{B}_{\mathbf{MI}}^+ = (\mathbf{B}^*\mathbf{MB})^{-1}\mathbf{B}^*\mathbf{M}, \quad (\mathbf{C}^*)_{\mathbf{IN}}^+ = \mathbf{N}^{-1}\mathbf{C}(\mathbf{C}^*\mathbf{N}^{-1}\mathbf{C})^{-1}$$
$$\mathbf{A}_{\mathbf{MN}}^+ = (\mathbf{C}^*)_{\mathbf{IN}}^+\mathbf{B}_{\mathbf{MI}}^+. \tag{11.3.1}$$

In particular

$$\mathbf{B}^+ = (\mathbf{B}^*\mathbf{B})^{-1}\mathbf{B}^*, \quad (\mathbf{C}^*)^+ = \mathbf{C}(\mathbf{C}^*\mathbf{C})^{-1}$$
$$\mathbf{A}^+ = \mathbf{C}(\mathbf{C}^*\mathbf{C})^{-1}(\mathbf{B}^*\mathbf{B})^{-1}\mathbf{B}^* = (\mathbf{C}^*)^+\mathbf{B}^+. \tag{11.3.2}$$

(b) Diagonal Reduction. Given a matrix $\mathbf{A}$ of order $m \times n$, there exist nonsingular (square) matrices $\mathbf{B}$ and $\mathbf{C}$ of orders m and n, respectively, such that $\mathbf{BAC} = \mathbf{\Delta}$. Then

$$\mathbf{A}^- = \mathbf{C\Delta}^-\mathbf{B}, \tag{11.3.3}$$

is a reflexive g-inverse of $\mathbf{A}$.

(c) Hermite Canonical Form. Given a square matrix $\mathbf{A}$ of order m, there exists a nonsingular matrix $\mathbf{C}$ of order m such that $\mathbf{CA}$ is in Hermite canonical form and idempotent (Rao, 1965, p. 18). Then $\mathbf{C}$ is $\mathbf{A}^-$. If $\mathbf{A}$ is $m \times n$ matrix, it can be made into a square matrix by adding $(n - m)$ zero rows [or $(m - n)$ zero columns]. We compute the corresponding $\mathbf{C}$ matrix and omit the last $(n - m)$ columns [or $(m - n)$ rows] to obtain a g-inverse of $\mathbf{A}$.

(d) Singular Value Decomposition. Given a matrix $\mathbf{A}$ of order $m \times n$, it can be written in the form

$$\mathbf{A} = \lambda_1\mathbf{V}_1\mathbf{U}_1^* + \cdots + \lambda_r\mathbf{V}_r\mathbf{U}_r^*, \tag{11.3.4}$$

where $\lambda_1^2, \ldots, \lambda_r^2$ are the nonzero eigenvalues of $\mathbf{AA}^*$ or $\mathbf{A}^*\mathbf{A}$ and the vectors $\mathbf{V}_1, \ldots, \mathbf{V}_r$ are orthonormal eigenvectors of $\mathbf{AA}^*$ and $\mathbf{U}_1, \ldots, \mathbf{U}_r$ are orthonormal eigenvectors of $\mathbf{A}^*\mathbf{A}$ corresponding to the eigenvalues $\lambda_1^2, \ldots, \lambda_r^2$. Then it is easy to verify that

$$\mathbf{A}^+ = \lambda_1^{-1}\mathbf{U}_1\mathbf{V}_1^* + \cdots + \lambda_r^{-1}\mathbf{U}_r\mathbf{V}_r^*. \tag{11.3.5}$$

More generally, if $\mathbf{M}$ and $\mathbf{N}$ are given p.d. matrices, $\mathbf{A}$ can be expressed in the form

$$\mathbf{A} = \mathbf{M}^{-1}[\mu_1\boldsymbol{\xi}_1\boldsymbol{\eta}_1^* + \cdots + \mu_r\boldsymbol{\xi}_r\boldsymbol{\eta}_r^*]\mathbf{N}, \tag{11.3.6}$$

where $\mu_1^2, \ldots, \mu_r^2$ are the nonzero eigenvalues of $\mathbf{AN}^{-1}\mathbf{A}^*$ with respect to $\mathbf{M}^{-1}$ or of $\mathbf{A}^*\mathbf{MA}$ with respect to $\mathbf{N}$. The vectors $\boldsymbol{\xi}_i$ are eigenvectors of $\mathbf{AN}^{-1}\mathbf{A}^*$ with respect to $\mathbf{M}^{-1}$ and $\boldsymbol{\eta}_i$ are eigenvectors of $\mathbf{A}^*\mathbf{MA}$ with respect to $\mathbf{N}$. Then

$$\mathbf{A}_{\mathbf{MN}}^+ = \mu_1^{-1}\boldsymbol{\eta}_1\boldsymbol{\xi}_1^* + \cdots + \mu_r^{-1}\boldsymbol{\eta}_r\boldsymbol{\xi}_r^*. \tag{11.3.7}$$

11.4 SPECIAL TECHNIQUES

Orthogonalization Method

Let $\mathbf{A}$ be a $m \times n$ matrix of rank r. Let $\boldsymbol{\xi}_i$, $i = 1, 2, \ldots, r$ be the orthonormal set of vectors obtained by Gram-Schmidt orthogonalization (see Rao 1965, p. 9) applied to columns of $\mathbf{A}$. Then we have (see Pyle 1964)

$$\mathbf{AA}^+ = \sum_{i=1}^{r} \boldsymbol{\xi}_i\boldsymbol{\xi}_i^*. \tag{11.4.1}$$

Note that $\mathbf{AA}^+$ is the projection operator projecting vectors onto $\mathscr{M}(\mathbf{A})$ and, hence, $\mathbf{AA}^+$ is hermitian and idempotent. So the eigenvalues of $\mathbf{AA}^+$ are 1 and 0 with algebraic multiplicities r and $m - r$, respectively. Further, observe that $\mathbf{AA}^+\boldsymbol{\xi}_i = \boldsymbol{\xi}_i$, $i = 1, 2, \ldots, r$ and, hence, $\boldsymbol{\xi}_i$, $i = 1, 2, \ldots, r$ are the eigenvectors corresponding to the eigenvalue 1 of $\mathbf{AA}^+$. Then the result (11.4.1) follows from the spectral decomposition theorem for hermitian matrices. Let $\mathbf{x}_i$ be a solution of the equation $\mathbf{Ax}_i = \mathbf{d}_i$ obtained by a gradient projection algorithm, where $\mathbf{d}_i$ is the i-th column of $\mathbf{AA}^+$ as defined in (11.4.1). Then $\mathbf{A}^+$ is the matrix whose i-th column is $\mathbf{x}_i$, $i = 1, \ldots, m$.

A computer program in FORTRAN-II for this method, written by Bhimasankaram, is available at the Indian Statistical Institute.

Iterative Method

Let $\mathbf{A}$ be a $m \times n$ matrix. Let $\mathbf{P}_{\mathbf{A}}$ and $\mathbf{P}_{\mathbf{A}^*}$ denote the orthogonal projection operators onto $\mathscr{M}(\mathbf{A})$ and $\mathscr{M}(\mathbf{A}^*)$, respectively. Then as shown by

Ben-Israel (1965b, 1966b) the sequence of matrices defined by

$$\mathbf{X}_{n+1} = \mathbf{X}_n(2\mathbf{P}_\mathbf{A} - \mathbf{A}\mathbf{X}_n) \qquad (n = 0, 1, 2, \ldots),$$

where $\mathbf{X}_o$ is a $n \times m$ matrix satisfying

(a) $\mathbf{X}_o = \mathbf{A}^*\mathbf{B}_o$ for some nonsingular matrix $\mathbf{B}_o$,
(b) $\mathbf{X}_o = \mathbf{C}_o\mathbf{A}^*$ for some nonsingular matrix $\mathbf{C}_o$,
(c) $\|\mathbf{A}\mathbf{X}_o - \mathbf{P}_\mathbf{A}\| < 1$, and (d) $\|\mathbf{X}_o\mathbf{A} - \mathbf{P}_{\mathbf{A}*}\| < 1$

converges to $\mathbf{A}^+$, where $\|\mathbf{A}\|$ is any valid matrix norm.

Remarks. For applying the above method one needs $\mathbf{P}_\mathbf{A}$ and $\mathbf{P}_{\mathbf{A}^*}$ which may imply that $\mathbf{A}^+$ should itself be known. Thus for practical purpose this does not seem to be of much value. This shortcoming is rectified by the following (also due to Ben-Israel, 1966b).

Let $\lambda_1 \geq \lambda_2 \geq \cdots \geq \lambda_r$ be the nonnull eigenvalues of $\mathbf{A}\mathbf{A}^*$. If the real scalar α satisfies, $0 < \alpha < 2/\lambda_1$, then the sequence defined by $\mathbf{X}_o = \alpha\mathbf{A}^*$ and $\mathbf{X}_{n+1} = \mathbf{X}_n(2\mathbf{I} - \mathbf{A}\mathbf{X}_n)$, $n = 0, 1, 2, \ldots$, converges to $\mathbf{A}^+$.

Singular Value Decomposition

First $\mathbf{A}$ is reduced to a bidiagonal form by using Householder's transformations on $\mathbf{A}$ to the left and right in a suitable way. Let $\mathbf{P}^*\mathbf{A}\mathbf{Q} = \mathbf{J}_o$, where $\mathbf{P}$ and $\mathbf{Q}$ are unitary matrices and $\mathbf{J}_o$ is in bidiagonal form. Then $\mathbf{J}_o$ is reduced to diagonal form by an algorithm similar to Q-R. algorithm (Francis, 1961, 1962) by making unitary transformation on the left and right of $\mathbf{J}_o$, iteratively. Let $\mathbf{G}\boldsymbol{\Sigma}\mathbf{H}* = \mathbf{J}_o$ where $\boldsymbol{\Sigma}$ is diagonal and $\mathbf{G}$ and $\mathbf{H}^*$ are unitary. Then we have

$$\mathbf{A} = \mathbf{P}\mathbf{G}\boldsymbol{\Sigma}\mathbf{H}^*\mathbf{Q}^* \text{ and } \mathbf{A}^+ = \mathbf{Q}\mathbf{H}\boldsymbol{\Sigma}^+(\mathbf{P}\mathbf{G})^*.$$

A program for the singular value decomposition, written in ALGOL, is available in Golub and Reinsch (1970).

11.5 LEAST SQUARES SOLUTION

A least-squares solution of $\mathbf{Y} = \mathbf{X}\boldsymbol{\beta}$ is of special interest in statistical and other applications. In this section we discuss the computational devices available for this purpose.

In particular we consider two techniques. One is traditional, involving the construction of the normal equation $\mathbf{X}'\mathbf{X}\boldsymbol{\beta} = \mathbf{X}'\mathbf{Y}$ and solving it by Gauss-Doolittle or square root methods; another is the approach advocated by Golub (1965), Businger and Golub (1965), and Björck and Golub (1968) of directly reducing the observational equation $\mathbf{X}\boldsymbol{\beta} = \mathbf{Y}$ by a series of Householder transformations (see Rao, 1965, p. 20). Computationally, the latter method has some advantages, but in problems where the normal equation is well conditioned the former method offers some simplicity. We give brief descriptions of both these methods. The computations involved in the normal equation approach were developed in Rao (1955), where the pseudoinverse was first introduced.

11.5.1 Normal Equation Approach

First we obtain the normal equation $\mathbf{X}'\mathbf{X}\boldsymbol{\beta} = \mathbf{X}'\mathbf{Y}$, which we represent as $\mathbf{S}\boldsymbol{\beta} = \mathbf{q}$, and reduce the matrix $(\mathbf{S} \vdots \mathbf{q} \vdots \mathbf{I})$, with a unit matrix appended to $(\mathbf{S} \vdots \mathbf{q})$, by Gauss-Doolittle or square root methods. We do this to obtain not only a solution of $\mathbf{S}\boldsymbol{\beta} = \mathbf{q}$, but a g-inverse of $\mathbf{S}$ which is needed in further statistical work, such as determining standard errors of estimated regression coefficients, etc., we illustrate the method by a numerical example. Let the normal equation be

$$\begin{pmatrix} 4 & 8 & 12 \\ 8 & 16 & 24 \\ 12 & 24 & 48 \end{pmatrix} \begin{pmatrix} \beta_1 \\ \beta_2 \\ \beta_3 \end{pmatrix} = \begin{pmatrix} 16 \\ 32 \\ 72 \end{pmatrix}$$

TABLE 1
GAUSS–DOOLITTLE TECHNIQUE

Row		S		0′	1′	2′	3′
1	4	8	12	16	1	.	.
2	8	16	24	32	.	1	.
3	12	24	48	72	.	.	1
1.0	4	8	12	16	1	.	.
1.1	1	2	3	4	1/4	.	.
2.0		0	0	0	−2	1	.
2.1		0	0	0	0	0	.
3.0			12	24	−3	0	1
3.1			1	2	−1/4	0	1/12

The method of reduction by Gauss–Doolittle pivotal condensation is well known (see Dwyer, 1951, p. 170). At each stage we have a reduced row $(i \cdot 0)$ and its normalized form $(i \cdot 1)$ with the pivotal coefficient reduced to unity.

When $\mathbf{S}$ is singular, we encounter a zero pivotal element at some stage. In such a case the elements in all columns under $\mathbf{S}$ and (0′) will be zero and not under the columns (1′, 2′, 3′). In the normalized form of such a row, we put zeros in all positions as in row (2.1) in Table 1. We define the product of two columns $(x)\cdot(y)$ in Table 1 as the sum of products of elements one chosen from row (i · 0) from col. (x) and another chosen from row $(i \cdot 1)$ from col. (y), $i = 1, 2, 3$. With this convention we have the following

Solution of Normal equation.

$$\beta_1 = (0')\cdot(1') = 16 \times (1/4) + (0 \times 0) + 24(-1/4) = -2$$

$$\beta_2 = (0')\cdot(2') = (16 \times 0) + (0 \times 0) + 24 \times 0 = 0$$

$$\beta_3 = (0')\cdot(3') = 16 \times 0 + (0 \times 0) + 24(1/12) = 2$$

g-inverse of $\mathbf{S}$ *denoted by* (c_{ij})

$$c_{11} = (1')\cdot(1') = 1(1/4) + (-2)(0) + (-3)(-1/4) = 1$$

$$c_{12} = (1')\cdot(2') = 1(0) + (-2)(0) + (-3)(0) = 0$$

$$\cdots \quad \cdots \quad \cdots \quad \cdots \quad \cdots \quad \cdots \quad \cdots \quad \cdots \quad \cdots$$

$$c_{33} = (3')(3') = (0)(0) + (0)(0) + (1)(1/12) = 1/12.$$

In order to obtain a g-inverse we need not reduce the entire unit matrix. It is sufficient to compute only the underlined elements in Table 1. In such a case the elements c_{ij} are obtained by a series of back solutions. First we solve the equation $\mathbf{Sx} = \mathbf{e}_3$, where $\mathbf{e}_3$ is the unit vector under (3′). The back solution gives

$$c_{33} = \tfrac{1}{12} \quad c_{23} = 0 \quad c_{13} = -\tfrac{1}{4}.$$

Next we solve the equation $\mathbf{Sx} = \mathbf{e}_2$, where $\mathbf{e}_2$ is the unit vector under (2′), observing that the solution (c_{32}) for x_3 is the same as that for x_2 in the previous equation (c_{23}). Now back solutions for c_{22} and c_{21} give

$$(c_{32} = c_{23} = 0) \quad c_{22} = 0 \quad c_{12} = 0.$$

Now we solve the equation $\mathbf{Sx} = \mathbf{e}_1$, where $\mathbf{e}_1$ is the unit vector under (1′), observing that $c_{31} = c_{13}$ and $c_{21} = c_{12}$. Now back solution for c_{11} is

$$(c_{31} = -\tfrac{1}{4} \quad c_{21} = 0) \quad c_{11} = 1.$$

The matrix (c_{ij}) is the same as that obtained earlier.

The expression for the residual sum of squares is $\mathbf{Y'Y} - (0')\cdot(0')$ where

$$(0')\cdot(0') = (16 \times 4) + (0 \times 0) + (24 \times 2) = 112.$$

Check that $(0')\cdot(0')$ is numerically the same as

$$\mathbf{q}'\hat{\boldsymbol{\beta}} = [16 \times (-2)] + [32 \times (0)] + [72 \times 2] = 112,$$

where $\hat{\boldsymbol{\beta}}$ is the least-squares solution of $\boldsymbol{\beta}$.

We shall now demonstrate the square root method. The reduction by this method is set out in Table 2.

TABLE 2
SQUARE ROOT METHOD

Row		S		0′	1′	2′	3′
1	4	8	12	16	1	.	.
2	8	16	24	32	.	1	.
3	12	24	48	72	.	.	1
1.1	2	4	6	8	1/2	.	.
2.1		0	0	0	0	0	0
3.1			$\sqrt{12}$	$24/\sqrt{12}$	$-3/\sqrt{12}$	0	$1/\sqrt{12}$

We define the product of two columns $(x)\cdot(y)$ in Table 2 as the sum of products of elements in columns (x) and (y) taken from rows $(i\cdot 1)$, $i = 1, 2, 3$. With this convention

$$\beta_1 = (0')\cdot(1') = 8\left(\frac{1}{2}\right) + (0)(0) + \left(\frac{24}{\sqrt{12}}\right)\left(\frac{-3}{\sqrt{12}}\right) = -2$$

$$\beta_2 = (0')\cdot(2') = 8(0) + (0)(0) + \left(\frac{24}{\sqrt{12}}\right)(0) = 0$$

$$\beta_3 = (0')\cdot(3') = 8(0) + (0)(0) + \left(\frac{24}{\sqrt{12}}\right)\left(\frac{1}{\sqrt{12}}\right) = 2$$

The elements c_{ij} of a g-inverse of **S** are

$$c_{11} = (1')\cdot(1') = \left(\frac{1}{2}\right)\left(\frac{1}{2}\right) + (0)(0) + \left(\frac{-3}{\sqrt{12}}\right)\left(\frac{-3}{\sqrt{12}}\right) = 1$$

$$c_{12} = (1')\cdot(2') = \left(\frac{1}{2}\right)(0) + (0)(0) + \left(\frac{-3}{\sqrt{12}}\right)(0) = 0$$

$$\cdots \quad \cdots \quad \cdots \quad \cdots \quad \cdots \quad \cdots \quad \cdots \quad \cdots \quad \cdots \quad \cdots$$

$$c_{33} = (3')(3') = (0)(0) + (0)(0) + \left(\frac{1}{\sqrt{12}}\right)\left(\frac{1}{\sqrt{12}}\right) = \frac{1}{12}$$

The residual sum of squares is $Y'Y - (0')\cdot(0')$, where

$$(0')\cdot(0') = (8)(8) + (0)(0) + \left(\frac{24}{\sqrt{12}}\right)\left(\frac{24}{\sqrt{12}}\right) = 112$$

11.5.2 Reduction by Householder Transformation

This method is useful when the normal equations are not well conditioned. We consider the matrix $(\mathbf{X} \vdots \mathbf{Y})$ and reduce it by a series of Householder transformations (HT).

Consider the first column of $\mathbf{X}$. If it has a nonzero element, apply HT to have a nonzero value in the first position and zero elsewhere in the first column (see Rao, 1965, p. 20). If all the elements of the first column are zero, move to the nearest column which has a nonzero value, say the i-th, and apply HT to have a nonzero value in the *first position* and zero elsewhere in the i-th column. Now omit the first row and repeat the process stated on the reduced matrix, and so on, till all the columns of $\mathbf{X}$ are covered. Then by a rearrangement of columns (i.e., by renaming the parameters), if necessary, the reduced matrix is of the form

$$\begin{pmatrix} \mathbf{T} & \mathbf{U} & \mathbf{Q}_1 \\ \mathbf{0} & \mathbf{0} & \mathbf{Q}_2 \end{pmatrix},$$

where $\mathbf{T}$ is an upper triangular and nonsingular matrix of rank r equal to that of $\mathbf{X}$, $\mathbf{Q}_1$ is r-vector, $\mathbf{Q}_2$ is $(n - r)$-vector and $\mathbf{U}$ is a $r \times (m - r)$ matrix. A least-squares solution is

$$(\beta_1, \ldots, \beta_r)' = \mathbf{T}^{-1}\mathbf{Q}_1, \beta_{r+1} = \cdots = \beta_m = 0$$

and the residual sum of squares is

$$R_o^2 = \mathbf{Q}_2'\mathbf{Q}_2.$$

A g-inverse of $\mathbf{S} = \mathbf{X}'\mathbf{X}$ is

$$\begin{pmatrix} \mathbf{T}^{-1}(\mathbf{T}^{-1})' & \mathbf{0} \\ \mathbf{0} & \mathbf{0} \end{pmatrix}.$$

Note. In actual practice, where approximations have to be made in numerical computations, it may be difficult to decide whether a particular value, at any stage of reduction, is a real deviation from zero or is a rounding off error in the place of zero. In many practical problems, it may be known in advance as to which columns are dependent on the previous ones, in which case the detection of a zero column at any stage does not present any problem. When such information is not available, one has to be careful and we recommend in this connection the computer program given by Golub and Reinsch (1970). Sometimes we have to consider least-squares problems with constraints on the parameters such as $\mathbf{F}\boldsymbol{\beta} = \boldsymbol{\xi}$. In such case the method of reduction by HT is as follows:

We consider the matrix

$$\begin{pmatrix} \mathbf{F} & \boldsymbol{\xi} \\ \mathbf{X} & \mathbf{Y} \end{pmatrix}.$$

Choose a nonzero element in the first row of $\mathbf{F}$, say in the i-th column, and sweep out the other elements in the i-th column. Omitting the first row and the swept-out columns, the same procedure is applied on the remaining matrix, that is, by choosing a nonzero element in the first row of the reduced $\mathbf{F}$ matrix and sweeping out an appropriate column. If a row consists of all zeroes at any stage, the next row is considered. Such a process of sweep-out is continued, till no row with a nonzero element is left in any reduced $\mathbf{F}$ matrix. For further operations, which start with the first row of the reduced $\mathbf{X}$ matrix we apply HT as in the case of no constraints. The final form of the matrix, after a rearrangement of columns, if necessary, and after omitting zero rows in the reduced form of $\mathbf{F}$, is

$$\begin{pmatrix} \mathbf{U}_1 & \mathbf{U}_2 & \mathbf{U}_3 & \mathbf{Q}_0 \\ \mathbf{0} & \mathbf{T}_1 & \mathbf{T}_2 & \mathbf{Q}_1 \\ \mathbf{0} & \mathbf{0} & \mathbf{0} & \mathbf{Q}_2 \end{pmatrix},$$

where $\mathbf{U}_1$ and $\mathbf{T}_1$ are upper triangular and nonsingular matrices. Let the order of

$$\begin{pmatrix} \mathbf{U}_1 & \mathbf{U}_2 \\ \mathbf{0} & \mathbf{T}_1 \end{pmatrix}$$

be $h \times h$. Then a solution to normal equations is

$$(\beta_1, \ldots, \beta_h)' = \begin{pmatrix} \mathbf{U}_1 & \mathbf{U}_2 \\ \mathbf{0} & \mathbf{T}_1 \end{pmatrix}^{-1} \begin{pmatrix} \mathbf{Q}_0 \\ \mathbf{Q}_1 \end{pmatrix}$$

$$\beta_{h+1} = 0 = \cdots = \beta_m,$$

and the residual sum of squares is

$$R_o^2 = \mathbf{Q}_2'\mathbf{Q}_2.$$

Bibliography on Generalized Inverses and Applications

Not all references are cited in the text.

Aitken, A. C. (1934): On least squares and linear combinations of observations, *Proc. Roy. Soc., Edinburgh*, **55**, 42–48.

Afriat, S. N. (1957): Orthogonal and oblique projectors and the characteristics of pairs of vector spaces, *Proc. Cambridge Philos. Soc.*, **LIII**, 800–816.

Albert, A. (1964): *An Introduction and Beginner's Guide to Matrix Pseudo Inverses*. ARCON—Advanced Research Consultants, Lexington, Mass.

——— (1965): *Real Time Computation of Constrained Least Squares Estimators*. ARCON—Advanced Research Consultants, Lexington, Mass.

Albert, A. and R. W. Sittler (1965): A method of computing least squares estimators that keep up with the data, *SIAM J. Control*, Ser A: **3**, 384–417.

Altman, M. (1957): On the approximate solution of linear algebraic equations, *Bull. Acad. Polon. Sci. Ser. Sci. Math. Astronom. Phys., Cl. 3*, **5**, 365–370.

Amburgey, J. K., T. O. Lewis, and T. L. Boullion (1968): On computing generalized characteristic vectors and values for a rectangular matrix, *Proc. Symp. Theo. Appl. Generalized Inverses of Matrices*, Mathematics Ser. No. 4, Texas Technological College, Lubbock.

Amir-Moez, A. R. (1967): *Singular Values of Linear Transformations*. Mathematics Ser. No. 2, Texas Technological College, Lubbock.

Anderson, W. N., Jr. and R. J. Duffin (1969): Series and parallel addition of matrices, *J. Math. Anal. Appl.*, **26**, 576–594.

Anderson, W. N., Jr., G. B. Kleindorfer, P. R. Kleindorfer, and M. B. Woodroofe (1969): Consistent estimates of parameters of a linear system, *Ann. Math. Statist.*, **40**, 2064–2075.

Arghiriade, E. (1963): Sur les matrices qui sont permutables avec leur inverse généralisée, *Atti Accad. Naz. Lincei Rend. Cl. Sci. Fis. Mat. Natur*, **35**, 244–251.

——— (1967): Remarques sur l'inverse généralisée d'un produit de matrices, *Atti Accad. Naz. Lincei Rend. Cl. Sci. Fis. Mat. Natur*, **42**, 621–625.

Arghiriade, E., and A. Dragomir (1963): Une nouvelle définition de l'inverse généralisée d'une matrices, *Atti Accad. Naz. Lincei Rend. Cl. Sci. Fis. Mat. Natur*, **35**, 158–163.

Autonne, L. (1917): Sur les matrices hypohermitiennes et sur les matrices unitaires, *Ann. Univ. Lyon Sect. A*, **38**, 1–77.

Banerji, K. S. (1966): Singularity in Hotelling's weighing designs and a generalized inverse, *Ann. Math. Statist.*, **37**, 1021–1032.

——— (1966): On non-randomized fractional weighing designs, *Ann. Math. Statist.*, **37**, 1836.

Banerji, K. S., and W. T. Federer (1968): On the structure and analysis of singular fractional replicates, *Ann. Math. Statist.*, **39**, 657.

Baskett, T. S., and I. J. Katz (1969): Theorems on products of EPr matrices, *J. Lin. Alg. Appl.*, **20**, 87–103.

Bauer, F. L. (1965): Elimination with weighted row combinations for solving linear equations and least squares problems, *Numer. Math.*, **7**, 338–352.

Ben-Israel, A. (1964): On direct sum decompositions of Hestenes algebras, *Israel J. Math.*, **2**, 50–54.

——— (1965a): A modified Newton–Raphson method for the solution of systems of equations, *Israel J. Math.*, **3**, 94–98.

——— (1965b): An iterative method for computing the generalized inverse of an arbitrary matrix, *Math. Comp.*, **19**, 452–455.

——— (1966a): A Newton–Raphson method for the solution of systems of equations, *J. Math. Anal. Appl.*, **15**, 243–252.

——— (1966b): A note on an iterative method for generalized inversion of matrices, *Math. Comp.*, **20**, 439–440.

——— (1966c): A note on the Cayley transform, *Notices Amer. Math. Soc.*, **13**, 599.

——— (1966d): On error bounds for generalized inverses, *SIAM J. Numer. Anal.*, **3**, 585–592.

——— (1966e): Application of Newton–Raphson method to the special eigenvalue problem. Unpublished manuscript.

——— (1967a): On the geometry of subspaces in Euclidean n-spaces, *SIAM J. Appl. Math.*, **15**, 1189–1198.

——— (1967b): On iterative methods for solving nonlinear least squares problems over convex sets, *Israel J. Math.*, **5**, 211–224.

———: *Lectures on generalized inverses.* To be published.

———: On Newton's method in nonlinear programming. Unpublished report.

——— (1968): On application of generalized inverses in nonlinear analysis. *Proc. Symp. Theo. Appl. Generalized Inverses of Matrices*, Mathematics Ser. No. 4, Texas Technological College, Lubbock.

Ben-Israel, A., and A. Charnes (1961): Projection properties and the Neumann–Euler expansion for the Moore–Penrose inverse of an arbitrary matrix. *ONR Research Memo.*, **40**, The Technological Institute, Northwestern University, Evanston, Ill.

——— (1963a): Contributions to the theory of generalized inverses, *SIAM J. Appl. Math.*, **11**, 667–699.

——— (1963b): Generalized inverses and the Bott–Duffin network analysis, *J. Math. Anal. Appl.*, **7**, 428–435, Erratum: ibid.

——— (1968a): An explicit solution of a special class of linear programming problems. *Operations Research*, **16**, 1166–1175.

——— (1968b): On the intersection of cones and subspaces, *Bull. Amer. Math. Soc.*, **74**, 541–544.

Ben-Israel, A., and P. D. Robers (1968): On generalized inverses and interval linear programming, *Proc. Symp. Theo. Appl. Generalized Inverses of Matrices*, Mathematics Ser. No. 4, Texas Technological College, Lubbock.

Ben-Israel, A., and D. Dohen (1966): On iterative computation of generalized inverses and associated projections, *SIAM J. Numer. Anal.*, **3**, 410–419.

Ben-Israel, A., and Y. Ijiri (1963): A report on the machine calculation of the generalized inverse of an arbitrary matrix. *ONR Research Memo.*, **110**, Carnegie Institute of Technology, Pittsburgh, Pa.

Ben-Israel, A. and S. J. Wersan (1962): A least square method for computing the generalized inverse of an arbitrary complex matrix, *ONR Research Mem.*, **61**, The Technological Institute, Northwestern University, Evanston, Ill.

——— (1963): An elimination method for computing the generalized inverse of an arbitrary complex matrix, *J. Assoc. Comput. Mach.*, **10**, 532–537.

Beutler, F. J. (1965): The operator theory of the pseudoinverse, *J. Math. Anal. Appl.*, **10**, 451–493.

Bhimasankaram, P. (1971): On generalized inverses of partitioned matrices, to appear in *Sankhyā*, Series A, **33**.

Bhimasankaram, P. and S. K. Mitra (1969): On a theorem of Rao on g-inverses of matrices, *Sankhyā*, Ser. A, **31**, 365–368.

Bjerhammer, A. (1951): Rectangular reciprocal matrices with special reference to geodetic calculations. *Bull. Géodésique*, **52**, 188–220.

——— (1957): Application of calculus of matrices to method of least squares with special reference to geodetic calculations, *Kungl. Tekn. Högsk. Handl. Stockholm*, **49**, 1–86.

——— (1958): A generalized matrix algebra, *Kungl. Tekn. Högsk. Handl. Stockholm*, **124**, 1–32.

Björck, A. and G. Golub (1968): Iterative refinements of linear least squares solutions by Householder transformations, Tech. Report No. CS **83**, Stanford University.

Blattner, J. W. (1962): Bordered matrices, *SIAM J. Appl. Math.*, **10**, 528–536.

Bleick, W. E. (1940): A least-square accumulation theorem, *Ann. Math. Statist.*, **11**, 225–226.

Boot, J. C. G. (1963): The computation of the generalized inverse of singular or rectangular matrices, *Amer. Math. Monthly*, **70**, 302–303.

——— (1965): *Projection Matrices and the Generalized Inverse*. State University of New York, Buffalo.

Boros, E. (1964): On the generalized inverse of an EPr matrix, *An. Univ. Timisoara Ser. Sti. Mat. Fiz.*, **2**, 33–38.

Boros, E. and I. Sturz (1963): On quasi-inverse matrices, *An. Univ. Timisoara Ser. Sti. Mat. Fiz.*, **1**, 59–66.

Bose, R. C. (1959): Analysis of variance. Unpublished lecture notes, University of North Carolina, Chapel Hill.

Bott, R. and R. J. Duffin (1953): On the algebra of networks, *Trans. Amer. Math. Soc.*, **74**, 99–109.

Boullion, T. L. and P. L. Odell (1969): A generalization of the Wielandt inequality, *Texas J. Sci.*, **20**, 255–260.

——— (1969): A note on the Scroggs–Odell pseudoinverse, *SIAM J. Appl. Math.*, **17**, 7–10.

den Broeder, C. G., Jr. and A. Charnes (1957): Contributions to the theory of generalized inverses for matrices, *Res. Memo ONR Project on Planning and Control of Industrial Operations*, Purdue University.

Brand, L. (1962): The solution of linear algebraic equations, *Math. Gaz.*, **46**, 203–207.

Businger, P. and G. H. Golub (1965): Linear least square solutions by Householder transformations, *Numer. Math.*, **7**, 269–276.

Charnes, A. and M. Kirby (1963): A linear programming application of a left inverse of a basis matrix, *ONR Research Memo.*, **91**, The Technological Institute, Northwestern University, Evanston, Ill.

——— (1965): Modular design, generalized inverses and convex programming, *Operations Res.*, **13**, 836–847.

Chernoff, H. (1953): Locally optimal designs for estimating parameters, *Ann. Math. Statist.*, **24**, 586–602.

Chipman, J. S. (1964): On least-squares with insufficient observations, *J. Amer. Statist. Assoc.*, **59**, 1078–1111.

——— (1969): Specification problems in regression analysis, *Proc. Symp. Theo. Appl. Generalized Inverses of Matrices*, Mathematics Ser. No. 4, Texas Technological College, Lubbock.

Chipman, J. S. and M. M. Rao (1964): Projections, generalized inverse and quadratic forms, *J. Math. Anal. Appl.*, **9**, 1–11.

——— (1964): On the treatment of linear restrictions in regression analysis, *Econometrica*, **32**, 198–209.

Cho, C. Y. (1966): Talks on generalized inverses and solutions of large, approximately singular linear systems, *MRC Tech. Summary Report*, **644**, Mathematics Research Center, U.S. Army, University of Wisconsin, Madison.

Clifford, A. H. (1941): Semigroups admitting relative inverses, *Ann. of Math.*, **42**, 1037–1049.

Clifford, A. H. and G. B. Preston (1961): The algebraic theory of semigroups, I, *Mathematical Surveys*, **7**, American Math. Soc., Providence, R.I.

Cline, R. E. (1958): On the computation of the generalized inverse A^+, of an arbitrary matrix A, and the use of certain associated eigenvectors in solving the allocation problem. Preliminary report, Statistical and Computing Laboratory, Purdue University.

——— (1964): Note on the generalized inverse of the product of matrices, *SIAM Rev.*, **6**, 57–58.

——— (1964): Representations for the generalized inverse of a partitioned matrix, *SIAM J. Appl. Math.*, **12**, 588–600.

——— (1965): Representations of the generalized inverse of sums of matrices, *SIAM J. Numer. Anal.*, Ser. B, **2**, 99–114.

——— (1965): An application of representations for the generalized inverse of a matrix, *MRC Tech. Summary Report*, **592**, Mathematics Research Center, U.S. Army, University of Wisconsin, Madison.

——— (1968): Inverses of rank invariant powers of a matrix, *Proc. Symp. Theor. Appl. Generalized Inverses of Matrices*, Mathematics Ser. No. 4, Texas Technological College, Lubbock.

Cochran, W. G. (1934): The distribution of quadratic forms in a normal system with applications to analysis of covariance, *Proc. Cambridge Philos. Soc.*, **30**, 178–191.

Craig, A. T. (1943): Note on the independence of certain quadratic forms, *Ann. Math. Statist.*, **14**, 195–197.

Croisot, R. (1953): Demi-groups inversifs et demi-groups réunions de semi-groups simples, *Ann. Sci. École Norm. Sup.*, **70**, 361–379.

David, F. N. and J. Neyman (1938): Extension of the Markoff theorem on least squares, *Statist. Res. Mem.*, **2**, 105–116.

Decell, H. P., Jr. (1965a): An alternative form of the generalized inverse of an arbitrary complex matrix, *SIAM Rev.*, **7**, 356–358.

——— (1965b): An application of the Cayley–Hamilton theorem to generalized matrix inversion, *SIAM Rev.*, **7**, 526–528.

——— (1965c): A special form of the generalized inverse of an arbitrary complex matrix, *NASA TN D-2784*, Washington, D.C.

——— (1965d): An application of generalized matrix inversion to sequential least squares parameter estimation, *NASA TN D-2830*, Washington, D.C.

——— (1968): A characterization of the maximal subgroups of the semigroup of $m \times n$ complex matrices, *Proc. Symp. Theo. Appl. Generalized Inverses of Matrices*, Mathematics Ser. No. 4, Texas Technological College, Lubbock.

Decell H. P., Jr. and C. L. Wiginson (1968): A characterization of the maximal subgroups of the semigroup of $n \times n$ complex matrices, *Czechoslovak Math. J.*, **18**, 675–677.

Decell, H. P., Jr. and S. W. Kahng (1966): An iterative method for computing the generalized inverse of a matrix, *NASA TN D-3464*, Washington, D.C.

Decell, H. P., Jr. and P. L. Odell (1966): A note concerning a generalization of the Gauss–Markov theorem, *Texas J. Sci.*, **27**, 21–24.

——— (1967): On computing the fixed point probability vector of regular or ergodic transition matrices, *J. Assoc. Comput. Mach.*, **14**, 765–768.

——— (1967): On the fixed point probability vector of regular or ergodic transition matrices, *J. Amer. Statist. Assoc.*, **62**, 600–602.

Delaney, J. C. and G. G. Gaffney: Efficiency of generalized matrix inversion methods. *NASA–MSC Internal Note.*

Delaney, J. C. and F. M. Speed: A new algorithm for calculating the generalized inverse of an arbitrary $m \times m$ matrix. *NASA–MSC Internal Note.*

Desoer, C. A. and B. H. Whalen (1963): A note on pseudoinverses, *SIAM J. Appl. Math.*, **11**, 442–447.

Doković, D. (1965): On the generalized inverse for matrices, *Glasnik Mat.-Fiz. Astron. Ser. II, Društvo Mat. Fiz. Hrvatske*, **20**, 51–55.

Dommanget, J. (1963): L'inverse d'un cracovien rectangulaire: Son emploi dans la résolution des systèmes d'équations linéaires, *Publ. Sci. Tech. Ministère de l'Air (Paris) Notes Tech.*, **128**, 11–41.

Doust, A. and V. E. Price (1964): The latent roots and vectors of a singular matrix, *Comput. J.*, **7**, 222–227.

Dragomir, P. (1963): On the Greville–Moore formula for calculating the generalized inverse matrix, *An. Univ. Timisoara Ser. Sti. Math-Fiz.*, **1**, 115–119.

——— (1964): The generalized inverse of a bilinear form, *An. Univ. Timisoara Ser. Sti. Mat.-Fiz.*, **2**, 71–76.

Drazin, M. P. (1958): Pseudoinverses in associative rings and semigroups, *Amer. Math. Monthly*, **65**, 506–514.

Dück, Werner (1964): Einzelschrittverfahren zur matrizeninversion, *Z. Angew Math. Mech.*, **44**, 401–403.

Eckart, C. and G. Young (1936): The approximation of one matrix by another of lower rank, *Psychometrika*, **I**, 211–218.

——— (1939), A principal axis transformation for nonhermitian matrices, *Bull. Amer. Math. Soc.*, **45**, 118–121.

Egervary, E. (1959): Über eine konstruktive methode zur reduktion einer matrix auf die Jorkansche normalform, *Acta Math. Acad. Sci. Hungar*, **10**, 31–54.

——— (1960): On rank diminishing operations and their applications to the solution of linear equations, *Z. Angew Math. Phys.*, **11**, 376–386.

Englefield, M. J. (1966): The commuting inverse of a square matrix, *Proc. Cambridge Philos. Soc.*, **62**, 667–671.

Erdelyi, I.: A generalized group inverse of square matrices. Unpublished report.

——— (1966a): On partial isometries in finite dimensional Euclidean spaces, *SIAM J. Appl. Math.*, **14**, 453–467.

——— (1966b): On the reverse order law related to the generalized inverse of matrix products, *J. Assoc. Comput. Mach.*, **13**, 439–443.

——— (1967a): The quasi-commuting inverses for a square matrix, *Atti. Accad. Naz. Lencei Rend. Cl. Sci. Fis. Math. Natur*, **42**, 626–633.

——— (1967b): On the matrix equation $Ax = \lambda Bx$. *J. Math. Anal. Appl.*, **17**, 119–132.

——— (1967c): On normal partial isometries in finite dimensional Euclidean spaces, *Technical Report*, **6**, Kansas State University, Manhattan.

——— (1968a): Partial isometries closed under multiplication on Hilbert spaces, *J. Math. Anal. Appl.*, **22**, 546–551.

——— (1968b): Partial isometries defined by a spectral property on unitary spaces, *Technical Report*, **12**, Kansas State University, Manhattan.

——— (1968c): Normal partial isometries closed under multiplication on unitary spaces, *Atti. Accad. Lincei Rend. Cl. Sci. Fis. Mat. Natur*, **43**,

——— (1968d): Partial isometrics and generalized inverses, *Proc. Symp. Theo. Appl. Generalized*

Inverses of Matrices, Mathematics Ser. No. 4, Texas Technological College, Lubbock.

Faddeev, D. K. and V. N. Faddeeva (1963): *Computational Methods of Linear Algebra*. Freeman, San Francisco.

Faddeev, D. K., V. N. Kublanovskaya, and V. N. Faddeeva (1966): On linear algebraic systems with rectangular and ill conditioned matrices, *Colloque International de Mathématiques du CNRS, Besancon*.

Fisher, A. G. (1967): On construction and properties of the general inverse, *SIAM J. Appl. Math.*, **15**, 269–272.

Foster, Manus (1961): An application of the Wiener–Kolmogorov smoothing theory to matrix inversion, *SIAM J. Appl. Math.*, **9**, 387–392.

Foulis, D. J. (1963): Relative inverses in Baer *-semigroups, *Michigan Math. J.*, **10**, 65–84.

Frame, J. S. (1964): Matrix operations and generalized inverses, *IEEE Spectrum*, 209–220.

Francis, J. (1961): The Q.R. transformation: a Unitary analogue to the L.R. transformation—Part I, *Computer J.*, **4**, 265–271.

——— (1962): The Q.R. transformation—Part II, *Computer J.*, **4**, 332–345.

Franck, P. (1962): Sur la distance minimale d'une matrice régulière donnée au lieu des matrices singulières, *Deux. Cong. Assoc. Française Calcul. et Traitement Information*, Paris, Gauthier-Villars, 55–60.

Gabrial, R. (1965): Estinderea complementilor algebrici generalizati la matrici oarecare, *Stud. Cerc. Mat.*, **10**,

——— (1967): Über die verallgemeinerte inverse einer matrix. Unpublished manuscript.

Gaches, J., J. L. Rigal, and X. Rousset De Pina (1965): Distance Euclidienne d'une application linéaire σ au lieu des applications de rang r donné. Détermination d'une meilleure approximation de rang r. *C. R. Acad. Sci.*, Paris, **260**, 5672–5674.

Gainer, P. A.: A method for computing the effect of an additional observation on a previous least squares estimate. *NASA TN D-1599*.

Gantmacher, F. R. (1959): *Applications of the Theory of Matrices*, Wiley-Interscience, New York.

Gavurin, M. K. (1964–65): Ill-conditioned systems of linear algebraic equations, *Internat. J. Comput. Math.*, **I**, 36–50.

Giurescu, C. and R. Gabriel (1964): Some properties of the generalized matrix inverse and semi-inverse, *An. Univ. Timisoara Ser. Sti. Mat.-Fiz.*, **2**, 103–111

Glassey, C. R. (1966): An orthogonalization method of computing the generalized inverse of a matrix, *Report ORC*, **66-10**, University of California College of Engineering, Operations Research Center, Berkeley.

Goldberger, A. S. (1961): Stepwise least squares residual analysis and specification error. *J. Amer. Statist. Assoc.*, **56**, 998–1000.

Golub, G. (1965): Numerical methods for solving linear least squares problems, *Numer. Math.*, **7**, 206–216.

——— (1968): Least squares, singular values and matrix approximations, *Appl. Math.*, **13**, 44–51.

Golub, G. and W. Kahan (1965): Calculating the singular values and pseudoinverse of a matrix, *SIAM J. Numer. Anal.*, Ser. B, **2**, 205–224.

Golub, G. and J. H. Wilkinson (1966): Note on iterative refinements of least squares solution, *Numer. Math.*, **9**, 139–148.

Golub, G. and C. Reinsch (1969): Singular value decomposition and least squares solutions, *Technical Report*, **CS133**, Computer Science Department, Stanford University.

Good, I. J. (1969): Some applications of the singular decomposition of a matrix, *Technometrics*, **11**, 823–831.

Graybill, F. A. (1961): *An Introduction to Linear Statistical Models, I*. McGraw-Hill, New York.

——— (1969): *Introduction to Matrices with Applications in Statistics*. Wadsworth, Belmont, Calif.

Graybill, F. A., C. D. Meyer, and R. J. Painter (1966): Note on the computation of the generalized inverse of a matrix, *SIAM Rev.*, **8**, 522–524.

Graybill, F. A. and G. Marsaglia (1957): Idempotent matrices and quadratic forms in the linear hypothesis, *Ann. Math. Statist.*, **28**, 678–686.

Greville, T. N. E. (1956): E. H. Moore's generalization of the concept of inverse of a matrix to include rectangular and singular matrices. Unpublished manuscript.

——— (1957): The pseudoinverse of a rectangular or singular matrix and its application to the solution of systems of linear equations, *SIAM News Letter*, **5**, 3–6.

——— (1957): On smoothing a finite table: a matrix approach, *SIAM J. Appl. Math.*, **5**, 137–154.

——— (1959): The pseudoinverse of a rectangular or singular matrix and its application to the solution of systems of linear equations, *SIAM Rev.*, **1**, 38–43.

——— (1960): Some applications of the pseudoinverse of a matrix, *SIAM Rev.*, **2**, 15–22.

——— (1961): Note on fitting of functions of several independent variables, *SIAM J. Appl. Math.*, **9**, 109–115, 317

——— (1962a): Notes on matrix pseudoinverses. Lecture Notes, Lehigh University.

——— (1962b): Further remarks on the pseudoinverse of a matrix. Unpublished manuscript, presented to Mathematics Club, University of Michigan.

——— (1963): A product characterization of the generalized inverse of a singular square matrix, The University of Wisconsin, Madison.

——— (1966): Note on the generalized inverse of a matrix product, *SIAM Rev.*, **8**, 518–521. Erratum, ibid., **9**, 249.

——— (1968a): Spectral generalized inverses of singular square matrices. 'Abstract' in *Notices Amer. Math. Soc.*, **15**, 111.

——— (1968b): Some new generalized inverses with spectral properties, *Proc. Symp. Theo. Appl. Generalized Inverses of Matrices*, Mathematics Ser. No. 4, Texas Technological College, Lubbock, Texas.

———: Spectral generalized inverses of square matrices. To appear as *MRC Tech. Summary Report*, **823**.

———: Generalized inverse of finite matrices. To be submitted to *SIAM Rev.*

Guedj, R. (1965): L'utilisation d'inverse généralisés dans la résolution de systèmes linéaires de rang quelconque, *Troisième Congr. de Calcul et de Traitement de l'Information AFCALTI*, 137–143.

Gura, I. A. (1967): Notes on the pseudoinverse of a matrix. Unpublished manuscript.

Halmos, P. R. and J. E. Mclaughlin (1963): Partial isometrics, *Pacific J. Math.*, **13**, 585–596.

Harris, W. A. and T. V. Helvig (1966a): Applications of the pseudoinverse to modelling, *Technometrics*, **8**, 351–357.

——— (1966b): Marginal and conditional distributions of singular distributions, *Publications of the Res. Inst. for Math. Sci.*, Kyoto University, Ser. A, **1**, 199–204.

Hawkins, J. B. and A. Ben-Israel (1968): On generalized matrix functions, *Systems Research Memo*, **193**, Northwestern University, Evanston, Ill.

Hearon, J. Z. (1967a): Construction of EPr generalized inverses by inversion of nonsingular matrices, *J. Res. Nat. Bur. Standards*, Sec. B, **71**, 57–60.

——— (1967b): A generalized matrix version of Rennie's inequality, *J. Res. Nat. Bur. Standards*, Sec. B, **71**, 61–64.

——— (1967c): Partially isometric matrices, *J. Res. Nat. Bur. Standards*, Sec. B, **71**, 225–228.

——— (1967d): Polar factorization of a matrix, *J. Res. Nat. Bur. Standards*, Sec. B, **71**, 65–67.

——— (1967e): Symmetrizable generalized inverses of symmetrizable matrices, *J. Res. Nat. Bur. Standards*, Sec. B, **71**, 229–231.

——— (1968): On the singularity of a certain bordered matrix, *SIAM J. Appl. Math.*, **15**, 1413–1421.

Hearon, J. Z. and J. W. Evans: Differentiable generalized inverses. Submitted for publication, Mathematical Research Branch, NIAMD, National Institute of Health, Bethesda, Md.

———: On spaces and maps of generalized inverses. To be published, Mathematical Research Branch, NIAMD, National Institute of Health, Bethesda, Md.

Henderson, C. R. (1953): Estimation of variance and covariance components, *Biometrics*, **9**, 226–252.

Herring, G. P. (1967): A note on generalized interpolation and the pseudoinverse, *SIAM J. Numer. Anal.*, **4**, 548–556.

Hestenes, M. R. (1958): Inversion of matrices by biorthogonalization and related results, *SIAM J. Appl. Math.*, **6**, 51–60.

——— (1961): Relative Hermitian matrices, *Pacific J. Math.*, **11**, 225–245.

——— (1961): Relative self-adjoint operators in Hilbert space, *Pacific J. Math.*, **11**, 1315–1357.

——— (1962): A ternary algebra with applications to matrices and linear transformations, *Arch. Rational Mech. Anal.*, **11**, 138–194.

Hogg, R. V. and A. T. Craig (1958): On the decomposition of certain chisquare variables, *Ann. Math. Statist.*, **29**, 608–610.

Householder, A. S. (1964): *The Theory of Matrices in Numerical Analysis*. Blasidell, New York.

Householder, A. S. and G. Young (1938): Matrix approximation and latent roots, *Amer. Math. Monthly*, **45**, 165–171.

Hsu, P. U. (1946): On a factorization of pseudoorthogonal matrices, *Quart. J. Math. Oxford Ser.*, **17**, 162–165.

Ijiri, Yuji (1965): On the generalized inverse of an incidence matrix, *SIAM J. Appl. Math.*, **13**, 827–836.

John, P. W. M. (1964): Pseudoinverses in the analysis of variance, *Ann. Math. Statist.*, **35**, 895–896.

Kalman, R. E. (1960): A new approach to linear filtering and prediction problems, *Jour. Basic Engineering*, 35–44.

Katz, I. J.: On the generalized inverse of a product and EPr matrices. To be published.

——— (1965): Weigmann type theorems for EPr matrices, *Duke Math. J.*, **32**, 423–427.

Katz, I. J. and M. H. Pearl (1966): On EPr and normal EPr matrices, *J. Res. Nat. Bur. Standards*, Sec. B, **70**, 47–77.

Keller, H. B. (1965): On the solution of singular and semi-definite linear systems by iteration, *SIAM J. Numer. Anal.*, **2**, 281–290.

Kellog, R. B. and J. Spanier (1965): On optimal alternating direction parameters for singular matrices, *Math. Comp.*, **19**, 448–452.

Khatri, C. G. (1963): Further contributions to Wishartness and independence of second degree polynomials in normal vectors, *J. Indian Statist. Assoc.*, **1**, 61–70.

——— (1968): Some results for the singular multivariate regression models, *Sankhyā*, *Ser. A*, **30**, 267–280.

——— (1969): A note on some results on generalized inverse of a matrix, *J. Indian Statist. Assoc.*, **7**, 38–45.

——— (1970): A note on a commutative g-inverse of a matrix. *Sankhyā*, *Ser. A*, **32**, 299–310.

Khatri, C. G. and C. R. Rao (1968): Solution to some functional equations and their applications to characterization of probability distribution, *Sankhyā*, *Ser. A*, **30**, 167–180.

Kim, J. B. (1966): On singular matrices, *J. Korean Math. Soc.*, **3**, 1–2.

Koop, J. C. (1963): Generalized inverse of a singular matrix, *Nature*, **198**, 1019–1020 and **200**, 716.

Korganoff, A. (1961): Functions of a normed vector space applied to the iterative solution of rectangular and square matrix non-linear equations of any given form, *Nordisk Symposium*, Oslo, August 1961, 18–22.

——— (1962): Les polynômes d'interpolation de matrices carrées à coefficients matriciels et les méthodes iteratives de résolution numérique des équations de matrices carrées de forme quelconque, *Proc. IFIP Congress* 1962, 102–106.

——— (1964): The inversion of rectangular matrices in the resolution of ill-conditioned linear systems, *Proc. Nordsam Congress*, Helsinki, **2**, 179–190.

Korganoff, A. and M. Pavel-Parvu (1964): Interprétation à l'aide des pseudo-inverses de la solution d'équations matricielles linéaires provenant de la discretisation d'opérateurs différentiels et intégraux, *83e Congrés de l'Association Française pour l'avancement des Sciences, Lille* (July, 1964).

——— (1967): *Éléments de théorie des matrices carrées et rectangles en analyse numérique*. Dunod, Paris.

Kruskal, W. (1968): When are Gauss–Markoff and least square estimates identical? A coordinate-free approach, *Ann. Math. Statist.*, **39**, 70–75.

Kublanovskaya, V. N. (1966): On the computation of the generalized matrix inverse and the projection, *USSR J. Comp. Math. Phys.*, **6**, 326–332.

Lanczos, C. (1958): Linear systems in self-adjoint form, *Amer. Math. Monthly*, **65**, 665–679.

——— (1961): *Linear Differential Operators*. Van Nostrand, Princeton, 124–129.

Langenhop, C. E. (1967): On generalized inverses of matrices, *SIAM J. Appl. Math.*, **15**, 1239–1246.

Lewis, T. O. and P. L. Odell (1967a): A generalization of the Gauss–Markov theorem, *J. Amer. Statist. Assoc.*, **61**, 1063–1066.

——— (1967b): A theory of linear estimation, *Texas Center for Research*.

Lewis, T. O., T. L. Boullion, and P. L. Odell (1968): A bibliography on generalized matrix inverses, *Proc. Symp. Theo. Appl. Generalized Inverses of Matrices*, Mathematics Ser. No. 4, Texas Technological College, Lubbock.

Lewis, T. O. and T. Newman (1968): Pseudoinverses of positive semidefinite matrices, *SIAM J. Appl. Math.*, **16**, 701–703.

Ljapin, E. S. (1958): Inversion of elements in semigroups, *Leningrad Gos. Ped. Inst. Ucen. Zap.*, **166**, 65–74.

Loud, W. S. (1966): Generalized inverses and generalized Green's functions, *SIAM J. Appl. Math.*, **14**, 342–369.

Magness, T. A. and T. B. Mcguire (1962): Comparison of least squares and minimum variance estimates on regression parameters, *Ann. Math. Statist.*, **33**, 462–470.

Marquardt, D. W. (1969): Generalized inverses, ridge regression, biased linear estimation and nonlinear estimation (mimeo report). Engineering Services Division, Engineering Department, E. I. du Pont de Nemours and Co., Wilmington, Delaware.

Marsaglia, G. (1964): Conditional means and variances of normal variables with singular covariance matrix, *J. Amer. Statist. Assoc.*, **59**, 1203–1204.

Mathai, A. M. (1966): Pseudoinverses in normal equations, *Estadistica*, **24**, 620–628.

Mayne, D. Q. (1966): An algorithm for calculation of the pseudoinverse of a singular matrix, *Comput. J.*, **9**, 312–317.

——— (1969): On the calculation of pseudoinverses, *IEEE Trans. Automatic Control*, **14**, 204–205.

Meicler, M. and P. L. Odell (1967a): Weighted generalized inverses, *NASA NAS 9-5384*, 1–13.

——— (1967b): Report *p–q* generalized inverses, *NASA NAS 9-5384*, 1–19.

Milne, R. D. (1968): An oblique matrix pseudoinverse, *SIAM J. Appl. Math.*, **16**, 931–944.

Milnes, H. W. (1968): Characteristic vectors for rectangular matrices, *Proc. Symp. Theo. Appl.*

Generalized Inverses of Matrices, Mathematics Ser. No. 4, Texas Technological College, Lubbock.

Milnes, H. W., J. Amburgey, T. O. Lewis, and T. L. Boullion (1968): Spectral eigen value property of A^+ for rectangular matrices, *Proc. Symp. Theo. Appl. Generalized Inverses of Matrices*, Mathematics Ser. No. 4, Texas Technological College, Lubbock.

Mitra Sanjit, K. (1969): *On the Analysis and Synthesis of Linear Active Networks*, Wiley, New York.

Mitra, S. K. (1968a): On a generalized inverse of a matrix and applications, *Sankhyā, Ser. A*, **30**, 107–114.

——— (1968b): A new class of g-inverse of square matrices, *Sankhyā, Ser. A*, **30**, 323–330.

——— (1970): A density free approach to matrix variate beta distribution, *Sankhyā, Ser. A*, **32**, 81–88.

——— (1971a): Restricted generalized inverse and the theory of minimum bias estimation in a Gauss–Markoff model, Tech. Report No. Math-Stat/2/71, Research and Training School, Indian Statistical Institute, Calcutta.

——— (1971b): Another look at the MINQUE of variance and covariance components, Tech. Report No. Math-Stat/3/71, Research and Training School, Indian Statistical Institute, Calcutta.

Mitra, S. K. and P. Bhimasankaram (1970a): Some results on idempotent matrices and a matrix equation connected with the distribution of quadratic forms, *Sankhyā, Ser. A*, **32**, 353–356.

——— (1970b): Generalized inverse of partitioned matrices and recalculation of least squares estimates for data or model changes. Research and Training School, Indian Statistical Institute, Technical Report No. Math. Stat./11/70.

——— (1970c): A characterisation of the Moore–Penrose Inverse and related results, submitted to *Sankhyā*.

Mitra, S. K. and C. R. Rao (1968a): Simultaneous reduction of a pair of quadratic forms, *Sankhyā, Ser. A*, **30**, 313–322.

——— (1968b): Some results in estimation and tests of linear hypothesis under the Gauss–Markoff model, *Sankhyā, Ser. A*, **30**, 281–290.

——— (1969): Conditions for optimality and validity of simple least squares theory, *Ann. Math. Statist.*, **40**, 1617–1624.

Mizel, V. J. and M. M. Rao (1962): Nonsymmetric projections in Hilbert space, *Pacific J. Math.*, **12**, 343–357.

Moore, E. H. (1920): On the reciprocal of the general algebraic matrix (abstract), *Bull. Amer. Math. Soc.*, **26**, 394–395.

——— (1935): *General Analysis*. American Philosophical Society, Philadelphia.

Morris, G. L. and P. L. Odell (1968a): A characterization for generalized inverses of matrices, *SIAM Rev.*, **10**, 208–211.

——— (1968b): Common solutions for n matrix equations with applications, *J. Assoc. Comput. Mach.*, **15**, 272–274.

Munn, W. D. (1961): Pseudoinverses in semigroups, *Proc. Cambridge Philos. Soc.*, **57**, 247–250.

Munn, W. D. and R. Penrose (1955): A note on inverse semigroups, *Proc. Cambridge Philos. Soc.*, **51**, 396–399.

Murray, F. J. and J. Von Neumann (1936): On rings of operators, *Ann. of Math.*, **37**, 116–229. (Vol. III, No. 2.)

Newcomb, Robert W. (1960): On the simultaneous diagonalization of two semidefinite matrices, *Quart. J. App. Math.*, **19**, 144–146.

Newman, T. G., P. L. Odell, and M. Meicler (1968): The concept of a p–q generalized inverse, *Proc. Symp. Theo. Appl. Generalized Inverses of Matrices*, Mathematics Ser. No. 4, Texas Technological College, Lubbock.

Noble, B. (1966): A method for computing the generalized inverse of a matrix, *SIAM J. Numer. Anal.*, **3**, 582–584.

Ogasawara, T. and M. Takahashi (1951): Independence of quadratic forms in normal system, *J. Sci. Hiroshima University*, **15**, 1–9.

Ogawa, J. (1949): On the independence of linear and quadratic forms of a random sample from a normal population, *Ann. Inst. Statist. Math.*, **1**, 83–108.

Osborne, E. E. (1961): On least squares solutions of linear equations, *J. Assoc. Comput. Mach.*, **8**, 628–636.

——— (1965): Smallest least squares solutions of linear equations, *SIAM J. Numer. Anal.*, **2**, 300–307.

Pearl, M. H. (1959a): On normal and EPr matrices, *Michigan Math. J.*, **6**, 1–5.

——— (1959b): On normal and EPr matrices, *Michigan Math. J.*, **6**, 89–94.

——— (1966): On generalized inverse of matrices, *Proc. Cambridge Philos. Soc.*, **62**, 673–677.

Pease, M. C. (1965): *Methods of Matrix Algebra*, Academic Press, New York and London.

Penrose, R. (1955): A generalized inverse for matrices, *Proc. Cambridge Philos. Soc.*, **51**, 406–413.

——— (1956): On best approximate solutions of linear matrix equations, *Proc. Cambridge Philos. Soc.*, **52**, 17–19.

Pereyra, V. (1969): Stability of general systems of linear equations, *Aequationes Mathematicae*, **2**, 194–206.

Pereyra, V. and J. B. Rosen (1964): Computation of the pseudoinverse of a matrix of unknown rank, *Tech. Rep.*, **CS13**, Computer Science Division, Stanford University.

Perlis, S. (1952): *Theory of Matrices*. Addison-Wesley, Cambridge, Mass.

Petryshyn, W. V. (1965): On the inversion of matrices and linear operators, *Proc. Amer. Math. Soc.*, **16**, 893–901.

——— (1967): On generalized inverse and on the uniform convergence of $(I - \beta\kappa)^n$ with application to iterative methods, *J. Math. Anal. Appl.*, **18**, 417–439.

Popa, C. (1962): Note on the inversion of singular matrices, *Lucrăr Sti. Inst. Ped. Timisoara Mat.-Fiz.*, 149–153.

Presić, S. B. (1963): Certaines équations matricielles, *Publications de la faculté d'électrotechnique de l'université à Belgrade Serie. Math. et Phys.*, **121**, 31–32.

Preston, G. B. (1954): Inverse semigroups, *J. London Math. Soc.*, **29**, 396–403.

Price, C. (1964): Matrix pseudoinverse and minimal variance estimates, *SIAM Rev.*, **6**, 115–120.

Pringle, R. M. and A. A. Rayner (1970): Expressions for generalized inverses of a bordered matrix with application to the theory of constrained linear models, *SIAM Rev.*, **12**, 107–115.

Pyle, L. D. (1964): Generalized inverse computations using the gradient projection method, *J. Assoc. Comput. Mach.*, **11**, 422–428.

——— (1965): The generalized inverse in linear programming—Basic theory. *Tech. Report*, **1**, Computer Science Department, Purdue University, Lafayette, Ind.

——— (1967): A generalized inverse ε-algorithm for construction of intersection projection matrices with applications, *Numer. Math.*, **10**, 86–102.

——— (1968): Remarks on a generalized inverse epsilon-algorithm for matrices, *Proc. Symp. Theo. Appl. Generalized Inverses of Matrices*, Mathematics Ser. No. 4, Texas Technological College, Lubbock.

Rado, R. (1956): Note on generalized inverses of matrices, *Proc. Cambridge Philos. Soc.*, **52**, 600–601.

Rainbolt, M. B.: Sequential least-squares parameter estimation. NASA–MSC unpublished report.

Rao, C. R. (1955): Analysis of dispersion for multiply classified data with unequal numbers in cells, *Sankhyā*, **15**, 253–280.

——— (1961): A study of large sample test criteria through properties of efficient estimates, *Sankhyā*, **23**, 25–40.

——— (1962): A note on a generalized inverse of a matrix with applications to problems in mathematical statistics, *J. Roy. Statist. Soc.*, Ser. B, **24**, 152–158.

——— (1965a): *Linear Statistical Inference and its Applications*. Wiley, New York.

——— (1965b): On the theory of least squares when parameters are stochastic and its application to analysis of growth curves, *Biometrika*, **52**, 447–458.

——— (1966): Generalized inverse for matrices and its applications in mathematical statistics, *Research Papers in Statistics*, Festschrift for J. Neyman, Wiley, New York.

——— (1967a): 'Least squares theory using an estimated dispersion matrix and its application to measurement of signals,' in *Proc. Fifth Berkeley Symposium on Math. Stat. and Prob.*, **1**, 355–372, University of California Press.

——— (1967b): Calculus of generalized inverse of matrices, Part I: General theory, *Sankhyā, Ser. A*, **29**, 317–342.

——— (1968): A note on a previous lemma in the theory of least squares and some further results, *Sankhyā, Ser. A*, **30**, 245–252.

——— (1970): Estimation of heteroscedastic variances in linear models, *J. Amer. Statist. Assoc.*, **65**, 161–172.

——— (1970): Estimation of variance and covariance components, Tech. Report No. Math-Stat/44/70, Research and Training School, Indian Statistical Institute, Calcutta.

——— (1971): Estimation of variance and covariance components—MINQUE theory, Discussion paper No. 58, Indian Statistical Institute, New Delhi.

Rao, C. R. and S. K. Mitra (1970): Generalized inverse of matrices and applications. To appear in *Proc. Sixth Berkeley Symposium on Math. Stat. and Probability*.

——— (1971): Further contributions to the theory of generalized inverse of matrices and its applications, to appear in *Sankhyā*, Ser. A, **33**.

Rao, K. K. (1966): A simpler proof of Gauss–Markov theorem when the regression matrix is of less than full rank, *Amer. Math. Monthly*, **73**, 394–395.

Rayner, A. A. and D. Livingstone (1965): On the distribution of quadratic forms in singular normal variates, *South African J. Agricultural Sci.*, **8**, 357–370.

Rayner, A. A. and R. M. Pringle (1967): A note on generalized inverses in the linear hypothesis not of full rank, *Ann. Math. Statist.*, **38**, 271–277.

Robinson, D. W. (1962): On the generalized inverse of an arbitrary linear transformation, *Amer. Math. Monthly*, **69**, 412–416.

Rohde, C. A. (1965): Generalized inverses of partitioned matrices, *SIAM J. Appl. Math.*, **13**, 1033–1035.

——— (1966): Some results on generalized inverses, *SIAM Rev.*, **8**, 201–205.

——— (1968): Special application of the theory of generalized matrix inversion to statistics, *Proc. Symp. Theo. Appl. Generalized Inverses of Matrices*, Mathematic Ser. No. 4, Texas Technological College, Lubbock.

Rohde, C. A. and J. R. Harvey (1965): Unified least squares analysis, *J. Amer. Statist. Assoc.*, **60**, 523–527.

Rohde, C. A. and G. M. Tallis (1969): Exact first and second order moments of estimates of components of covariance, *Biometrika*, **56**, 517–525.

Rosen, J. B. (1960): The gradient projection method for nonlinear programming, Part I: Linear constraints, *SIAM J. Appl. Math.*, **8**, 181–217.

——— (1961): The gradient projection method for nonlinear programming, Part II: Nonlinear constraints, *SIAM J. Appl. Math.*, **9**, 514–532.

——— (1964): Minimum and basic solutions to singular linear systems, *SIAM J. Appl. Math.*, **12**, 156–162.

Rust, B., W. R. Burrus, and C. Schnuberger (1966): A simple algorithm for computing the generalized inverse of a matrix, *Comm. ACM*, **9**, 381–387.

Sakamoto, H. (1944): On independence of statistics (in Japanese), *Res. Memoirs Instit. Stat. Math.*, **1**, 1–25.

Schneider, H. (1956): A matrix problem concerning projections, *Proc. Edinburgh Math. Soc.*, **10**, 129–130.

Schulz, G. (1963): Iterative berechnung der reziproken matrix, *Z. Angew Math. Mech.*, **13**, 57–59.

Scroggs, J. E. and P. L. Odell (1966): An alternative definition of the pseudoinverse of a matrix, *SIAM J. Appl. Math.*, **14**, 796–810.

Searle, S. R. (1965): Additional results concerning estimator function and generalized inverse matrices, *J. Roy. Statist. Soc.*, Ser. B, **27**, 486–490.

——(1966): *Matrix algebra for the Biological Sciences* (including applications in statistics), Wiley, New York.

—— (1968): Another look at Henderson's methods of estimating variance components, *Biometrics*, **24**, 749–778.

Seshu, S. and M. B. Reed (1961): *Linear Graphs and Electrical Networks*. Addison-Wesley, Reading, Mass.

Shanbag, D. N. (1968): Some remarks on Khatri's result in quadratic forms, *Biometrika*, **55**, 593–595.

—— (1970): On the distribution of a quadratic form, *Biometrika*, **57**, 222–223.

Sharpe, G. E. and B. Spain (1960): On the solution of networks by equicofactor matrix, *IRE Transactions on Circuit Theory*, **CT-7**, 230–239.

Sharpe, G. E. and G. P. H. Styan (1965a): A note on the generalized network inverse, *IEEE Transactions on Circuit Theory*, **12**, 632–633.

—— (1965b): Circuit duality and the general network inverse, *IEEE Transactions on Circuit Theory*, **12**, 22–27.

—— (1967): A note on equicofactor matrices, *Proc. IEEE*, **15**, 1226–1227.

Sheffield, R. D. (1958): A general theory of linear systems, *Amer. Math. Monthly*, **65**, 109–111.

Showalter, D. (1967): Representation and computation of the pseudoinverse, *Proc. Amer. Math. Soc.*, **18**, 584–587.

Shurbet, G. L., T. O. Lewis and H. W. Milnes (1969): Recovery of linear transformations using collinear invariant points and pseudoinverses, *Texas Journal of Science*, **20**, 361–366.

Sibuya, Masaki (1969): Generalized inverses of matrices. Part I (Japanese) *Instit. Stat. Math.*, **17**, 109–131.

—— (1970): Generalized inverse of mappings, to appear *Sankhyā*.

—— (1970): Subclasses of generalized inverses of matrices, *Ann. Inst. Statist. Math.* (in press).

Siegel, C. L. (1937): Über die analytische Theorie der quadratischen Formen III, *Annals of Mathematics*, **38**, 212–291.

Speed, F. M. and A. Feiveson: Testing hypotheses and generalized inverse. Unpublished report.

Stewart, G. W., III (1966): Perturbation bounds for the linear least squares problem. Computing Technological Center, Oak Ridge, Tenn.

—— (1969): On the continuity of the generalized inverse, *SIAM J. Appl. Math.*, **17**, 33–45.

Styan, G. P. H. (1969): Hadamard products and multivariate analysis (abstract), *Ann. Math. Statist.*, **40**, 1149–1150.

—— (1970): Notes on the distribution of quadratic forms in singular normal variables. Unpublished manuscript.

Tewarson, R. P. (1967): A direct method for generalized matrix inversion, *SIAM J. Numer. Anal.*, Ser. B, **4**, 499–507.

Tseng, Y. (1949): Generalized inverses of unbounded operators between two unitary spaces, *Dokl. Akad. Nauk, SSSR (N.S.)*, **67**, 431–434.

———(1949): Properties and classifications of generalized inverses of closed operators, *Dokl. Akad. Nauk, USSR*, **67**, 607–610.

—— (1956): Virtual solutions and generalized inversions, *Uspehi Math. Nauk (N.S.)*, **11**, 213–215.

Urquhart, N. S. (1968): Computation of generalized inverse of matrices which satisfy specified conditions, *SIAM Rev.*, **10**, 216–218.

Von Neumann, J. (1936): On regular rings, *Proc. Nat. Acad. Sci., U.S.A.*, **22**, 707–713.

Watson, G. S. (1967): Linear least square regression. *Ann. Math. Statist.*, **38**, 1679–1699.

Wedderburn, J. H. M. (1934): *Lectures on Matrices*, American Math. Soc., Providence, R.I.

Wedin, P. (1967): Results on pseudoinverses. Unpublished manuscript.

Wilkinson, G. N. (1958): Estimation of missing values for the analysis of incomplete data, *Biometrics*, **14**, 257–286.

Willner, L. B. (1967): An elimination method for computing the generalized inverse, *Math. Comp.*, **21**, 227–229.

Zacks, S. (1964): Generalized least squares estimators for random fractional replication designs, *Ann. Math. Statist.*, **35**, 696–704.

Zadeh, L. A. and C. A. Desoer (1963): *Linear System Theory*. McGraw-Hill, New York.

Zelen, M. (1962): The role of constraints in the theory of least squares. MRC Technical Summary Report, 312, University of Wisconsin, Math. Res. Center, U.S. Army. Madison, Wisconsin.

Zelen, M. and A. J. Goldman (1964): Weak generalized inverses and minimum variance linear unbiased estimation, *J. Res. Nat. Bur. Standards, Sect. B*, **68B**, 151–172.

Zimmule, D.: Techniques for computing the pseudoinverse of a matrix. Unpublished manuscript.

Zyskind, G. (1967): On canonical forms, negative covariance matrices and best and simple, least square linear estimator in linear models, *Ann. Math. Statist.*, **38**, 1092–1110.

Unpublished Master's Theses and Doctoral Dissertations

Anderson, C. L. (1967): A geometric theory of pseudoinverses and some applications in statistics. Master's thesis, Southern Methodist University.

Barton, C. P. (1966): Pseudoinverses of rectangular matrices. Master's thesis, Southern Methodist University.

Bhimasankaram, P. (1971): Some contributions to the theory, application and computation of generalized inverses of matrices. Doctoral dissertation, Indian Statistical Institute.

Boullion, T. L. (1966): Contributions to the theory of pseudoinverses. Doctoral dissertation, University of Texas.

Cline, R. E. (1963): Representations for the generalized inverse of matrices with applications in linear programming. Doctoral dissertation, Purdue University.

Dekerlegand, R. J. (1967): Analysis of generalized inverse computation schemes. Master's thesis, University of Southwestern Louisiana.

Gatley, W. Y. (1962): Application of the generalized inverse concept to the theory of linear statistical models. Doctoral dissertation, Oklahoma State University.

Kirby, M. J. L. (1965): Generalized inverses and chance-constrained programming. Doctoral dissertation, Northwestern University.

Lewis, T. P. (1966): Application of the theory of generalized matrix inversion to statistics. Doctoral dissertation, University of Texas.

Manherz, R. K. (1968): Applications of the generalized matrix inverse in network theory. Doctoral dissertation, Northwestern University.

Meicler, M. (1966): Weighted generalized inverses with minimal p and q norms. Doctoral dissertation, University of Texas.

Morris, G. L. (1967): Characterizations of generalized inverses for matrices. Doctoral dissertation, Texas Technological College.

Pyle, L. D. (1960): The generalized inverse in linear programming. Doctoral dissertation, Purdue University.

Rohde, C. A. (1964): Contributions to the theory, computation and applications of generalized inverses. Doctoral dissertation, North Carolina State University.

Vortuba, C. (1963): Generalized inverses and singular equations in functional analysis. Doctoral dissertation, University of Michigan.

Additional References

Bickart, T. A. (1968): On the compensation of multivariate control systems, *Proc. IEEE*, **58**, 1258–1259.

Manherz, R. K. and S. L. Hakimi (1969): The generalized inverse in network analysis and quadratic-error minimization problems, *IEEE Transactions on Circuit Theory*, **CT-16**, 559–562.

Olivares, J. E. (1968): Generalized inverse of the tie-set and cut-set matrix for networks with complete graphs, *Proc. Hawaii International Conference on System Sciences*, 583–586.

Wee, W. G. (1971): A generalized inverse approach to clustering pattern selection and classification, *IEEE Transactions on Information Theory*, **17**.

Author Index

Subject Index